Lecture Notes in Economics and Mathematical Systems

T0250574

Lecture Notes in Economics
and Mathematical Systems

588

Founding Editors:

M. Beckmann
H.P. Künzi

Managing Editors:

Prof. Dr. G. Fandel
Fachbereich Wirtschaftswissenschaften
Fernuniversität Hagen
Feithstr. 140/AVZ II, 58084 Hagen, Germany

Prof. Dr. W. Trockel
Institut für Mathematische Wirtschaftsforschung (IMW)
Universität Bielefeld
Universitätsstr. 25, 33615 Bielefeld, Germany

Editorial Board:

A. Basile, A. Drexl, H. Dawid, K. Inderfurth, W. Kürsten, U. Schittko

Don Grundel · Robert Murphey
Panos Pardalos · Oleg Prokopyev
(Editors)

Cooperative Systems

Control and Optimization

With 173 Figures and 17 Tables

 Springer

Dr. Don Grundel
AAC/ENA
Suite 385
101 W. Eglin Blvd.
Eglin AFB, FL 32542
USA
don.grundel@eglin.af.mil

Dr. Robert Murphey
Guidance, Navigation and
Controls Branch
Munitions Directorate
Suite 331
101 W. Eglin Blvd.
Eglin AFB, FL 32542
USA
robert.murphey@eglin.af.mil

Dr. Panos Pardalos
University of Florida
Department of Industrial and
Systems Engineering
303 Weil Hall
Gainesville, FL 32611-6595
USA
pardalos@ufl.edu

Dr. Oleg Prokopyev
University of Pittsburgh
Department of Industrial Engineering
1037 Benedum Hall
Pittsburgh, PA 15261
USA
prokopyev@engr.pitt.edu

Library of Congress Control Number: 2007920269

ISSN 0075-8442

ISBN 978-3-540-48270-3 Springer Berlin Heidelberg New York

Springer is part of Springer Science+Business Media

springer.com

© Springer-Verlag Berlin Heidelberg 2007

Production: LE-TEX Jelonek, Schmidt & Vöckler GbR, Leipzig
Cover-design: WMX Design GmbH, Heidelberg

SPIN 11916222   /3100YL - 5 4 3 2 1 0   Printed on acid-free paper

Preface

Cooperative systems are pervasive in a multitude of environments and at all levels. We find them at the microscopic biological level up to complex ecological structures. They are found in single organisms and they exist in large sociological organizations. Cooperative systems can be found in machine applications and in situations involving man and machine working together. While it may be difficult to define to everyone's satisfaction, we can say that cooperative systems have some common elements: 1) more than one entity, 2) the entities have behaviors that influence the decision space, 3) entities share at least one common objective, and 4) entities share information whether actively or passively.

Because of the clearly important role cooperative systems play in areas such as military sciences, biology, communications, robotics, and economics, just to name a few, the study of cooperative systems has intensified. That being said, they remain notoriously difficult to model and understand. Further than that, to fully achieve the benefits of manmade cooperative systems, researchers and practitioners have the goal to optimally control these complex systems. However, as if there is some diabolical plot to thwart this goal, a range of challenges remain such as noisy, narrow bandwidth communications, the hard problem of sensor fusion, hierarchical objectives, the existence of hazardous environments, and heterogeneous entities.

While a wealth of challenges exist, this area of study is exciting because of the continuing cross fertilization of ideas from a broad set of disciplines and creativity from a diverse array of scientific and engineering research. The works in this volume are the product of this cross-fertilization and provide fantastic insight in basic understanding, theory, modeling, and applications in cooperative control, optimization and related problems. Many of the chapters of this volume were presented at the 5th International Conference on "Cooperative Control and Optimization," which took place on January 20-22, 2005 in Gainesville, Florida. This 3 day event was sponsored by the Air Force Research Laboratory and the Center of Applied Optimization of the University of Florida.

We would like to acknowledge the financial support of the Air Force Research Laboratory and the University of Florida College of Engineering. We are especially grateful to the contributing authors, the anonymous referees, and the publisher for making this volume possible.

Don Grundel
Rob Murphey
Panos Pardalos
Oleg Prokopyev

December 2006

Contents

Optimally Greedy Control of Team Dispatching Systems

Venkatesh G. Rao[1] and Pierre T. Kabamba[2]

[1] Mechanical and Aerospace Engineering, Cornell University
    Ithaca, NY 14853
    E-mail:vr47@cornell.edu
[2] Aerospace Engineering, University of Michigan
    Ann Arbor 48109
    E-mail: kabamba@engin.umich.edu

Summary. We introduce the team dispatching (TD) problem arising in coopera-tive control of multiagent systems, such as spacecraft constellations and UAV fleets. The problem is formulated as an optimal control problem similar in structure to queuing problems modeled by restless bandits. A near-optimality result is derived for greedy dispatching under oversubscription conditions, and used to formulate an approximate deterministic model of greedy scheduling dynamics. Necessary condi-tions for optimal team configuration switching are then derived for restricted TD problems using this deterministic model. Explicit construction is provided for a spe-cial case, showing that the most-oversubscribed-first (MOF) switching sequence is optimal when team configurations have low overlap in their processing capabilities. Simulation results for TD problems in multi-spacecraft interferometric imaging are summarized.

1 Introduction

In this chapter we address the problem of scheduling multiagent systems that accomplish tasks in teams, where a *team* is a collection of agents that acts as a single, *transient* task processor, whose capabilities may partially overlap with the capabilities of other teams. When scheduling is accomplished using *dispatching* [1], or assigning tasks in the temporal order of execution, we re-fer to the associated problems as TD or *team dispatching* problems. A key characteristic of such problems is that two processes must be controlled in parallel: *task sequencing* and *team configuration switching*, with the associ-ated control actions being *dispatching* and *team formation and breakup events* respectively. In a previous paper [2] we presented the class of MixTeam dis-patchers for achieving simultaneous control of both processes, and applied it to a multi-spacecraft interferometric space telescope. The simulation results in [2] demonstrated high performance for *greedy* MixTeam dispatchers, and provided the motivation for this work. A schematic of the system in [2] is in Figure 1, which shows two spacecraft out of four cooperatively observing a target along a particular line of sight. In interferometric imaging, the resolution of the virtual telescope synthesized by two spacecraft depends on their separation. For our purposes, it is sufficient to note that features such as this distinguish the capabilities of different teams in team scheduling domains. When such features are present, team configuration switching must be used in order to fully utilize system capabilities.

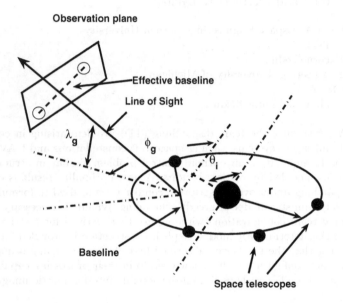

Fig. 1. Interferometric Space Telescope Constellation

The scheduling problems handled by the MixTeam schedulers are NP-hard in general [3]. Work in empirical computational complexity in the last decade [4, 5] has demonstrated, however, that worst-case behavior tends to be confined to small regions of the problem space of NP-hard problems (suitably-parameterized), and that average performance for good heuristics outside this region can be very good. The main analytical problem of interest, therefore, is to provide performance guarantees for specific heuristic approaches in specific parts of problem space, where worst-case behavior is rare and local structure may be exploited to yield good average performance. In this work we are concerned with *greedy* heuristics in *oversubscribed* portions of the problem space.

TD problems are structurally closest to *multi-armed bandit* problems [6] (in particular, the sub-class of *restless* bandit problems [7, 8, 9]), and in [2] we utilized this similarity to develop exploration/exploitation learning methods inspired by the multi-armed bandit literature. Despite the broad similarity of TD and bandit problems, however, they differ in their detailed structure, and decision techniques for bandits cannot be directly applied. In this chapter we seek *optimally greedy* solutions to a special case of TD called RTD (Resricted Team Dispatching). Optimally greedy solutions use a greedy heuristic for dispatching (which we show to be asymptotically optimal) and an optimal team configuration switching rule.

The results in this chapter are as follows. First, we develop an input-output representation of switched team systems, and formulate the TD problem. Next we show that greedy dispatching is asymptotically optimal for a single static team under oversubscription conditions. We use this to develop a deterministic model of the scheduling process, and then pose the restricted team dispatching (RTD) problem of finding optimal switching sequences with respect to this deterministic model. We then show that switching policies for RTD must belong to the class OSPTE (one-switch-persist-till-empty) under certain realistic constraints. For this class, we derive a necessary condition for the optimal configuration switching functions, and provide an explicit construction for a special case. A particularly interesting result is that when the task processing capabilities of possible teams overlap very little, then the *most oversubscribed first* (MOF) switching sequence is optimal for minimizing total cost. Qualitatively, this can be interpreted as the principle that when team capabilities do not overlap much, *generalist* team configurations should be instantiated before *specialist* team configurations.

The original contribution of this chapter comprises three elements. The first is the development of a systematic representation of TD systems. The second is the demonstration of asymptotic optimality properties of greedy dispatching under oversubscription conditions. The third is the derivation of necessary conditions and (for a special case) constructions for optimal switching policies under realistic assumptions.

In Section 2, we develop the framework and the problem formulation. In Sections 3 and 4, we present the main results of the chapter. In Section 5 we summarize the application results originally presented in [2]. In Section 6 we present our conclusions.The appendix contains sketches of proofs. Full proofs are available in [3].

2 Framework and Problem Formulation

Before presenting the framework and formulation for TD problems in detail, we provide an overview using an example.

Figure 2 shows a 4-agent TD system, such as Figure 1, represented as a queuing network. A set of tasks $G(t)$ is waiting to be processed (in general tasks may arrive continuously, but in this chapter we will only consider tasks sets where no new jobs arrive after $t = 0$). If we label the agents a, b, c and d, and legal teams are of size two, then the six possible teams are ab, ac, ad, bc, *bd* and *cd*. Legal *configurations* of teams are given by *ab-cd, ac-bd* and *ad-bc* respectively. These are labeled C_1, C_2 and C_3 in Figure 1. Each configuration, therefore, may be regarded as a set of processors corresponding to constituent teams, each with a queue capable of holding the next task. At any given time, only one of the configurations is in existence, and is determined by the configuration function $\bar{C}(t)$. Whenever a team in the current configuration is free, a trigger is sent to the dispatcher, *d*, which releases a waiting *feasible* task from the unassigned task set $G(t)$ and assigns it to the free team, which then executes it. The *control* problem is to determine the signal $\bar{C}(t)$ and the dispatch function *d* to optimize a performance measure. In the next subsection, we present the framework in detail.

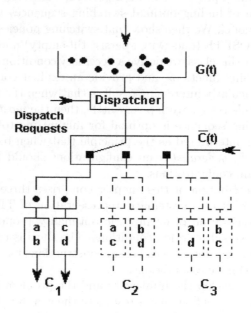

Fig. 2. System Flowchart

2.1 System Description

We will assume that time is discrete throughout, with the discrete time index t ranging over the non-negative integers **N**. There are three agent-based entities in TD systems: individual agents, teams, and configurations of teams. We define these as follows.

Agents and Agent Aggregates

1. Let $\mathcal{A} \triangleq \{A_1, A_2, \ldots, A_q\}$ be a set of q distinguishable agents.

2. Let $\mathcal{T} \triangleq \{T_1, T_2, \ldots, T_r\}$ be a set of r teams that can be formed from members of $\mathcal{A}$, where each team maps to a fixed subset of $\mathcal{A}$. Note that multiple teams may map to the same subset, as in the case when the ordering of agents within a team matters.

3. Let $\mathcal{C} \triangleq \{C_1, C_2, \ldots, C_m\}$ be a set of m *team configurations*, defined as a set of teams such that the subsets corresponding to all the teams constitute a partition of $\mathcal{A}$. Note that multiple configurations can map to the same set partition of $\mathcal{A}$. It follows that an agent A must belong to exactly one team in any given configuration C.

Switching Dynamics

We describe formation and breakup by means of a switching process defined by a *configuration function*.

1. Let a *configuration function* $\bar{C}(t)$ be a map $\bar{C} : \mathbf{N} \to \mathcal{C}$ that assigns a configuration to every time step t. The value of $\bar{C}(t)$ is the element with index i_t in $\mathcal{C}$, and is denoted C_{i_t}. The set of all such functions is denoted **C**.

2. Let time t be partitioned into a sequence of half-open intervals $[t_k, t_{k+1})$, $k = 0, 1, \ldots$, or *stages*, during which $\bar{C}(t)$ is constant. The t_k are referred to as the *switching times* of the configuration function $\bar{C}(t)$.

3. The configuration function can be described equivalently with either time or stage, since, by definition, it only changes value at stage boundaries. We therefore define $C(k) = \bar{C}(t)$ for all $t \in [t_k, t_{k+1})$. We will refer to both $C(k)$ and $\bar{C}(t)$ as the configuration function. The sequence $C(0), C(1), \ldots$ is called the *switching sequence*

4. Let the *team function* $\bar{T}(C, j)$ be the map $T : \mathcal{C} \times \mathbf{N} \to \mathcal{T}$ given by team j in configuration C. The maximum allowable value of j among all configurations in a configuration function represents the maximum number of logical teams that can exist simultaneously. This number is referred to as the number of *execution threads* of the system, since it is the maximum number of parallel task execution processes that can exist at a given time. In this chapter we will only analyze single-threaded TD systems, but present simulation results for multi-threaded systems.

Tasks and Processing Capabilities

We require notation to track the status of tasks as they go from unscheduled to executed, and the capabilities of different teams with respect to the task set. In particular, we will need the following definitions:

1. Let X be an arbitrary collection of teams (note that any configuration C is by definition such a collection). Define $G(X, t) \triangleq \{g_r :$ the set of all tasks that are available for assignment at time t, and can be processed by some team in $X\}$.

$$\bar{G}(C,t) = G(C,t) - \bigcup_{C_i \neq C} G(C_i,t)$$

$$\bar{G}(T,t) = G(T,t) - \bigcup_{T_i \neq T} G(T_i,t). \qquad (1)$$

If $X = \mathcal{T}$, then the set $G(X,t) = G(\mathcal{T},t)$ represents all unassigned tasks at time t. For this case, we will drop the first argument and refer to such sets with the notation $G(t)$. A task set $G(t)$ is by definition feasible, since at least one team is capable of processing it. Team capabilities over the task set are illustrated in the Venn diagram in Figure 3.

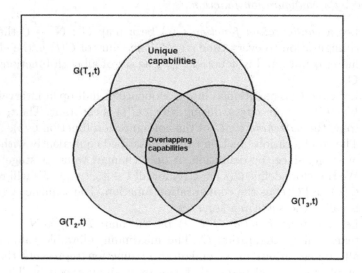

Fig. 3. Processing capabilities and task set structure

2. Let X be a set of teams (which can be a single team or configuration as in the previous definition). Define

$$n_X(t) = \left| \bigcup_{T_i \in X} G(T_i,t) \right|, \text{ and}$$

$$\bar{n}_X(t) = \left| \bigcup_{T_i \in X} G(T_i,t) - \bigcup_{T_i \notin X} G(T_i,t) \right|. \qquad (2)$$

If X is a set with an index or time argument, such as $C(k)$, $\bar{C}(t)$ or C_i, the index or argument will be used as the subscript for n or $\bar{n}$, to simplify the notation.

Dispatch Rules and Schedules

The scheduling process is driven by a dispatch rule that picks tasks from the unscheduled set of tasks, and assigns them to free teams for execution. The schedule therefore evolves forward in time. Note that this process does not backtrack, hence assignments are irrevocable.

1. We define a *dispatch rule* to be a function $d : \mathcal{T} \times \mathbf{N} \rightarrow G(t)$ that *irrevocably* assigns a free team to a feasible unassigned task as follows,

$$d(T, t) = g \in G(T, t), \tag{3}$$

   where $t \in \{t_d^i\}$ the set of *decision points*, or the set of end times of the most recently assigned tasks for the current configuration. d belongs to a set of available dispatch rules D.
2. A dispatch rule is said to be *complete* with respect to the configuration function $\bar{C}(t)$ and task set $G(0)$ if it is guaranteed to eventually assign all tasks in $G(0)$ when invoked at all decision points generated starting from $t = 0$ for all teams in $\bar{C}(t)$.
3. Since a configuration function and a dispatch rule generate a schedule, we *define* a schedule[3] to be the ordered pair $(\bar{C}(t), d)$, where $\bar{C}(t) \in \mathbf{C}$, and $d \in D$ is complete with respect to $G(0)$ and $\bar{C}(t)$.

Cost Structure

Finally, we define the various cost functions of interest that will allow us to state propositions about optimality properties.

1. Let the real-valued function $c(g, t) : G(t) \times \mathbf{N} \rightarrow \mathbf{R}$ be defined as the cost incurred for *assigning*[4] task g at time t_g. We refer to c as the *instantaneous cost function*. c is a random process in general. Let $\mathcal{J}(\bar{C}(t), d)$ be the *partial cost function* of a schedule $(\bar{C}(t), d)$. The two are related by:

$$\mathcal{J}(\bar{C}(t), d) = \sum_{g \in G(0)} c(g, t_g), \tag{4}$$

   where t_g is the actual time at which g is assigned. This model of costs is defined to model the specific instantaneous cost of *slack time* in processing a task in [2], and the overall cost of *makespan* [1]. Other interpretations are possible.

---

[3] Strictly speaking, $(\bar{C}(t), d)$ is insufficient to uniquely define a schedule, but sufficient to define a schedule up to interchangeable tasks, defined as tasks with identical parameters. Sets of schedules that differ in positions of interchangeable tasks constitute an equivalence class with respect to cost structure. These details are in [3].

[4] Task costs are functions of *commitment* times in general, not just the start times. See [3] for details.

2. Let a configuration function $C(k) = C_{i_k} \in \mathcal{C}$ have $k_{\max}$ stages. The *total cost function* $\mathcal{J}^T$ is defined as

$$\mathcal{J}^T(\bar{C}(t), d) = \mathcal{J}(\bar{C}(t), d) + \sum_{k=1}^{k_{\max}} J^S(i_k, i_{k-1}), \tag{5}$$

where $J^S(i_k, i_{k+1})$ is the *switching* cost between configurations i_k and i_{k+1}, and is finite. Define $J_{\min}^S = \min J^S(i, j)$, $J_{\max}^S = \max J^S(i, j)$, i, $j \in 1, \ldots, m,$.

2.2 The General Team Dispatching (TD) Problem

We can now state the general team dispatching problem as follows:
General Team Dispatching Problem (TD) Let $G(0)$ be a set of tasks that must be processed by a finite set of agents $\mathcal{A}$, which can be partitioned into team configurations in $\mathcal{C}$, comprising teams drawn from $\mathcal{T}$. Find the schedule $(\bar{C}^*(t), d^*)$ that achieves

$$(\bar{C}^*(t), d^*) = \operatorname{argmin} E(\mathcal{J}^T(\bar{C}(t), d)), \tag{6}$$

where $\bar{C}(t) \in \mathbf{C}$ and $d \in D$.

3 Performance Under Oversubscription

In this section, we show that for the TD problem with a set of tasks $G(0)$, whose costs $c(g, t)$ are bounded and randomly varying, and a static configuration comprising a single team, a greedy dispatch rule is asymptotically optimal when the number of tasks tends to infinity. We use this result to justify a simplified *deterministic oversubscription model* of the greedy cost dynamics, which will be used in the next section.

Consider a system comprising a single, static team, T. Since there is only a single team, $C(t) = C = \{T\}$, a constant. Let the value of the instantaneous cost function $c(g, t)$, for any g and t, be given by the random variable X, as follows,

$$c(g, t) = X \in \{c_{\min} = c_1, c_2, \ldots, c_k = c_{\max}\},$$
$$P(X = c_i) = 1/k, \tag{7}$$

such that the finite set of equally likely outcomes, $\{c_{\min} = c_1, c_2, \ldots, c_k = c_{\max}\}$ satisfies $c_i < c_{i+1}$ for all $i < k$. The index values $j = 1, 2, \ldots k$ are referred to as *cost levels*. Since there is no switching cost, the total cost of a schedule is given by

$$\mathcal{J}^T(\bar{C}(t), d) \equiv \mathcal{J}(\bar{C}(t), d) \equiv \sum_{g \in G(0)} c(g, t_g), \tag{8}$$

where t_g are the times tasks are assigned in the schedule.

Definition 1: We define the *greedy dispatch rule*, d_m, as follows:

$$d_m(T, t) = g^* \in G(T, t),$$

$$c(g^*, t) \leq c(g, t) \ \forall g \in G(T, t), \ g \neq g^*. \tag{9}$$

We define the *random dispatch rule* $d_r(T, t)$ as a function that returns a randomly chosen element of $G(T, t)$. Note that both greedy and random dispatch rules are complete, since there is only one team, and any task can be done at any time, for a finite cost.

Theorem 1: *Let $G(0)$ be a set of tasks such that (7) holds for all $g \in G(0)$, for all $t > 0$. Let j_m be the lowest occupied cost level at time $t > 0$. Let $n \triangleq |G(t)|$. Then the following hold:*

$$\lim_{n \to \infty} E(c(d_m(T, t), t)) = c_{\min}, \tag{10}$$

$$\lim_{n \to \infty} E(j_m) = 1, \tag{11}$$

$$E(\mathcal{J}_m) < E(\mathcal{J}_r) \text{ for large } n, \tag{12}$$

$$\lim_{n \to \infty} \frac{E(\mathcal{J}_m) - \mathcal{J}^*}{\mathcal{J}^*} = 0, \tag{13}$$

where $\mathcal{J}_m \equiv \mathcal{J}^T(\bar{C}(t), d_m)$ and $\mathcal{J}_r \equiv \mathcal{J}^T(\bar{C}(t), d_r)$ are the total costs of the schedules $(\bar{C}(t), d_m)$ and $(\bar{C}(t), d_r)$ computed by the greedy and random dispatchers respectively, and $\mathcal{J}^*$ is the cost of an optimal schedule.

Remark 1: Theorem 1 essentially states that if a large enough number of tasks with randomly varying costs are waiting, we can nearly always find one that happens to be at $c_{\min}$.[5] All the claims proved in Theorem 1 depend on the behavior of the probability distribution for the lowest occupied cost level j_m as n increases. Figure 4 shows the change in $E(j_m)$ with n, for $k = 10$, and as can be seen, it drops very rapidly to the lowest level. Figure 5 shows the actual probability distribution for j_m with increasing n and the same rapid skewing towards the lowest level can be seen. Theorem 1 can be interpreted as a *local optimality* property that holds for a single execution thread between switches (a single stage).

Theorem 1 shows that for a set of tasks with randomly varying costs, the expected cost of performing a task picked with a greedy rule varies inversely with the size of the set the task is chosen from. This leads to the conclusion that the cost of a schedule generated with a greedy rule can be expected to converge to the optimal cost in a relative sense, as the size of the initial task set increases.

Remark 2: For the spacecraft scheduling domain discussed in [2], the sequence of cost values at decision times are well approximated by a random sequence.

---

[5] Theorem 1 is similar to the idea of 'economy of scale' in that more tasks are cheaper to process on average, except that the economy comes from probability rather than amortization of fixed costs.

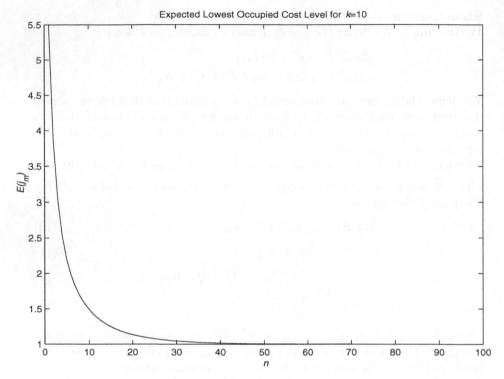

Fig. 4. Change in expected value of j_m with n

3.1 The Deterministic Oversubscription Model

Theorem 1 provides a relation between the degree of oversubscription of an agent or team, and the performance of the greedy dispatching rule. This relation is stochastic in nature and makes the analysis of optimal switching policies extremely difficult. For the remainder of this chapter, therefore, we will use the following model, in order to permit a *deterministic* analysis of the switching process.

Deterministic Oversubscription Model: The costs $c(g,t)$ of all tasks is bounded above and below by $c_{\max}$ and $c_{\min}$, and for any team T, if two decision points t and t' are such that $n_T(t) > n_T(t')$ then

$$c(d_m(t),t) \equiv c(n_T(t)) < c(d_m(t'),t') \equiv c(n_T(t)). \qquad (14)$$

The model states that the cost of processing the task picked from $G(T,t)$ by d_m is a deterministic function that depends *only* on the size of this set, and decreases monotonically with this size. Further, this cost is bounded above and below by the constants $c_{\max}$ and $c_{\min}$ for all tasks. This model may be regarded as a deterministic approximation of the stochastic correlation between degree of oversubscription and performance that was obtained in Theorem 1. We now use this to define a *restricted* TD problem.

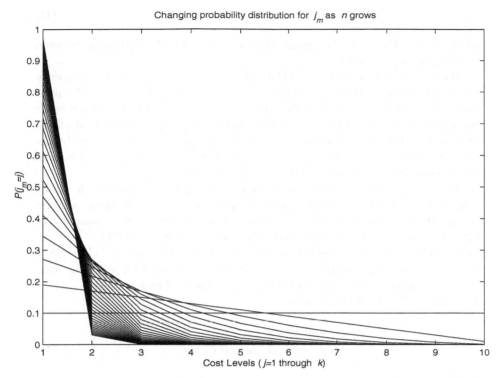

Fig. 5. Change in distribution of j_m with n. The distributions with the greatest skewing towards $j = 1$ are the ones with the highest n

4 Optimally Greedy Dispatching

In this section, we present the main results of this chapter: necessary conditions that optimal configuration functions must satisfy for a subclass, RTD, of TD problems, under reasonable conditions of high switching costs and decentralization. We first state the restricted TD problem, and then present two lemmas that demonstrate that under conditions of high switching costs and information decentralization, the optimal configuration function must belong to the well-defined *one-switch, persist-till-empty* (OSPTE) dominance class. When Lemmas 1 and 2 hold, therefore, it is sufficient to search over the OSPTE class for the optimal switching function, and in the remaining results, we consider RTD problems for which Lemmas 1 and 2 hold.

Restricted Team Dispatching Problem (RTD) Let $G(0)$ be a feasible set of tasks that must be processed by a finite set of agents $\mathcal{A}$, which can be partitioned into team configurations in $\mathcal{C}$, comprising teams drawn from $\mathcal{T}$. Let there be a one to one map between the configuration and team spaces, $\mathcal{C} \leftrightarrow \mathcal{T}$ and $C_i = \{T_i\}$, i.e., each configuration comprises only one team. Find the schedule $(\bar{C}^*(t), d_m)$ that achieves

$$(\bar{C}^*(t), d_m) = \text{argmin } \mathcal{J}^T(\bar{C}(t), d_m),\tag{15}$$

where $\bar{C}(t) \in \mathbf{C}$, d_m is the greedy dispatch rule, and the deterministic over-subscription model holds.

RTD is a specialization of TD in three ways. First, it is a *deterministic* optimization problem. Second, it has a *single execution thread*. For team dispatching problems, such a situation can arise, for instance, when every configuration consists of a team comprising a unique permutation of all the agents in A. For such a system, only one task is processed at a time, by the current configuration. Third, the dispatch function is fixed ($d = d_m$) so that we are only optimizing over configuration functions.

We now state two lemmas that show that under the reasonable conditions of *high switching cost* (a realistic assumption for systems such as multi-spacecraft interferometric telescopes) and *decentralization*, the optimal configuration function for greedy dispatching must belong to OSPTE.

Definition 2: For a configuration space $\mathcal{C}$ with m elements, the class OS of *one-switch* configuration functions comprises all configuration functions, with exactly m stages, with each configuration instantiated exactly once.

Lemma 1: *For an RTD problem, let*

$$|G(0)| = n$$
$$\bar{G}(C_i, 0) \neq \emptyset, \text{ for all } C_i \in \mathcal{C},\tag{16}$$

and let

$$mJ_{\min}^S - (m-1)J_{\max}^S > n\left(c_{\max} - c_{\min}\right).\tag{17}$$

Under the above conditions, the optimal configuration function $\bar{C}^*(t)$ is in OS.

Lemma 1 provides conditions under which it is sufficient to search over the class of schedules with configuration functions in OS. This is still a fairly large class. We now define OSPTE defined as follows:

Definition 3: A *one-switch persist-till-empty* or OSPTE configuration function $\bar{C}(t) \in$ OS is such that every configuration in $\bar{C}(t)$, once instantiated, persists until $G(C_k, t) = \emptyset$.

Constraint 1: (*Decentralized Information*) Define the *local knowledge set* $K_i(t)$ to be the set of truth values of the membership function $g \in G(C_i, t)$ over $G(t)$ and the truth value of Equation 17. The *switching time* t_{k+1} is only permitted to be a function of $K_i(t)$.

Constraint 2: (*Decentralized Control*): Let $C(k) = C_i$ where C_i comprises the single team T_i. For stage k, the switching time t_{k+1} is only permitted to take on values such that $t_k \geq t_C$, where t_C is the *earliest time* at which

$$K_i(t) \Rightarrow \square \ \exists (t' < \infty) : (G(T_i, t') = \emptyset)\tag{18}$$

is true

Lemma 2: *If Lemma 1 and constraints 1 and 2 hold, then the optimal configuration function is OSPTE.*

Remark 3: Constraint 1 says that the switching time can only depend on information concerning the capabilities of the *current* configuration. This captures the case when each configuration is a decision-making agent, and once instantiated, determines its own dissolution time (the switching time t_{k+1}) based only on knowledge of its *own* capabilities, i.e., it does not know what other configurations can do.[6] Constraint 2 uses the modal operator □ ("In all possible future worlds") [10] to express the statement that the switching time cannot be earlier than the earliest time at which the knowledge set K_i is *sufficient* to guarantee completion of all tasks in $G(C(k))$ at some future time. This means a configuration will only dissolve itself when it *knows* that there is a time t', when all tasks within its range of capabilities will be done (possibly by another configuration with overlapping capabilities). Lemma 2 essentially captures the intuitive idea that if an agent is required to be sure that tasks will be done by some other agent in the future in order to stop working, it must necessarily *know* something about what other agents *can do*. In the absence of this knowledge, it must do everything it can possibly do, to be safe.

We now derive properties of solutions to RTD problems that satisfy Lemmas 1 and 2, which we have shown to be in OSPTE.

4.1 Optimal Solutions to RTD Problems

In this section, we first construct the optimal switching sequence for the simplest RTD problems with two-stage configuration functions (Theorem 2), and then use it to derive a necessary condition for optimal configuration functions with an arbitrary number of stages (Theorem 3). We then show, in Theorem 4, that if a dominance property holds for the configurations, Theorem 3 can be used to construct the optimal switching sequence, which turns out to be the most-oversubscribed-first (MOF) sequence.

Theorem 2 *Consider a RTD problem for which Lemmas 1 and 2 hold. Let* $\mathcal{C} \triangleq \{C_1, C_2\}$. *Assume, without loss of generality, that* $|C_1| \geq |C_2|$. *For this system, the configuration function* $(C(0) = C_1, C(1) = C_2)$ *is optimal, and unique when* $|C_1| > |C_2|$.

Theorem 2 simply states that if there are only two configurations, the one that can do more should be instantiated first. Next, we use Theorem 2 to derive a necessary condition for arbitrary numbers of configurations.

Theorem 3: *Consider an RTD system with* m *configurations and task set* $G(0)$. *Let Lemmas 1 and 2 hold. Let* $C(k) = C(0), \ldots, C(m-1)$ *be an optimal configuration function. Then any subsequence* $C(k), \ldots, C(k')$ *must be the optimal configuration function for the RTD with task set* $G(t_k) - G(t_{k'+1})$. *Furthermore, for every pair of neighboring configurations* $C(j)$, $C(j+1)$

$$n_j(t_j) > n_{j+1}(t_j). \tag{19}$$

---

[6] Parliaments are a familiar example of multiagent teams that dissolve themselves and do not know what future parliaments will do.

Theorem 3 is similar to the principle of optimality. Note that though it is merely necessary, it provides a way of improving candidate OSPTE configuration functions by applying Equation 19 locally and exchanging neighboring configurations to achieve local improvements. This provides a local optimization rule.

Definition 4: The *most-oversubscribed first* (MOF) sequence $C_D(k) \triangleq C_{i_0} \ldots C_{i_{m-1}}$ is a sequence of configurations such that $n_{i_0}(0) \geq n_{i_1}(0) \geq \ldots \geq n_{i_{m-1}}(0)$

Definition 5: The *dominance order relation* $\succ$ is defined as

$$C_i \succ C_j \iff \bar{n}_i(0) > n_j(0). \tag{20}$$

Theorem 4: *If every configuration in $C_D(k)$ dominates its successor, $C_D(k) \succ C_D(k+1)$, then the optimal configuration function is given by $(C_D(k), d_m)$.*

Theorem 3 is an analog of the principle of optimality, which provides the validity for the procedure of dynamic programming. For such problems, solutions usually have to be computed backwards from the terminal state. Theorem 4 can be regarded as a tractable special case, where a property that can be determined *a priori* (the MOF order) is sufficient to compute the optimal switching sequence.

Remark 4: The relation $\succ$ may be interpreted as follows. Since the relation is stronger than size ordering, it implies either a strong convergence of task set sizes for the configurations *or* weak overlap among task sets. If the number of tasks that can be processed by the different configurations are of the same order of magnitude, the only way the ordering property can hold is if the *intersections* of different task sets (of the form $G(C_i, t) \bigcap G(C_j, t)$ are all very small. This can be interpreted qualitatively as the prescription: *if capabilities of teams overlap very little, instantiate generalist team configurations before specialist team configurations.*

Theorem 3 and Theorem 4 constitute a basic pair of analysis and synthesis results for RTD problems. General TD problems and the systems in [2] are much more complex, but in the next section, we summarize simulation results from [2] that suggest that the provable properties in this section may be preserved in more complex problems.

5 Applications

While the abstract problem formulation and main results presented in this chapter capture the key features of the multi-spacecraft interferometric telescope TD system in [2] (greedy dispatching and switching team configurations), the simulation study had several additional features. The most important ones are that the system in [2] had multiple parallel threads of execution, arbitrary (instead of OSPTE) configuration functions and, most importantly, *learning* mechanisms for discovering good configuration functions automatically. In the following, we describe the system and the simulation results obtained. These demonstrate that the fundamental properties of greedy dispatching and optimal switching deduced analytically in this chapter are in fact present in a much richer system.

The system considered in [2] was a constellation of 4 space telescopes that operated in teams of 2. Using the notation in this chapter, the system can be described by $\mathcal{A} = \{a, b, c, d\}$, $\mathcal{T} = \{T_1, \ldots, T_6\} \triangleq \{ab, ac, ad, bc, bd, cd\}$ and $\mathcal{C} = \{C_1, C_2, C_3\} \triangleq \{ab-cd, ac-bd, ad-bc\}$ (Figure 2). The goal set $G(0)$ comprised 300 tasks in most simulations. The dispatch rule was greedy (d_m). The local cost c_j was the *slack* introduced by scheduling job j, and the global cost was the *makespan* (the sum of local costs plus a constant). The switching cost was zero. The relation of oversubscription to dispatching cost observed empirically is very well approximated by the relation derived in Theorem 1. For this system, the greedy dispatching performed approximately 7 times better than the random dispatching, even with a random configuration function. The MixTeam algorithms permit several different exploration/exploitation learning strategies to be implemented, and the following were simulated:

1. *Baseline Greedy:* This method used greedy dispatching with random configuration switching.
2. *Two-Phase:* This method uses reinforcement learning to identify the effectiveness of various team configurations during an exploration phase comprising the first k percent of assignments, and preferentially creates these configurations during an exploitation phase.
3. *Two-Phase with rapid exploration:* this method extends the previous method by forcing rapid changes in the team configurations during exploration, to gather a larger amount of effectiveness data.
4. *Adaptive:* This method uses a continuous learning process instead of a fixed demarcation of exploration and exploitation phases.

Table 1 shows the comparison results for the the three learning methods, compared to the basic greedy dispatcher with a random configuration function. Overall, the most sophisticated scheduler reduced makespan by 21% relative to the least sophisticated controller. An interesting feature was that the preference order of configurations learned by the learning dispatchers approximately matched the MOF sequence that was proved to be optimal under the conditions of Theorem 4. Since the preference order determines the time fraction assigned to each configuration by the MixTeam schedulers, the *dominant* configuration during the course of the scheduling approximately followed the MOF sequence. This suggests that the MOF sequence may have optimality or near-optimality properties under weaker conditions than those of Theorem 4.

Table 1. Comparison of methods

Method	Best Makespan (hours)	Best $\mathcal{J}_m/\mathcal{J}^*$	% change (w.r.t greedy)
1.	54.41	0.592	0%
2.	48.42	0.665	-11%
3.	47.16	0.683	-13.3%
4.	42.67	0.755	-21.6%

6 Conclusions

In this chapter, we formulated an abstract team dispatching problem and demonstrated several basic properties of optimal solutions. The analysis was based on first showing, through a probabilistic argument, that the *greedy* dispatch rule is asymptotically optimal, and then using this result to motivate a simpler, *deterministic* model of the oversubscription-cost relationship. We then derived properties of optimal switching sequences for a restricted version of the general team dispatching problem. The main conclusions that can be drawn from the analysis are that greed is asymptotically optimal and that a most-oversubscribed-first (MOF) switching rule is the optimal greedy strategy under conditions of small intersections of team capabilities. The results are consistent with the results for much more complex systems that were studied using simulation experiments in [2].

The results proved represent a first step towards a complete analysis of dispatching methods such as the MixTeam algorithms, using the greedy dispatch rule. Directions for future work include the extension of the stochastic analysis to the switching part of the problem, derivation of optimality properties for multi-threaded execution, and demonstrating the *learnability* of near-optimal switching sequences, which was observed in practice in simulations with Mix-Team learning algorithms.

References

1. Pinedo, M., *Scheduling: theory, algorithms and systems*, Prentice Hall, 2002.
2. Rao, V. G. and Kabamba, P. T., "Interferometric Observatories in Circular Orbits: Designing Constellations for Capacity, Coverage and Utilization," *2003 AAS/AIAA Astrodynamics Specialists Conference*, Big Sky, Montana, August 2003.
3. Rao, V. G., *Team Formation and Breakup in Multiagent Systems*, Ph.D. thesis, University of Michigan, 2004.
4. Cook, S. and Mitchell, D., "Finding Hard Instances of the Satisfiability Problem," *Proc. DIMACS workshop on Satisfiability Problems*, 1997.
5. Cheeseman, P., Kanefsky, B., and Taylor, W., "Where the Really Hard Problems Are," *Proc. IJCAI-91*, Sydney, Australia, 1991, pp. 163–169.

6. Berry, D. A. and Fristedt, B., *Bandit Problems: Sequential Allocation of Experiments*, Chapman and Hall, 1985.
7. Whittle, P., "Restless Bandits: Activity Allocation in a Changing World," *Journal of Applied Probability*, Vol. 25A, 1988, pp. 257–298.
8. Weber, R. and Weiss, G., "On an Index Policy for Restless Bandits," *Journal of Applied Probability*, Vol. 27, 1990, pp. 637–348.
9. Papadimitrou, C. H. and Tsitsiklis, J. N., "The Complexity of Optimal Queuing Network Control," *Math and Operations Research*, Vol. 24, No. 2, 1999, pp. 293–305.
10. Weiss, G., *Multiagent Systems: A Modern Approach to Distributed Artificial Intelligence*, MIT Press, Cambridge, MA, 2000.

A Proofs

In this appendix we present a sketch of the proof of Theorem 1, and briefly outline the main arguments of the other proofs. Full proofs are available in [3].

Proof of Theorem 1: To prove the first and second claims we first derive expressions for $E(c(d_m(t), t))$ and $E(j_m)$,

$$E(j_m) = \sum_{j=1}^{j=k} j P(j_m = j),$$

$$E(c(d_m(t), t)) = \sum_{j=1}^{j=k} c_j P(j_m = j). \tag{21}$$

Define the $\phi(j)$, the *occupancy* of cost-level j, as the number of waiting tasks for which $c(g, t) = c_j$. We write $\alpha = (j-1)/k$ and $\beta = (1 - 1/(k - j + 1))$. It can be shown [3] that

$$E(j_m) = \sum_{j=1}^{j=k} j (1 - \alpha)^n (1 - \beta^n), \tag{22}$$

and similarly

$$E(c(d_m(t), t)) = \sum_{j=1}^{j=k} c_j (1 - \alpha)^n (1 - \beta^n). \tag{23}$$

By taking limits on the term inside the summand

$$P(j_m = j) = (1 - \alpha)^n (1 - \beta^n) \tag{24}$$

it can be shown that

$$\lim_{n \to \infty} E(c(d_m(t), t)) = c_{\min},$$

$$\lim_{n \to \infty} E(j_m) = 1, \tag{25}$$

which proves the first two claims. To prove 12, we first prove that the convergence for 10 and 11 is monotonic after a sufficiently high n for each of the summands. Specifically, we can show that for $n > \eta_j^*$, the jth summand decreases monotonically, where η_j^* is given by

$$\eta_j^* = \ln\left(\frac{\lambda}{1+\lambda}\right)/\ln\beta$$
$$= \ln\left(\frac{\ln(1-\alpha)/\ln\beta}{1+\ln(1-\alpha)/\ln\beta}\right)/\ln\beta. \tag{26}$$

Picking $n^* > n_j^*$ for all j, we can show that the cost approaches $c_{\min}$ monotonically for $n > n^*$. We can use this fact to bound the total cost of the schedule by partitioning it into the cost of the last n^* tasks and the first $n - n^*$ tasks to show that for arbitrary ϵ:

$$E(\mathcal{J}_m) < N(\epsilon)(c_{\max} - c_{\min} - \epsilon) + n(c_{\min} + \epsilon), \tag{27}$$

which yields

$$E(\mathcal{J}_r) - E(\mathcal{J}_m) > 0 \text{ as } n \to \infty. \tag{28}$$

Finally, 13 follows immediately from the fact that the schedule cost is bounded below by $nc_{\min}$, which yields, for sufficiently large n

$$\lim_{n\to\infty} \frac{(E(\mathcal{J}_m) - \mathcal{J}^*)}{\mathcal{J}^*} \leq O(\epsilon/c_{\min}). \tag{29}$$

Since we can choose ϵ arbitrarily small, the right-hand side cannot be bounded away from 0, therefore

$$\lim_{n\to\infty} \frac{(E(\mathcal{J}_m) - \mathcal{J}^*)}{\mathcal{J}^*} = 0. \tag{30}$$

□

Proof of Lemma 1: This lemma is proved by showing that with high enough switching costs, the worst case cost for a schedule with $m - 1$ switches is still better than the best-case cost for a schedule with m switches. Details are in [3] □

Proof of Lemma 2: Constraint 1 says that the switching time t_{k+1} out of stage k can only depend on information $K_i(t)$ about whether or not the current configuration $C(k) = C_i$ can do each of the remaining jobs. Constraint 2 specifies this dependence further, and says that the switching time cannot be less than the *earliest* time at which $K_i(t)$ is sufficient to guarantee that all jobs in $G(C_i, t)$ will eventually get done (in a finite time). Clearly, if $G(C_i, t_{k+1})$ is empty at the switching time t_{k+1}, then it will continue to be empty in all future worlds and constraints 1 and 2 are trivially satisfied.

To establish that $C(k)$ is OSPTE, it is sufficient to show that $G(C_i, t)$ must be empty at $t = t_k$. We show this by contradiction. Assume it is non-empty and let $g \in G(C(k), t_{k+1})$. Then by constraint 2, it must be that $K_i(t_k)$ is sufficient to establish the existence of $t' > t_{k+1}$ such that $G(C(k), t') = \emptyset$. This implies it is also sufficient to establish that there exists at least one configuration C' to be instantiated in the future, that can (and will) process g. Now, either $C' = C_i$ or $C' \neq C_i$. By assumption it is known that Equation 15 holds, and by Constraint 1, this is part of $K_i(t)$. Therefore $K_i(t_k)$ is sufficient information to conclude that C_i will not be instantiated again in the future. Therefore $C' \neq C$. But this means something is known about the truth value of membership relation $g \in G(C', t')$, for a $C' \neq C_i$, which is impossible by Constraint 1. Therefore, by contradiction, $G(C(k), t_{k+1}) = \emptyset$ and the configuration function must be in OSPTE. $\square$

Proof of Theorem 2: This theorem is a consequence of the deterministic oversubscription model which leads to lower marginal costs for doing tasks when they are assigned to the more capable configuration. See [3] for details.

Proof of Theorem 3: Theorem 3 is a straightforward generalization of Theorem 2 and hinges on the fact that each task is done by the first configuration that can process it, which implies that the tasks processed by a subsequence of configurations do not depend on the ordering within that subsequence. Therefore the state of the task sets before and after the subsequence are not changed by changing the subsequence, implying that each subsequence must be the optimal permutation among all permutations of the constituent configurations. This principle does not hold in general. For details see [3]. $\square$

Proof of Theorem 4: This theorem hinges on the fact that the relation $C_i \succ C_j$ cannot be changed by any possible processing by configurations instantiated before either C_i or C_j is instantiated, since the relation depends on the number of tasks each is *uniquely* capable of processing. This relation, *a fortiori*, allows us to use reasoning similar to Theorems 2 and 3 to recover a construction of the optimal sequence. For details see [3]. $\square$

Heuristics for Designing the Control of a UAV Fleet With Model Checking

Christopher A. Bohn*

Department of Systems and Software Engineering
Air Force Institute of Technology
Wright-Patterson AFB OH 45385, USA
E-mail: christopher.bohn@afit.edu

Summary. We describe a pursuer-evader game played on a grid in which the pursuers can move faster than the evaders, but the pursuers cannot determine an evader's location except when a pursuer occupies the same grid cell as that evader. The pursuers' object is to locate all evaders, while the evader's object is to prevent collocation with any pursuer indefinitely. The game is loosely based on autonomous unmanned aerial vehicles (UAVs) with a limited field-of-view attempting to locate enemy vehicles on the ground, where the idea is to control a fleet of UAVs to meet the search objective. The requirement that the pursuers move without knowing the evaders' locations necessitates a model of the game that does not explicitly model the evaders. This has the positive benefit that the model is independent of the number of evaders (indeed, the number of evaders need not be known); however, this has the negative side-effect that the time and memory requirements to determine a pursuer-winning strategy is exponential in the size of the grid. We report significant improvements in the available heuristics to abstract the model further and reduce the time and memory needed.

1 Introduction

The challenge of an airborne system locating an object on the ground is a common problem for many applications, such as tracking, search and rescue, and destroying enemy targets during hostilities. If the target is not facilitating the search, or is even attempting to foil it by moving to avoid detection, the difficulty of the search effort is greater than when the target aids the search. Our research is intended to address a technical hurdle for locating moving targets with certainty. We have abstracted this problem of controlling a fleet of UAVs to meet some search objective into a pursuer-evader game played on

---

* The views expressed in this article are those of the author and do not necessarily reflect the official policy of the Air Force, the Department of Defense, or the US Government.

a finite grid. The pursuers can move faster than the evaders, but the pursuers cannot ascertain the evaders' locations except by the collocation of a pursuer and evader. Further, not only can the evaders determine the pursuers' past and current locations, they have an oracle providing them with the pursuers' future moves. The pursuers' objective is to locate all evaders eventually, while the evaders' objective is to prevent indefinitely collocation with any pursuer.

We previously [5] described how and why we modeled this game as a system of concurrent finite automata, and the use of symbolic model checking to extract pursuer-winning search strategies for games involving single- and multiple-pursuers, games with rectilinear and hexagonal grids, games with and without terrain features, and games with varying pursuer-sensor footprints. We further outlined the state-space explosion problem essential to our approach and suggested heuristics that may be suitable to cope with this problem.

Here we present the results of our investigation into these heuristics. In Section 2, we reiterate the technique of using model checking to discover pursuer-winning search strategies. In Section 3, we describe our heuristics and demonstrate their utility. In Section 4, we establish necessary pursuer qualities for a pursuer-winning search strategy to exist. Finally, in Section 5 we consider directions for future work.

2 Background

We begin by describing model checking, an automatic technique to verify properties of systems composed of concurrent finite automata. After examining model checking, we review the model of the pursuer-evader game and how model checking can be used to discover pursuer-winning search strategies.

2.1 Model Checking

Model checking is a software engineering technique to establish or refute the correctness of a finite-state concurrent system relative to a formal specification expressed using a temporal logic. Originally, model checking involved the explicit representation of an automaton's states, which placed a considerable constraint on the size of models that could be checked. With the advent of symbolic model checking, checking models with greater state spaces was possible. Symbolic model checking differs from explicit-state model checking in that the models are represented by reduced, ordered binary decision diagrams, which are canonical representations of boolean formulas. Examples of symbolic model checkers are SMV [2] and its re-implementation, NuSMV [1]; Spin [3] is an examplar explicit-state model checker. Should a model fail to satisfy its specification, SMV, NuSMV, and Spin all provide computation traces that serve as witnesses to the falsehood of the specification; these counterexamples are often used to identify and correct errors the model.

The computational complexity of model checking is not unreasonable. For example, consider a model M consisting of the set of states S and the transition relation $\mathcal{R}$ and the formula f. Let $|S|$ and $|\mathcal{R}|$ be the cardinalities of S and $\mathcal{R}$, respectively. Then we define $|M| = |S| + |\mathcal{R}|$, and we further define $|f|$ as the number of atomic propositions and operators in f. The model-checking complexity of Computation Tree Logic, a temporal logic used by SMV and NuSMV, is $\mathcal{O}\left(|M| \cdot |f|\right)$; that is, it is linear in the size of the model and in the size of the specification. On the other hand, the model-checking complexity of Linear Temporal Logic, a logic used by Spin and NuSMV, is $\mathcal{O}\left(|M| \cdot 2^{\mathcal{O}(|f|)}\right)$ [7].

2.2 Modeling the Game

In our model, each pursuer is represented by a nondeterministic finite automaton. If a pursuer can move *speed* times faster than the evaders, then in each round of movement, the automaton modeling that pursuer will make *speed* nondeterministic moves, each move being either a transition into an adjacent grid cell or remaining in-place. While we directly model the pursuers, we do not explicitly include evaders. Instead, each grid cell has a single boolean state variable *cleared* that indicates whether it is possible for an undetected evader to occupy that cell. *Cleared* is TRUE if and only if no undetected evader can occupy that cell, and *cleared* is FALSE if it is *possible* for an undetected evader to occupy that cell. Trivially, cells occupied by pursuers are *cleared* – either there's no evader occupying that cell, or it has been detected. A cell that is not *cleared* becomes *cleared* when and only when a pursuer occupies it. A *cleared* cell ceases to be *cleared* when and only when it is adjacent to an un*cleared* cell during the evaders' turn to move; if all its neighboring cells are *cleared* then it remains *cleared*.

Consider Figure 1. In this hypothetical scenario, the pursuer has *cleared* a region of the southwest corner of the grid, as shown by the shaded portion of Figure 1(a), and can conclude that all the evaders must be outside that region. The pursuer moves four spaces north and west in Figure 1(b), increasing the *cleared* region by three cells (one of the visited cells was already cleared). Since the pursuer does not know where the evaders are located, the *cleared* region must shrink in accordance with the union of all possible moves by the evaders. A move by the evader south from the northeastern-most corner would not cause the evader to enter a previously-*cleared* cell, but Figure 1(c) shows there are six ways evaders *could* move from an un*cleared* cell into a *cleared* cell, and the five *cleared* cells that could now be occupied by evaders may no longer be considered *cleared*.

We now check whether, in the resulting system, invariably at least one cell is not *cleared*. If this specification holds, then there is no pursuer-winning search strategy: no matter what the pursuers do, the evaders will always be able to avoid detection. On the other hand, if the specification does not hold, then the model checker will provide a counterexample: a sequence of states

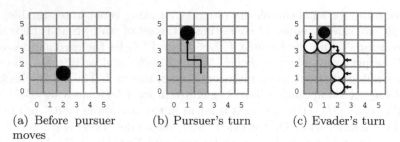

(a) Before pursuer moves   (b) Pursuer's turn   (c) Evader's turn

Fig. 1. Examples of changes in the possible locations for the evader. Evader is known to be in unshaded region.

that lead to a state in which every cell is *cleared*. If every cell is *cleared*, then there is no cell that contains an undetected evader; ergo, every evader has been detected. By examining the counterexample trace, we can infer the moves the pursuers made and use this as a pursuer-winning search strategy.

3 Heuristics

While the technique we have described works, the time and memory requirements grow exponentially with the size of the grid. Consider a game on an $m \times n$ grid with p pursuers moving at *speed* spaces/turn. The number of states, then, is:

$$\underbrace{(mn)^p}_{\substack{\text{pursuers'} \\ \text{locations}}} \cdot \underbrace{2^{mn}}_{\substack{\text{cells} \\ \textit{cleared?}}} \cdot \underbrace{(speed + 1)}_{\substack{\text{scheduling} \\ \text{counter}}} \tag{1}$$

That model checking can be accomplished in time that is linear is the number of states is of little comfort when the number of states grows exponentially in the size of the problem. This exponential growth is shown in Figure 2.

3.1 Heuristic Descriptions

To overcome this complexity, we turned to heuristics, three of which we describe here.

Clear-Column

The Clear-Column heuristic involves breaking the problem of clearing the grid into the smaller problem of clearing one column and ending up positioned to clear the next column, without permitting any undetected evaders to pass into previously-*cleared* columns; see Figure 3. If it is ever possible for the

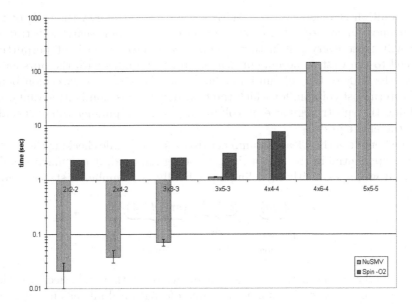

Fig. 2. Total mean execution times to generate winning search strategies for pursuer. Where no time is listed, the model checker exceeded available memory. Error bars indicate minimum and maximum values from the test data.

evader to enter the westernmost region, then the technique of clearing columns will not compose. However, if it is possible to accomplish this feat, repeated applications of this Clear-Column procedure can be composed to clear the whole grid by sweeping from one side of the grid to the other. Now we only need to model $w \times n$ cells explicitly (where w is the width of the subgrid we model; $2 \le w \ll m$), which can be a significant reduction in the size of the state space.

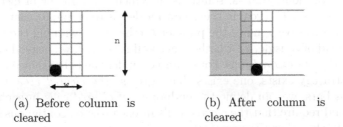

(a) Before column is cleared

(b) After column is cleared

Fig. 3. Abstraction of grid unbounded along the horizontal axis.

The general approach is inductive on the columns: assume the western region has been *cleared*; that is, any evaders to the west have already been detected. If the pursuer is in the westernmost column of the actual grid, then this condition is vacuously true. With the pursuer at one of the ends of the westernmost un*cleared* column, the pursuer executes some search substrategy that will cause every cell in that column to be *cleared* without permitting any cell to the west to become un*cleared* and terminates with the pursuer at one of the ends of the column immediately to the east (the exception being the easternmost column, for which the terminating position is irrelevant). By applying the substrategy at each column in turn, the pursuer will eventually clear the entire grid.

The benefit of the Clear-Column heuristic is that, while checking the model is still exponential in the size of the grid being modeled, it is a much smaller grid that we are explicitly modeling. Specifically, the number of states is now:

$$\underbrace{(wn)}_{\substack{\text{pursuers'} \\ \text{locations}}} \cdot \underbrace{2^{wn+1}}_{\substack{\text{cells} \\ \text{cleared?}}} \cdot \underbrace{(speed+1)}_{\substack{\text{"clock"} \\ \text{artifact}}} \tag{2}$$

The property to check is no longer an invariant; rather, we check whether the region to the west of column c remains *cleared* until all cells in column c and the region to the west are *cleared* when the pursuer is positioned to clear column $c+1$. The obvious downside to the Clear-Column heuristic is that if it is possible for a pursuer to win by a strategy that does not involve clearing the columns in sequence, and no comparable strategy exists which does involve column-clearing, then this heuristic would not reveal that pursuer-winning strategy.

Cleared-Bars

Besides composing subsolutions, we also consider changes to the manner in which we model the game. The alternate models we present here reflect our belief that when pursuer-winning solutions exist, there are pursuer-winning monotonic solutions; that is, solutions in which the number of *cleared* cells does not decrease. The goal in these new models is to eliminate many possible states that, intuitively, move the pursuer further from winning the game.

So instead of considering whether each cell is *cleared*, we instead can define sets of contiguous *cleared* cells. For example, under the belief that if a pursuer-winning strategy exists, one exists that "grows" the *cleared* area as a set of contiguous bars, we can define the endpoints of *cleared* cells in each row (or column) and require that the *cleared* cells in each row be contiguous from one endpoint to the other (Figure 4(a)).

The number of states in the Cleared-Bars model is:

$$\underbrace{(mn)^{2p}}_{\substack{\text{pursuers'} \\ \text{locations}}} \cdot \underbrace{(m+1)^{2n}}_{\substack{\text{endpoints} \\ \text{of bars}}} \cdot \underbrace{(speed+1)}_{\substack{\text{"clock"} \\ \text{artifact}}} \tag{3}$$

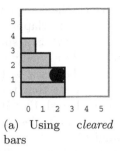

(a) Using   *cleared*
bars

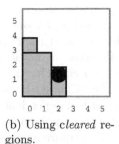

(b) Using *cleared* re-
gions.

Fig. 4. Alternate ways to describe the configuration of Figure 1(a).

The first term is raised to the power of $2p$ instead of p because, as we described above, there are conditions in which the pursuers' current and last locations are needed to update the bars correctly. The middle term is $m + 1$ instead of m to provide for "endpoints" when there are no *cleared* cells in a given row. The property to check is that invariantly there is a row whose left endpoint is not in the leftmost column or whose right endpoint is not in the rightmost column.

We earlier reported our preliminary performance results of the Cleared-Bars heuristic using the SMV model checker [5]. Unfortunately, that was the extent of our success with the SMV (or NuSMV) model checker. Describing the Cleared-Bars model with the SMV model description language is overly complex and difficult to reason about. The result was that generating each model was an error-prone process for even the simplest models, and the tendency toward insidious errors rapidly increased as the problem size grew. For this reason we re-implemented the model to be checked with Spin. Spin's model description language, Promela, uses guarded commands that made for a far simpler model description that was less amenable to implementation errors. The performance of Cleared-Bars using Spin is reported in Figure 6 along with our other results.

Cleared-Regions

Alternatively, we might instead define the *cleared* regions geometrically by possibly-overlapping convex polygons: for rectilinear grids, rectangles. Figure 4(b) shows how the *cleared* area in Figure 1(a) can be described using three rectangles. While this will dramatically increase the complexity of the model description, it will also dramatically decrease the number of states in the model because each rectangle can be fully characterized by two opposing corners.

We believe that when a pursuer-winning search strategy exists, it will have contiguous regions of *cleared* cells throughout the game, as opposed to isolated *cleared* cells scattered across the grid. Moreover, when a pursuer-winning search strategy exists, at least one exists for which these regions of *cleared* cells can be grouped into a small number of possibly-overlapping rectangles. In essence, the "Cleared Bars" heuristic detailed above is a special case of the "Cleared Regions" heuristic: there are potentially as many rectangles as there are rows. Our claim for the "Cleared Regions" heuristic is stronger than our claim for the "Cleared Bars" heuristic. We believe that the number of rectangles needed is independent of the size of the board, that it is in fact a small constant: for example, pursuer-winning search strategies on a rectangular rectilinear grid require at most three rectangles.

While we have proposed this heuristic before, we have now implemented the Cleared-Regions heuristic and can report its performance.

The critical issue to be addressed is how to determine the positions and dimensions of the rectangles. While we could take a brute-force approach and try to fit each possible selection of rectangles until all *cleared* cells and only *cleared* cells are enclosed by a rectangle, the time to do this would tend to offset any gain achieved by model checking the smaller state space. Instead, we shall use a fast and satisficing approach.

We define a total ordering on the grid cells in row-major order starting in the lower-left corner. Starting in the first cell, we examine the cells in order until we locate a *cleared* cell. This is the lower-left corner of a rectangle. We then continue searching the cells in order until we reach the right edge of the grid or until we encounter an un*cleared* cell; we now have the breadth of the rectangle. Now we examine all the cells in the next row within the columns touched by the rectangle; for example, if we begin the rectangle in row 2 and it stretches from column 5 to column 8, then we examine the cells in row 3, columns 5–8. If all those cells are *cleared*, then the rectangle's height grows by one. We continue to grow the rectangle's height until we reach a row in which at least one of the cells within the rectangle's breadth is not *cleared*.

Construction of the next rectangle begins by resuming the examination of the cells where we had stopped to adjust the previous rectangle's height. Again, we examine the cells in order until we locate a *cleared* cell that is not already in a previously-constructed rectangle. Once we have located such a cell, the rectangle is constructed as before. This process continues until all cells have been examined.

The algorithm we have described is suboptimal in that it may require more rectangles than are necessary for a particular arrangement of *cleared* cells. For example, consider the arrangement in Figure 5(a). The method presented here would require the three rectangles shown in Figure 5(b). The *cleared* region could in fact be covered by two rectangles, as shown in Figure 5(c). Indeed, the problem of covering the *cleared* cells is an instance of the the Minimal Set Cover Problem, which is known to be NP-complete [8]. This algorithm, though, runs in linear time: if we allow up to some constant k rectangles, then each cell will be examined at most k times. We are willing to accept using three rectangles to cover a configuration that could be covered with two, as we know of no pursuer-winning strategies for grids larger than 2×2 for which two rectangles are sufficient for all confingurations in the general case nor in the specific instances that we checked.

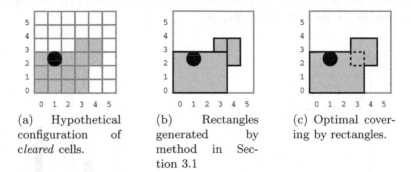

(a) Hypothetical configuration of *cleared* cells.

(b) Rectangles generated by method in Section 3.1

(c) Optimal covering by rectangles.

Fig. 5. Example game configuration demonstrating suboptimality of cell-covering algorithm.

With p pursuers moving *speed* spaces/turn and k rectangles describing the *cleared* regions, the size of the state space is:

$$\underbrace{(mn)^p}_{\substack{\text{pursuers'} \\ \text{locations}}} \cdot \underbrace{((m+1)(n+1))^{2k}}_{\substack{\text{diagonal} \\ \text{corners of} \\ \text{rectangles}}} \cdot \underbrace{(speed+1)}_{\substack{\text{"clock"} \\ \text{artifact}}} \tag{4}$$

And the property to check is that invariantly at least one grid cell is not covered by a rectangle. If this property does not hold, then a pursuer-winning search strategy exists and can be extracted from the counterexample witness.

3.2 Performance

The first question to be answered is whether the heuristics fail to find pursuer-winning search strategies for games which are known to have pursuer-winning search strategies. The answer is no. For every problem we checked using the basic approach, the heuristics' solutions did not require faster pursuers. Moreover, we have proven that there are no pursuer-winning search strategies permitting slower pursuers than those produced by our technique here; this proof is in Section 4.

We have demonstrated three heuristics that can be used to reduce the time to determine if and how the pursuers can locate the evaders. Clear-Columns was based on composing solutions to subproblems, whereas Cleared-Bars and Cleared-Regions were based on alternate ways to describe the arrangement of *cleared* and un*cleared* cells on the grid. Each of the three was able to provide a pursuer-winning search strategy for a single pursuer travelling at the slowest speed possible for it to win in the full model. This suggests the heuristics are effective. As Figure 6 shows, for sufficiently large grids — by the 4 × 4 grid for all three — the heuristics also provided solutions faster than the full model. This suggests the heuristics are efficient.

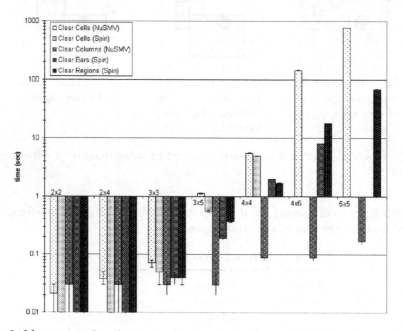

Fig. 6. Mean execution times to generate winning search strategies for pursuer, for the full model checked with NuSMV and with Spin, for the Clear-Columns model checked with NuSMV, and for the Cleared-Bars and Cleared-Regions models checked with Spin.

On a 933 MHz Pentium III workstation with 1 GB main memory, the Clear-Column heuristic is efficient enough to permit games with up to 15 × ∞ grids. The Cleared-Bars and Cleared-Regions permitted no larger than 4 × 6 and 5 × 5 grids, respectively, given the memory requirements for Spin. This is larger than possible with the full model with Spin, but no larger than is possible with the full model with NuSMV – though checking these models with Spin *is* faster than checking the full model with NuSMV. Should we implement these heuristics with a symbolic model checker, much larger grids should be manageable.

For the problem sizes we checked, despite its lower big-$\mathcal{O}$ complexity, Cleared-Regions did not provide a clear benefit over Cleared-Bars, other than permitting a 5 × 5 grid. This is can be explained in part by their constant factors; further, as shown in Figure 7, for these problem sizes, the Cleared-

Regions model has a larger state space than the Cleared-Bars model. For grids at least as large as 6×6, though, the ranking of the number of states among the four models is as we would expect, except that the size of the statespace of Clear-Columns will overtake that of Cleared-Regions at 32×32.

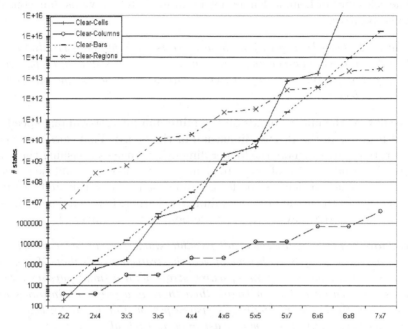

Fig. 7. Number of states for the four models as a function of grid size.

4 Necessary Pursuer Qualities for Simple Game Variants

We previously reported the sufficient pursuer qualities for a pursuer win the game [5, 6], though we were unable to prove the necessary conditions in general. We showed that a single pursuer moving at a rate of n spaces/turn is sufficient to detect all evaders on an $m \times n$ board (where n is the shorter dimension) when the evaders do not move diagonally, regardless of whether the pursuer moves diagonally. We also showed that when the evaders do move diagonally, a pursuer speed of $n + 1$ spaces/turn is sufficient for the pursuer to win. We now prove that, under a reasonable assumption, these speeds are also necessary; that is, a pursuer moving $n - 1$ spaces/turn cannot win the game, nor can a pursuer moving n spaces per turn when the evaders move diagonally. We begin with a lemma whose proof should be obvious; in the interest of space we do not reproduce the proof for Lemma 1 here, though can be found elsewhere [4].

Lemma 1. *Let s be a speed for which there is a pursuer-winning search strategy for a single pursuer on an m × n board. Then s is also a speed for which there is a pursuer-winning search strategy for a single pursuer on an $(m-1) \times n$ board.*

The relevance of Lemma 1 may not be immediately obvious, but consider its contrapositive:

Corollary 1. *Let s be a speed for which there is not a pursuer-winning search strategy for a single pursuer on an m × n board. Then s is a speed for which there is not a pursuer-winning search strategy for a single pursuer on an $(m + 1) \times n$ board.*

Recall that the upper bounds on the minimum puruser-winning speed are defined in terms of the shorter dimension of the board. That does not mean, however, that we can ignore the longer dimension when establishing the lower bounds. We shall use Corollary 1 to demonstrate that an insufficient speed does not become sufficient as the longer dimension grows. But first, we turn our attention to the assumption we alluded to earlier. Let us define a class of search strategies that have a property we believe to be universal:

Definition 1. *Let S be the set of all possible single-pursuer pursuer-winning search strategies. $A \subseteq S$ is the set of search strategies such that: if a search strategy $S \in A$ is a pursuer-winning search strategy for a single pursuer moving s spaces per turn on an m × n board, then there is a pursuer-winning search strategy for a single pursuer moving s spaces per turn on an m × n board such that the pursuer visits each row at least once in each of its turns.*

The most immediate consequence of Definition 1 is that no strategy in A has a pursuer speed less than $n - 1$, where n is the shorter dimension of the board. This does not, however, provide us with the tight bounds we seek.

Definition 2. *Let S be the set of all possible single-pursuer pursuer-winning search strategies. $B \subseteq S$ is the set of search strategies such that: if a search strategy $S \in B$ is a pursuer-winning search strategy for a single pursuer moving s spaces per turn on an m × n board, then there is a pursuer-winning search strategy for a single pursuer moving s spaces per turn on an m × n board such that the number of cells in which an undetected evader may be present never decreases when counted at the end of each round of movement. That is, there is a pursuer-winning search strategy such that the number of cleared cells is non-strictly monotonically increasing.*

For the proof of our next lemma, we require one more definition.

Definition 3. *The frontier is the set of cells from which an evader can enter a cell that is known not to contain an evader.*

Lemma 2. $B \subseteq A$

Proof. Consider an arbitrary pursuer-winning search strategy with speed s: $S_s \in \mathcal{B}$.

Ignoring for the moment the edges of the grid, then for any given number of *cleared* cells, the smallest frontier is realized by forming a contiguous region of *cleared* cells that is square. The frontier can be halved by placing this square in a corner such that only two sides of the square are exposed to the frontier. When the square is $\frac{n}{2} \times \frac{n}{2}$, the frontier will consist of n cells ($n + 1$ if the evader can move diagonally), and the pursuer must be able to cover at least this distance each turn to preserve monotonicity. Since $s \geq n$ ($s \geq n + 1$ if the evader can move diagonally), the pursuer has enough speed to execute the algorithms we used to prove the sufficient pursuer qualities [5], which are elements of $\mathcal{B}$ since thay are monotonic, but more importantly, are also elements of $\mathcal{A}$ since in each turn the pursuer visits each row at least once.

Alternatively, the *cleared* cells may be grown as a contiguous region contacting three edges of the board; in such a configuration, the frontier can never be fewer than $n - 1$ cells, and to preserve monotonicity, the pursuer must be able to visit each row in each turn; thus $S_s \in \mathcal{A}$. (Note that when there are more than $\frac{n^2}{4}$ cleared cells, a smaller frontier is realized by growing the *cleared* region contacting three edges than by growing the region as a square.)

As arbitrary strategy $S_s \in \mathcal{B}$ is also in $\mathcal{A}$, we conclude that $\mathcal{B} \subseteq \mathcal{A}$.

Conjecture 1. $\mathcal{A}$ is the set of all single-pursuer pursuer-winning search strategies; that is, $\mathcal{A} = \mathcal{S}$.

It is worth noting that it may not be possible to prove the correctness of any pursuer-winning search strategies which are not in $\mathcal{B}$. This, in part, is why we believe Conjecture 1.

Lemma 3. *Let s be a speed for which there is a pursuer-winning search strategy $S_1 \in \mathcal{A}$ for a single pursuer on an $m \times n$ board, where n is the shorter dimension. Then $s - 1$ is a speed for which there is a pursuer-winning search strategy $S_2 \in \mathcal{A}$ for a single pursuer on an $(m - 1) \times (n - 1)$ board.*

The proof of Lemma 3 may not be as obvious as that of Lemma 1; however, the intuition is that we have been defining pursuer-winning speeds in terms of the shorter dimension of the board; if the shorter dimension of the board is decreased by 1, then the pursuer's speed can also be decreased by 1. The full proof is available elsewhere [4]. Again, we consider the contrapositive of this lemma:

Corollary 2. *Let s be a speed for which there is not a pursuer-winning search strategy for a single pursuer on an $m \times n$ board, where n is the shorter dimension. Then $s + 1$ is not a speed for which there is a pursuer-winning search strategy for a single pursuer on an $(m + 1) \times (n + 1)$ board.*

We now intend to use an inductive argument. Before we do so, we need our base cases.

Lemma 4. *On a 2×2 board, a single pursuer with speed $s = 1$ does not have a pursuer-winning search strategy.*

Lemma 5. *On a 2×2 board, a single pursuer with speed $s = 2$ does not have a pursuer-winning search strategy when the evaders can move diagonally.*

We have demonstrated both of these base cases through model checking, and we have also proven the lemmas analytically [4]. We are now ready to prove the lower bounds on the minimum pursuer speed.

Theorem 1. *To catch all evaders on an $m \times n$ board, $\forall m, n \geq 2$, the pursuer's minimum speed is at least $\min(m, n)$ spaces/turn when using a search strategy $S \in \mathcal{A}$.*

Proof. The proof is inductive. For the basis, Lemma 4 states that the pursuer's minimum speed on a 2×2 board is at least $\min(2, 2) = 2$ spaces/turn. If $n - 2$ is insufficient for an $(n - 1) \times (n - 1)$ board, then by Corollary 2, $n - 1$ is insufficient for an $n \times n$ board. If $n - 1$ is insufficient for an $n \times n$ board, then by Corollary 1, $n - 1$ is insufficient for an $m \times n$ board, where $m \geq n$. Thus, for an $m \times n$ board, the minimum pursuer-winning speed is at least $\min(m, n)$.

Theorem 2. *To catch all evaders that can move diagonally on an $m \times n$ board, $\forall m, n \geq 2$, the pursuer's minimum speed is at least $\min(m, n) + 1$ spaces/turn when using a search strategy $S \in \mathcal{A}$.*

Proof. The proof is inductive. For the basis, Lemma 5 states that the pursuer's minimum speed on a 2×2 board is at least $\min(2, 2) + 1 = 3$ spaces/turn. If $n - 1$ is insufficient for an $(n - 1) \times (n - 1)$ board, then by Corollary 2, n is insufficient for an $n \times n$ board. If n is insufficient for an $n \times n$ board, then by Corollary 1, n is insufficient for an $m \times n$ board, where $m \geq n$. Thus, for an $m \times n$ board, the minimum pursuer-winning speed is at least $\min(m, n) + 1$.

Thus, we established the lower bounds on the minimum pursuer-winning speeds for a particular class of search strategies. Repeating our earlier sufficiency theorems:

Theorem 3. *To catch all evaders that cannot move diagonally on an $m \times n$ board, $\forall m, n \geq 2$, the pursuer's minimum speed is at most $\min(m, n)$ spaces/turn.*

Proof. Follows from Theorems 9 and 16 of Bohn & Sivilotti [6].

Theorem 4. *To catch all evaders that can move diagonally on an $m \times n$ board, $\forall m, n \geq 2$, the pursuer's minimum speed is at most $\min(m, n) + 1$ spaces/turn.*

Proof. Follows from Theorems 13 and 20 of Bohn & Sivilotti [6].

Theorem 5. *When the evaders cannot move diagonally, the minimum speed for which a single pursuer to be assured it can detect all evaders on an $m \times n$ board, $\forall m, n \geq 2$, is $\min(m, n)$ when using a search strategy $S \in \mathcal{A}$.*

Proof. Follows from Theorems 1 and 3.

Conjecture 2. When the evaders cannot move diagonally, the minimum speed for which a single pursuer to be assured it can detect all evaders on an $m \times n$, $\forall m, n \geq 2$, board is $\min(m, n)$.

Theorem 6. *When the evaders can move diagonally, the minimum speed for which a single pursuer to be assured it can detect all evaders on an $m \times n$ board, $\forall m, n \geq 2$, is $\min(m, n) + 1$ when using a search strategy $S \in \mathcal{A}$.*

Proof. Follows from Theorems 2 and 4.

Conjecture 3. When the evaders can move diagonally, the minimum speed for which a single pursuer to be assured it can detect all evaders on an $m \times n$ board, $\forall m, n \geq 2$, is $\min(m, n) + 1$.

We have shown the bounds are tight for all single-pursuer pursuer-winning search strategies in $\mathcal{A}$. We believe the bounds are tight for all single-pursuer pursuer-winning search strategies (that is, $\mathcal{A} = \mathcal{S}$) in large part because we have not found evidence to the contrary. We further believe the conjectures because they hold for smaller boards where the evaders have less freedom to move. Every pursuer-winning search strategy involves cornering the evaders; that is, reducing the places they can escape to. On an $m \times n$ board where $m, n \geq 2$, there will be four cells with two (or three) escapes, $2(m-2)+2(n-2)$ cells with three (or five) escapes, and $(m-2) \cdot (n-2)$ cells with four (or eight) escapes. If there is some board for which our conjecture does not hold, then on that board, the pursuer can detect all evaders at the same speed as on a smaller board, despite the greater number of ways the evaders can avoid detection.

5 Conclusions and Future Work

We have shown that heuristics exist that can reduce the size of the models and, by extension, reduce the time to obtain pursuer-winning search strategies. The heuristics we demonstrated generate pursuer-winning search strategies without loss in effectiveness. Clear-Columns demonstrated remarkable efficiency, though this is in large part because the structure of the solution is incorporated in the heuristic. Cleared-Bars and Cleared-Regions offer promise, but they are limited by Spin's memory reqauirements.

A necessary future direction of this research is obtaining an improved model checker. The nature of our technique requires a model checker that produces counterexample witnesses. The memory requirements of an explicit-state model checker demand that we use a symbolic model checker for our future research. SMV's model description language is ill-suited for the Cleared-Bars and Cleared-Regions heuristics (indeed, preparing models for these heuristics was very error-prone when using SMV — this was the reason we used Spin for these last two heuristics) and so we need a model checker that uses guarded commands in its description language. We can find many model checkers that satisfy any two of these requirements, but finding one satisfying all three remains elusive.

That problem dealt with, we shall then investigate other games. These may be variations of the current game, such as on-line coordination (for example, dealing with the loss of a pursuer, with a heterogeneous fleet of UAVs, or cooperation with a sensor net). Or these may be games on arbitrary graphs, as opposed to grids, such as might be found with sensor nets. Or they may even be completely new games; for example, we have a proof-of-concept model for a popular puzzle-solving board game.

References

1. NuSMV home page. http://nusmv.irst.itc.it/. Viewed 11 March 2005.
2. The SMV system. http://www.cs.cmu.edu/~modelcheck/smv.html. Viewed 11 March 2005.
3. Spin – formal verification. http://spinroot.com/. Viewed 11 March 2005.
4. C. Bohn. *In Pursuit of a Hidden Evader*. PhD thesis, The Ohio State University, 2004.
5. C. Bohn, P. Sivilotti, and B. Weide. Designing the control of a UAV fleet with model checking. In D. Grundel, R. Murphey, and P. Pardalos, editors, *Theory and Algorithms for Cooperative Systems*, chapter 2. World Scientific, 2004.
6. C. A. Bohn and P. A.G. Sivilotti. Upper bounds for pursuer speed in rectilinear grids. Technical Report OSU-CISRC-1/01-TR01, The Ohio State University, 2004.
7. E. M. Clarke, Jr., O. Grumberg, and D. A. Peled. *Model Checking*. The MIT Press, 1999.
8. M. R. Garey and D. S. Johnson. *Computers and Intractability: A Guide to the Theory of NP-Completeness*. W.H. Freeman and Company, 1979.

Unmanned Helicopter Formation Flight Experiment for the Study of Mesh Stability *

Elaine Shaw[1], Hoam Chung[1], J. Karl Hedrick[1] and Shankar Sastry[2]

[1] Department of Mechanical Engineering,
University of California Berkeley,
Berkeley, CA 94720, USA
E-mail: eshaw@vehicle.me.berkeley.edu, hachung@eecs.berkeley.edu,
khedrick@me.berkeley.edu
[2] Department of Electrical Engineering and Computer Science,
University of California Berkeley,
Berkeley, CA 94720, USA
E-mail: sastry@eecs.berkeley.edu

Summary. The authors have performed formation flights of UAVs using two of UC Berkeley's BEAR unmanned helicopters together in realtime with a simulated leader and six simulated helicopters. The goal of this experiment was to verify the mesh stability theory. The experimental results differ from the ideal theoretical results. Upon closer examination, this discrepancy is due to the effects of having a heterogeneous formation while the theory used is meant for homogeneous formations. While much work remains to be done, we can still show that using leader information in the control law is better than not using leader information. However, a new question arises regarding how one should define mesh stability for a heterogeneous mesh.

1 Introduction

Connective stability in one dimension, called string stability, has been studied by many sources including Chu [1], Eyre [3], Hedrick [5], and Swaroop [12, 13]. The generalization of string stability to two dimensions, called mesh stability, has been studied by Hedrick [4], Pant [7], and Seiler [8, 9]. A string/mesh stable interconnected system has the property of damping disturbances as the disturbances travel away from the source. Mesh stability is an important aspect of formation flight as it allows the possibility of large formations. Formation flight of UAVs is of great interest in recent years due to its application in areas such as surveillance and terrain mapping. In the near future, for-

* This work was supported by the Office of Naval Research's Autonomous Intelligent Network System Program under the grant N00014-99-10756.

mation flight theory can be applied to a swarm of micro-robotic insects for sensing applications.

To verify the theoretical mesh stability results, formation flight experiments are performed using the BErkeley AeRobot (BEAR) Rotorcraft-based Unmanned Aerial Vehicles (RUAVs) and virtual helicopters running simultaneously on computers in realtime.

The outline of the paper is as follows: We will briefly review some useful results in the analysis of string and mesh stability followed by the problem formulation. Then we will describe the experimental setup and scenario. Finally, we will present and discuss the results of the experiments.

Fig. 1. Snapshot of the formation flight experiment. The left one is Ursa Magna 2, and the right one is Ursa Magna 1.

2 Background

The theory and design methodology used for this experiment will be briefly summarized in this section.

2.1 String and Mesh Stability Theory

A summary of string and mesh stability theory can be found in Seiler [8] and is briefly presented below. We will use the norms $\|f(\cdot)\|_\infty = \sup_{t \geq 0} |f(t)|$ and $\|f(\cdot)\|_1 = \int_0^\infty |f(\tau)| d\tau$. A string of vehicles is called string stable if the ith error, ϵ_i, and the $(i+1)^{th}$ error, ϵ_{i+1}, in the chain satisfy $\|\epsilon_{i+1}\|_\infty \leq \|\epsilon_i\|_\infty$. For a linear system, if $y = h * u$, then Desoer [2] shows:

$$\|y(t)\|_\infty \leq \|h(t)\|_1 \|u(t)\|_\infty \tag{1}$$

Hedrick and Swaroop [5] found an LTI convolution kernel, $h(t)$, that relates the errors in a vehicle following chain by: $\epsilon_{i+1} = h * \epsilon_i$. The $\|h(t)\|_1$ represents the maximum amplification of any error as it propagates down the chain, and thus allowing us to determine the string stability of the chain of vehicles. If $\|h(t)\|_1 \leq 1$, then $\|\epsilon(t)_{i+1}\|_\infty \leq \|\epsilon(t)_i\|_\infty$. If $\|h(t)\|_1 > 1$, then the system is string unstable, and we can find an input error which will be amplified as it propagates down the chain. If $h(t)$ does not change sign, then the time domain string stability condition of $\|h(t)\|_1 \leq 1$ is equivalent to the frequency domain condition of $\|H(j\omega)\|_\infty \leq 1$. Hedrick and Swaroop [5] found that a system can be made string stable if reference vehicle information is used.

The SISO string stability results can be generalized to the MIMO case for mesh stability. We will use the norm $\|f(\cdot)\|_\infty = \max_i \sup_{t \geq 0} |f_i(t)|$. For an n-input, n-output linear system, if h(t) is the convolution kernel, and y=h*u, then Desoer [2] shows:

$$\|y(t)\|_\infty \leq (\max_i \sum_{j=1}^{n} \|h_{ij}(t)\|_1) \cdot \|u(t)\|_\infty \qquad (2)$$

As in the SISO case, if none of the $h_{ij}(t)$ changes sign, then

$$\|y(t)\|_\infty \leq (\max_i \sum_{j=1}^{n} \|H_{ij}(j\omega)\|_\infty) \cdot \|u(t)\|_\infty. \qquad (3)$$

2.2 Problem Formulation

Figure 2 shows the 3x3 grid formation used for the experiment. $X_{1,1}$ denotes the lead vehicle, while the others are follower vehicles. The goal is to force the

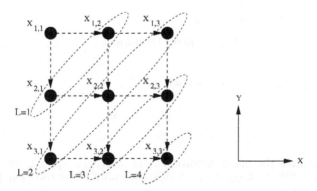

Fig. 2. 3×3 Mesh Schematic

mesh spacing errors to zero and to make sure that the disturbances on the mesh damp out instead of amplify. The analysis done in Seiler [8, 9] assume the following about the formation:

1. All the vehicles have the same model.

2. The vehicle model is linear.
3. The vehicle has the same number of inputs and outputs.
4. All vehicles use the same control law.
5. The desired spacings are constants independent of time and formation index.

Let $\mathbf{p}_{i,j}(t)$, a 4×1 vector, denote the outputs, $(x, y, z, \psi \equiv heading)$, of the $(i, j)^{th}$ vehicle. For the experiments without reference vehicle information, each follower vehicle gets the position information of its nearest top neighbor and its nearest left neighbor, as shown in Figure 2. For the experiments with reference vehicle information, each follower vehicle also gets the position information of the lead vehicle as well as from the two neighbors.

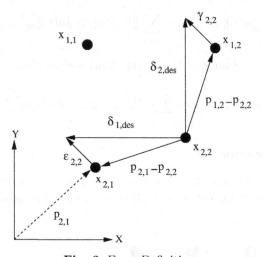

Fig. 3. Error Definitions

The mesh spacing errors are defined as:

$$\epsilon_{i,j} = \delta_{1,des} - (\mathbf{p}_{i,j-1} - \mathbf{p}_{i,j}) \tag{4}$$
$$\gamma_{i,j} = \delta_{2,des} - (\mathbf{p}_{i-1,j} - \mathbf{p}_{i,j}) \tag{5}$$

Figure 3 illustrates the error definitions. An averaged spacing error vector, $e_{i,j} = (\epsilon_{i,j} + \gamma_{i,j})/2$, is defined for analysis purposes. For the vehicles on the top and the left boundaries of the mesh, the averaged spacing error vector is defined as $e_{i,j} = \epsilon_{i,j}$ if $i = 1$ and $e_{i,j} = \gamma_{i,j}$ if $j = 1$.

By assumption 1) mentioned above, the formation exhibits symmetry. Performance "level sets", shown in Figure 2, can be defined by $L = i + j - 2$ as in Seiler [9]. Due to the formation symmetry, all vehicles in the same level set have the same error trajectories, and thus the same averaged spacing errors, i.e. $e_{1,2} = e_{2,1}$, $e_{1,3} = e_{2,2} = e_{3,1}$, and $e_{2,3} = e_{3,2}$. If a formation is mesh stable, then the errors for vehicles of level set $L = i$ are less than the errors for vehicles of level set $L = i - 1$. We will think in terms of level sets for the analyses presented in the next two sections in order to simplify the notations.

2.3 Design of Mesh Stability Controller

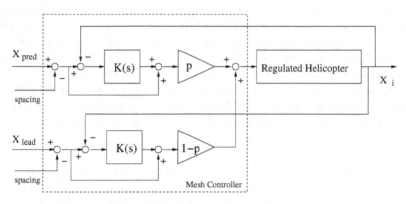

Fig. 4. Mesh Controller Structure

The design methodology for a mesh stability controller, as shown in Figure 4, can be found in Hedrick [4] and is briefly presented below. Assume that we are given a regulated helicopter model. Let $X_i \in \Re^4$ be the regulated outputs of this helicopter model and $X_i = H(s)X_{id}$.[3] Let X_{id}^p and X_{id}^l denote the desired positions of the ith vehicle with respect to the preceding and the lead vehicle, respectively. E_i^p and E_i^l are the corresponding errors. In order to have mesh stability, we must include the lead vehicle information, and so we implement:

$$X_{id} = p(K(s)E_i^p + X_{id}^p) + (1 - p)(K(s)E_i^l + X_{id}^l), \qquad (6)$$

where p is a scalar $\in [0, 1]$. This equation can be manipulated into the form:

$$E_i^p = (p[I + HK]^{-1}[H + HK])E_{i-1}^p \qquad (7)$$

Define:
$$G(s) \equiv [I + HK]^{-1}[H + HK] \qquad (8)$$

Let $g(t)$ be the impulse response of the individual transfer function entries of $G(s)$. Using the input-output relation described in section 2.1, we know that $p\|g(t)\|_1 \leq 1$ implies that $\|e(t)_i\|_\infty \leq \|e(t)_{i-1}\|_\infty$. The mesh controller can be designed by manipulating the parameters $K(s)$ and p. $K(s)$ should be designed to minimize $\|G(j\omega)\|_\infty$ while keeping $G(s)$ stable. The system will

---

[3] Uppercase letters represent the Laplace transform of time functions.

be mesh stable if we choose p such that $p \leq \frac{1}{\|g(t)\|_1}$. To avoid collisions, p should be kept as large as possible so that each vehicle is tightly coupled to its neighbors.

2.4 Robustness to Disturbances

Robustness analysis for the mesh controller in Hedrick [4] is briefly presented below. All the prior analysis assume linear regulated helicopter models, perfect tracking at steady state, and no external disturbances. In reality, helicopter dynamics are nonlinear, perfect tracking cannot be achieved, and external disturbances, such as wind gust, will act on the helicopters. We will show that a mesh stable system damps out the effects of all such disturbances.

Consider the given regulated helicopter model presented earlier with an added disturbance, giving $X_i = H(s)X_{id} + D_i(s)$. Using equation (6) and the same analysis as in the preceding section, we get the following equation:

$$E_i(s) = \hat{G}(s)E_{i-1}(s) + \bar{G}(s)(D_{i-1}(s) - D_i(s)), \qquad (9)$$

where $\hat{G}(s) = pG(s)$, $G(s)$ is given in equation (8), and $\bar{G}(s) = [I + HK]^{-1}$. Equation (9) shows us that disturbances are propagated via $\hat{G}(s)$. For example, suppose the second helicopter ($i = 2$) is affected by D_2. D_2 affects the error, E_2, through $\bar{G}$, i.e. $E_2 = \hat{G}E_1 + \bar{G}D_2$. Notice that D_2 also propagates to other errors through $\hat{G}$. For $i > 2$, $E_i = \hat{G}^{i-1}E_2 + \hat{G}^{i-3}\bar{G}D_3$. If a system is mesh stable, then $\|\hat{G}\|_\infty < 1$, and the effects of the disturbances will decay. If a system is mesh unstable, then $\|\hat{G}\|_\infty > 1$, and there exists disturbances which will cause the errors to grow.

3 Experimental Setup

The experimental system is made up of real and simulated helicopters working together in realtime.

3.1 Real Helicopters

BEAR UAV Testbed

The BEAR RUAVs used in the experiments are the Ursa Magna series, which are based on the Yamaha R-50 industrial helicopters. The two RUAVs have identical software and hardware structure, except for minor enhancements in Ursa Magna 1 and the control algorithms, which will be discussed later. MiLLennium RT-2 GPS from NovAtel and DQI-NP from Systron Donner make up the main navigation sensor system, and WaveLAN from Orinoco provides wireless communication links, one is between the vehicle and the monitoring station, and the other is for the differential GPS solution. For detailed descriptions about the hardware structure of BEAR RUAVs, please refer to Shim [10].

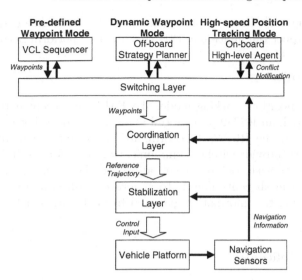

Fig. 5. Hierarchical flight management system of BEAR RUAVs

Hierarchical Flight Management System

A flight management system (FMS) for an intelligent UAV system should provide various flexible communication protocols for higher level agents, which can be on-board or off-board. In order to realize this feature, and to supply important functions for autonomous navigation of RUAVs, the BEAR group uses a hierarchical flight management system by Shim [11].

Figure 5 shows the current hierarchical structure of the FMS implemented on the Ursa Magna series. The hierarchical structure consists of high-level strategy planners, a coordination layer, a stabilization layer, and a physical vehicle platform. On the top level of the hierarchy, high-level strategy planners send waypoints to the lower layer, and the switching layer determines which planner takes the control. For this experiment, the on-board mesh stability controller is chosen as the high-level agent.

The stabilization layers of the two helicopters are different from each other. A multi-loop PID controller is used on Ursa Magna 2, see Shim [10], and a reinforcement learning controller is used on Ursa Magna 1, see Ng [6]. The multi-loop PID controller is implemented on the assumption of loosely coupled helicopter dynamics around hovering condition, while the reinforcement learning controller is more aggressively tuned and is trained for high velocity and complicated maneuvers. Here, "aggressive" does not mean only the size of the gains but also the consideration of the MIMO nature of helicopter dynamics. The multi-loop PID shows superior tracking performance at low velocities, while the reinforcement learning controller shows better performance when the reference trajectory requires simultaneous control of all channels, such as a circular trajectory. In the experiments, only slow reference trajectories are used, so we expect the multi-loop PID controller to perform better.

Mesh Controller

In high-speed position tracking mode, the FMS can accept external inputs at 10 Hz through an RS232 port. Stand alone mesh controller units are built and mounted on the RUAVs to coordinate the formation flight task. The mesh controllers provide wireless communication links for the RUAVs so they can communicate with each other and also with the virtual helicopters. The purpose of the mesh controllers is to generate mesh stable trajectories for the RUAVs using the information gathered from the other helicopters in the formation.

3.2 Virtual Helicopters

Since we only have access to two real helicopters, in order to perform experiments as shown in Figure 6, we will have to augment our fleet using virtual helicopters that are simulated in realtime. The lead vehicle, $X_{1,1}$, is simply a reference trajectory. Six additional virtual helicopters, with dynamics based on the real helicopters, will be used alongside the two real helicopters.

Realtime Helicopter Simulations

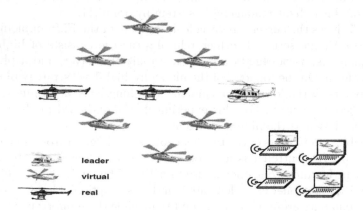

Fig. 6. Experimental Scenario

The helicopter model and controller described in Shim [10] is coded in C. This code can be run in realtime to simulate the virtual helicopters. Four laptops are used to simulate the virtual leader helicopter and six virtual follower helicopters (see Figure 6). One laptop acts as the virtual leader by broadcasting the desired trajectory to the rest of the formation. The remaining three laptops each simulates two virtual helicopters. Each helicopter simulation is composed of two processes. One process is in charge of simulating the helicopter dynamics, while the other process is in charge of the wireless communication.

Communication Between Helicopters

Each vehicle in the formation communicates wirelessly with the others at 10 Hz using an Orinoco 802.11b card. A token ring protocol is used to allow a large number of vehicles to communicate without packet collision.

3.3 Software Structure Outline

An outline of the software is shown in Figure 7. Each mesh controller is composed of two processes, and each virtual helicopter is also composed of two processes. The virtual leader is composed of only 1 process because it is just a trajectory with no dynamics. The whole setup for the mesh formation layer (i.e. not including the low level RUAVs regulation layer) is made up of 17 processes, each one running at 10Hz through the use of timer interrupts.

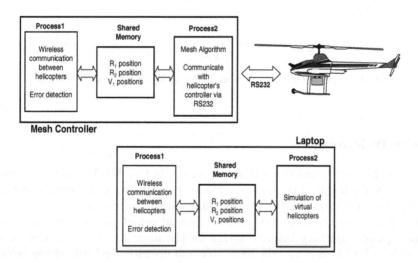

Fig. 7. Software Outline

3.4 Experimental Scenarios

In the experiments, $X_{2,2}$ and $X_{3,3}$ are the real helicopters, while the other ones are the simulated helicopters. $X_{1,1}$, also known as the virtual leader, traces out a trajectory, and the rest of the formation follows while trying to keep the spacings between them constant.

We performed two experiments. In the first experiment, no lead vehicle information is used, i.e. $p = 1.0$. In the second experiment, lead vehicle information is used, and the parameter p is chosen as described below in section 3.4. For comparison purposes, we first simulated the above scenarios using only the virtual helicopters.

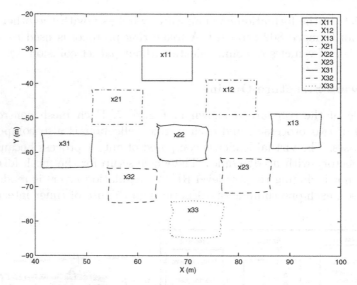

Fig. 8. Experimental Formation Trajectories: X versus Y, $p = 1.0$

Leader Trajectory

The virtual leader traces out a 10 meter square in the $X - Y$ plane, as shown in Figure 8. Each side of the square trajectory is constructed using a constant acceleration of $0.5m/s^2$, with the maximum velocity limited at $1m/s$, followed by a constant deceleration of $0.5m/s^2$. The lead vehicle pauses after each leg of the square trajectory before starting on the next leg. These pauses allow the vehicles farther down in the mesh to catch up with the lead vehicle before starting the next leg. This trajectory is chosen due to limited safe operating space.

Mesh Controller Parameters

A regulated helicopter model, $H(s)$, is constructed using the linear helicopter model and controller given by Shim [10]. Using the mesh stability controller design methodology described in section 2.3, we briefly checked what the p parameter should be. Since the stabilized helicopter model has reasonable frequency response, we decided not to design for the $K(s)$, see section 2.3. By setting $K(s) = 0_{4\times4}$, $G(s)$ is simply $H(s)$. Calculations show that $\|G(j\omega)\|_\infty = 1.217$ (achieved near 0.35 radians/second), so we know that if $p < 0.82$, then mesh stability is guaranteed. However, considering the nonlinearities and the uncertainties in the actual system, we simply choose $p = 0.5$ to make sure that the system will be mesh stable.

4 Results

4.1 All Virtual Helicopters Simulation Results

Table 1. Simulation: $\|e_{i,j}\|_\infty$ Without Leader Info ($p = 1$)

$\|e_{i,j}\|_\infty(meter)$	j=1	j=2	j=3
i=1	n/a	4.52	3.62
i=2	4.52	3.62	3.10
i=3	3.58	3.10	2.88

Table 2. Simulation: $\|e_{i,j}\|_\infty$ With Leader Info ($p = 0.5$)

$\|e_{i,j}\|_\infty(meter)$	j=1	j=2	j=3
i=1	n/a	4.53	1.82
i=2	4.48	1.83	0.80
i=3	1.82	0.79	0.37

For comparison purposes, we performed simulations using the identical experimental setup as in the actual experiments except with all virtual helicopters. These simulations are performed using all the assumptions presented in the theory section, i.e. constant spacings, identical vehicle models, and identical control laws. The infinity norm of the averaged spacing error vectors, as defined in section 2.2, are shown in Tables 1 and 2.

The results shown, which are performed on four laptops communicating wirelessly, are almost identical in structure to the simulation results obtained in Seiler [8]. The data exhibit performance "level sets", described in section

2.2, and the error for each $(i, j)^{th}$ vehicle is only a function of the level sets. The results simulated on the four laptops deviate slightly, i.e. $e_{1,2} \approx e_{2,1}$, $e_{1,3} \approx e_{2,2} \approx e_{3,1}$, and $e_{2,3} \approx e_{3,2}$, because the timing for the 17 processes, running at 10Hz each, across four laptops, cannot be completely synchronized.

Table 1 shows that, for the leader trajectory chosen, the spacing errors attenuate even if no reference vehicle information is used. The theory states that if a formation is mesh unstable, one can find an input that will cause the spacing errors to amplify, but not all inputs will cause the spacing errors to amplify. The very slow stabilized helicopter dynamics make it difficult to find a trajectory that will cause error amplification even though the formation without reference vehicle information is theoretically a mesh unstable formation.

Based on the simulation results, we expect the experimental errors to attenuate even if no lead vehicle information is used. However, we also expect that the addition of lead vehicle information will improve the performance of the mesh. Since the helicopters are not identical, we expect the actual results to be slightly different from the simulation results, especially with regards to the real helicopters $X_{2,2}$ and $X_{3,3}$. But our assumption is that the real helicopters' dynamics will be roughly the same as the simulated helicopters' dynamics since the helicopter model used is based on the real helicopters.

4.2 Real and Virtual Helicopters Experimental Results

Table 3. Experiment: $\|e_{i,j}\|_\infty$ Without Leader Info ($p = 1$)

$\|e_{i,j}\|_\infty (meter)$	j=1	j=2	j=3
i=1	n/a	4.57	3.65
i=2	4.52	3.73	3.20
i=3	3.64	3.23	1.69

Table 4. Experiment: $\|e_{i,j}\|_\infty$ With Leader Info ($p = 0.5$)

$\|e_{i,j}\|_\infty (meter)$	j=1	j=2	j=3
i=1	n/a	4.59	1.85
i=2	4.55	1.93	1.30
i=3	1.79	1.32	3.00

Tables 3 and 4 show the infinity norm of the averaged spacing errors obtained from the real helicopter experiments. When compared to the simulation results in Tables 1 and 2, we notice the loss in performance "level sets". While $e_{2,2}$ stays roughly the same as in the simulations, $e_{3,3}$ is quite different. We will attempt to explain this phenomenon in the next section.

Figures 9 and 10 show the positions of the helicopters as a function of time. Please note that the plots are shifted so that they all start at the same X (and Y) position, allowing us to compare the performances of the helicopters. Figures 11 and 12 show the averaged spacing errors as a function of time.

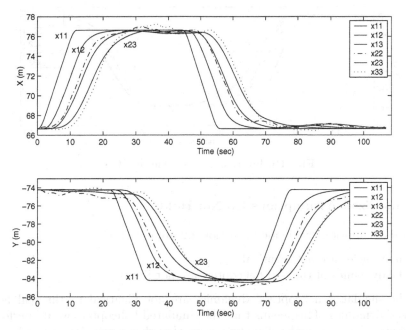

Fig. 9. Position versus Time, $p = 1.0$

5 Discussion of Results

5.1 Differences between Simulation and Actual Experiments

There are many reasons why the experimental results are different from the simulation results. Some of the reasons that intuitively come to mind are that the helicopters are not identical, there are external disturbances acting on the real helicopters, wireless communication packet loss, and imprecise GPS readings.

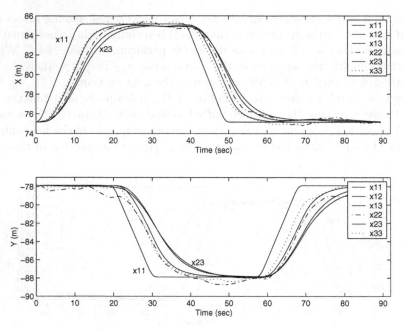

Fig. 10. Position versus Time, $p = 0.5$

The Idealized Assumptions Do Not Hold

Several assumptions we made in section 2.2 are not true.

- The vehicles are not identical.
- The dynamics of the real helicopters are not linear.

The simulated helicopters' dynamics are not identical to the real helicopters' dynamics. This means that the simulated helicopters would perform differently than the real helicopters, even though, as mentioned previously, our initial assumption is that their dynamics would roughly be the same. Due to our limited access to the experimental platform, we are not able to dictate which type of controllers the real helicopters use as different controllers are tried out on these helicopters constantly. As a result of this, the simulated helicopters use PD controllers as described by Shim [10], while the real helicopters use a PID controller and a reinforcement learning controller. Once again, we initially assume that this difference would not cause too much difference in the performance of the helicopters since only slow reference trajectories are used.

Looking at Figures 9 and 10, we noticed that both real helicopters have faster dynamics than the simulated helicopters. However, when we look at the infinity norm of the averaged spacing errors in Tables 3 and 4, $X_{2,2}$ have slightly larger errors than the ideal case in both experiments, while $X_{3,3}$ have

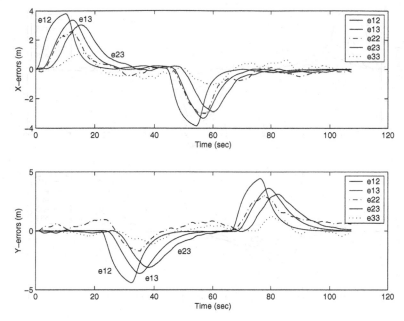

Fig. 11. Error versus Time, $p = 1.0$

errors that are smaller than the ideal without-leader-information case, and bigger than the ideal with-leader-information case.

We believe $X_{2,2}$'s errors can be explained by the external disturbances, imprecise GPS readings, and communication issues.

External Disturbances

The real helicopters are affected by external disturbances such as wind gusts, while the simulated helicopters are not affected. Suppose while the formation is traveling in the x direction, a wind gust pointing in the y direction blows on the real helicopters causing them to drift to one side, this alone can create the slightly larger errors that we noticed in $X_{2,2}$.

Wireless Communication Issues

Wireless communication packet losses also contribute to the deterioration in performance. If packets are not received by a particular helicopter, that helicopter will not move as much as if it had received the packets, thus causing errors. This effect increases if consecutive packets are lost. The simulations, performed in a laboratory environment with the laptops in close proximity to each other, have negligible packet loss rate. The actual experiments, performed outside in a field, have a worst case packet loss rate of 1.5 percent.

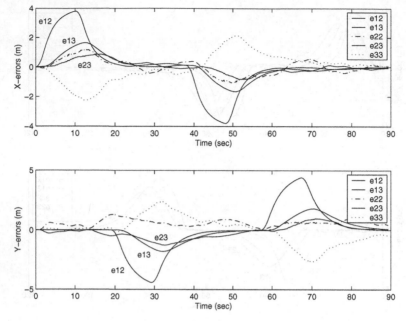

Fig. 12. Error versus Time, $p = 0.5$

Uncertainties in GPS Readings

The GPS does not produce good readouts at all times (see Shim [10]). Even when the GPS readings are good, the standard deviation of the readings is around $0.04m$ during the experiments.

While the errors of $X_{2,2}$ are quite similar to the ideal case and can be explained by the three issues mentioned above, the errors of $X_{3,3}$ are too different to have been caused by the above issues alone.

Effects of Heterogeneity

Even though we expect differences to exist, we did not expect the differences to be so large. The robustness analysis done in section 2.4 tells us that disturbances on a mesh stable formation will be damped out, so why does $X_{3,3}$ seem to perform worse when a mesh stable formation is used? Upon careful examination, it becomes obvious that $X_{3,3}$'s dynamics are much faster than the other helicopters (see Figures 10 and 12). The figures show that $X_{3,3}$ is moving so far ahead that its errors are opposite in sign from that of the other helicopters. $X_{3,3}$'s dynamics are sufficiently different from the rest of the helicopters, therefore the formation should be considered heterogeneous rather than homogeneous.

Using this information, new simulations are performed in MATLAB to imitate the experimental scenario, i.e. we improved the performance of the

Table 5. Simulation: $\|e_{i,j}\|_\infty$ Without Leader Info ($p = 1$), $X_{2,2}$ slower and $X_{3,3}$ much faster than others

$\|e_{i,j}\|_\infty(meter)$	j=1	j=2	j=3
i=1	n/a	3.00	2.80
i=2	3.00	2.91	2.60
i=3	2.80	2.60	0.97

Table 6. Simulation: $\|e_{i,j}\|_\infty$ ($p = 0.7$) $X_{2,2}$ slower and $X_{3,3}$ much faster than others

$\|e_{i,j}\|_\infty(meter)$	j=1	j=2	j=3
i=1	n/a	3.00	1.96
i=2	3.00	2.08	1.26
i=3	1.96	1.26	0.89

Table 7. Simulation: $\|e_{i,j}\|_\infty$ ($p = 0.4$) $X_{2,2}$ slower and $X_{3,3}$ much faster than others

$\|e_{i,j}\|_\infty(meter)$	j=1	j=2	j=3
i=1	n/a	3.00	1.12
i=2	3.00	1.25	0.38
i=3	1.12	0.38	1.67

$X_{3,3}$ helicopter and decreased the performance of $X_{2,2}$. Note that since we cannot simulate the real helicopters' exact dynamics, we cannot recreate the identical experimental scenario. The MATLAB simulation results, shown in Tables 5, 6, 7, 8, reaffirm the experiment results and fill in the missing gaps. Intuitively, if $X_{3,3}$ is much faster than the rest of the formation and $p = 1$, $X_{3,3}$'s mesh spacing errors will be much smaller than for the case when all the helicopters have identical dynamics. Even though the formation is already attenuating errors in the $p = 1$ case, adding leader information by using $p = 0.7$, making the formation mesh stable, causes the errors to damp out even faster. However, as more emphasis is placed on the leader's position information, as in the case of $p = 0.4$, $X_{3,3}$, in its attempt to follow the leader, starts to move much faster than its nearest neighbors and closes in on

Table 8. Simulation: $\|e_{i,j}\|_\infty$ ($p = 0.0$) $X_{2,2}$ slower and $X_{3,3}$ much faster than others

$\|e_{i,j}\|_\infty (meter)$	j=1	j=2	j=3
i=1	n/a	3.00	0.00
i=2	3.00	0.14	0.07
i=3	0.00	0.07	2.01

them, causing its errors to go up. In the limiting case, when all vehicles follow only the leader ($p = 0$), each vehicle is traveling as fast as it can to keep up with the leader without any clue as to where their neighbors are, $X_{3,3}$ shows its largest error because it has moved even closer to its neighbors, distorting the mesh.

For vehicles with different tracking performances, what we noticed is that even though using reference vehicle information can still improve the performance (in the sense that the maximum error magnitude decreases), placing too much emphasis on the lead vehicle information will eventually cause the performance to degrade again.

As previously discussed in section 2.3, even though a homogeneous formation is guaranteed to be mesh stable as long as we keep $p \leq \frac{1}{\|g(t)\|_1}$, in practice, one would want to make p as large as possible, i.e. $p = \frac{1}{\|g(t)\|_1}$, in order to keep each vehicle tightly coupled to its neighbors, decreasing the possibility of collisions if one of the vehicles fails to behave as expected. When one is simulating a homogeneous formation on a computer, it is easy to forget this fact since the more we decrease p, the better the overall performance (in the sense of smaller mesh spacing errors). However, in the real world, where vehicles can break down and fail in unexpected ways, and the formation may not be homogeneous, we quickly remember to follow this rule.

5.2 Mesh Stability Controller Improves Performance

Even though having a heterogeneous mesh makes it more difficult to design a mesh stability controller, the use of a mesh stability controller can still be beneficial, at least for the simple case we cover in this paper. By choosing the appropriate weight for the leader information, the magnitude of the mesh spacing errors can be decreased. Since these errors are the inputs into the mesh controller, this means that the control effort for a mesh stable formation is in turn smaller.

6 Conclusion

The results of the experiments are different than expected because the combination of real and simulated helicopters created a heterogeneous formation. Once the formation is not homogeneous, we must come up with a new definition for mesh stability because the definition involving "level sets" no longer applies. Despite having a heterogeneous formation, one can still show that the use of reference information improves the performance of the formation flight, although the balance between using the lead vehicle's information versus neighboring vehicles' information is much more delicate. The design of a mesh controller is no longer straight forward.

Obviously this is only one very simple heterogeneous formation scenario out of many possible scenarios. For further analysis on heterogeneous formations, please refer to Shaw [14].

Acknowledgments

The authors would like to thank Aniruddha Pant, Rosemary Huang, Marco Zennaro, Jusuk Lee, Xiao Xiao, Tony Mak, David Shim, Ron Tal, Perry Kavros, Richard Niedzwiecki, and Ted Phares.

References

1. Chu, Kai-ching. "Decentralized Control of High Speed Vehicular Strings." *Transportation Science,* p.361-384, 1974.
2. Desoer, C.A. and M. Vidyasagar. *Feedback Systems: Input-Output Properties.* Academic Press, 1975.
3. Eyre, J., D. Yanakiev, and I. Kanellakopoulos. "A Simplified Framework for String Stability Analysis of Automated Vehicles". *Vehicle System Dynamics,* 30(5):375-405, November, 1998.
4. Karl Hedrick, Aniruddha Pant, and Pete Seiler. "Mesh Stability of Helicopters." In *Proceedings of the 11th Yale Workshop on Adaptive and Learning Systems,* 2001.
5. Hedrick, J.K. and D. Swaroop. "Dynamic Coupling in Vehicles under Automatic Control." *13th IAVSD Symposium,* p.209-220, August 1993.
6. Andrew Y. Ng, H. Jin Kim, Michael I. Jordan, and Shankar Sastry. "Flying a Helicopter with Reinforcement Learning." (to be published).
7. Aniruddha G. Pant. *Mesh Stability of Formations of Unmanned Aerial Vehicles.* PhD thesis, University of California, Berkeley, 2002.
8. Pete Seiler, Aniruddha Pant, and J.K. Hedrick. "Preliminary Investigation of Mesh Stability for Linear Systems." In *Proceedings of the ASME: DSC Division,* volume 67, p.359-364, 1999.
9. Peter J. Seiler. *Coordinated Control of Unmanned Aerial Vehicles.* PhD thesis, University of California, Berkeley, 2001.

10. Hyunchul Shim. *Hierarchial Flight Control System Synthesis for Rotorcraft-based Unmanned Aerial Vehicles.* PhD thesis, University of California, Berkeley, 2001.
11. D.H. Shim, H.J. Kim, H. Chung, and S. Sastry, "Multi-functional Autopilot design and Experiments for Rotorcraft-based Unmanned Aerial Vehicles," *20th Digital Avionics Systems Conference,* Florida, 2001.
12. D. Swaroop. *String Stability of Interconnected Systems: An Application to Platooning in Automated Highway Systems.* PhD thesis, University of California, Berkeley, 1994.
13. D. Swaroop and J.K. Hedrick. "String Stability of Interconnected Systems." *IEEE Transactions on Automatic Control,* 41:349-357, March, 1996.
14. Elaine Shaw and J.Karl Hedrick. "String Stability Analysis for Heterogeneous Vehicle Strings." to be submitted to the *2007 American Control Conference.*

Cooperative Estimation Algorithms Using TDOA Measurements

Kenneth A. Fisher, John F. Raquet and Meir Pachter

Air Force Institute of Technology
E-mail: {Kenneth.Fisher,John.Raquet,Meir.Pachter}@afit.edu

Summary. A navigation algorithm using Time Difference of Arrival (TDOA) measurements obtained from signals of opportunity (SOPs) is developed. SOP-derived TDOA measurements are signals that are transmitted for purposes other than navigation (such as communication, telecasts, etc.) and are motived as appealing alternatives to GPS. In the scenario considered herein, the received SOPs are generated by asynchronous emitters at known locations. The measured TDOA are with reference to a reference receiver also at a known location. Although it would appear that the equations governing TDOA measurements, and consequently TDOA GDOP, are quite different from their GPS counterparts, it is shown that the TDOA measurement equations can be transformed into GPS pseudorange equations. This interpretation not only provides a direct evaluation of the TDOA GDOP and affords a characterization of the optimal measurement geometry, but, in addition, lets us fall back on well known GPS algorithms.

1 Introduction

Systems and applications have increasingly become dependent upon and rely on accurate position determination afforded by the Global Positioning System (GPS). GPS is a line-of-sight (LOS) system – that is, the satellites must be in "view" of the receiver antenna to receive the signal. Efforts are made to reduce this limitation, most notably in urban areas and indoors. Urban areas are characterized by tall buildings, which block satellites from view and create multipath signals. Indoors, the signals are present but greatly attenuated, so that it is difficult to get a reliable GPS position solution. The demand for accurate position information is growing. When GPS solutions are unattainable, another method with comparable accuracy is desired.

Alternatives to aid in navigation without GPS or to improve navigation in urban areas and indoors include inertial navigation [12], LORAN [5], and signals of opportunity [2, 3]. Signals of opportunity (SOP) are transmitted for non-navigation purposes; however, clever techniques can exploit these signals for navigation. SOPs are convenient sources of navigation for several reasons. First, SOPs are abundant, which is useful in ensuring sufficient signals are available for position determination and for reducing the position error. Second, the signal-to-noise ratio is often higher than for signals such as GPS [2]. Finally, there are no deployment costs or operating expenses related to the signal for the navigational user. (Of course, there are navigation equipment costs incurred by the user).

The use of signals of opportunity for navigation has been suggested by Hall in [2, 3]. In this paper, a navigation algorithm using Time Difference of Arrival (TDOA) measurements obtained from SOPs is developed. In the scenario considered herein the received SOPs are generated by asynchronous emitters at known locations. The measured TDOA are with reference to a reference receiver also at a known location. Using a stationary and a mobile receiver, a time difference of arrival (TDOA) measurement is formed as the difference in received time of a single SOP received at the stationary receiver compared to the received time at the mobile receiver. Multiple TDOA measurements can be formed using multiple SOP. In this paper, it is shown that TDOA measurements from SOP can be treated as GPS psuedorange measurements. The derived positioning algorithm is an elegant adaptation of the conventional positioning algorithm used in GPS positioning and which operates on pseudorange measurements. Subsequent position solution and geometry considerations follow directly from current GPS techniques.

The paper is organized as follows. Section 2 presents the *formation* of measurements from SOP. TDOA measurements are motivated, and the general measurement equation is derived. Section 3 presents the *use* of TDOA measurements from SOP for navigation. It is shown that the navigation algorithm parallels that of the GPS system with minor differences. Concluding remarks are made in Section 4.

2 Measurements from Signals of Opportunity

SOPs can be exploited for navigation in several ways. Hall [2, 3] used two receivers, one at a known location and one at the location to be determined, to track the phase of an incoming amplitude modulation (AM) radio station signal and was able to obtain GPS-like accuracy under certain restrictions. Rosum Corporation [8, 9] uses analog and digital television signals. Measurements are formed by comparing the incoming TV signal with a generated reference signal. Timing information is provided by a second receiver at a known location. GPS-like accuracy is obtained indoors in specific coverage areas. This paper proposes using two receivers and TDOA measurements. As shown in Fig. 1, one receiver, the base station or reference receiver, is at a known location while the other receiver, the rover, is at a location to be determined. The SOP transmitter location is assumed to be known. Fixed

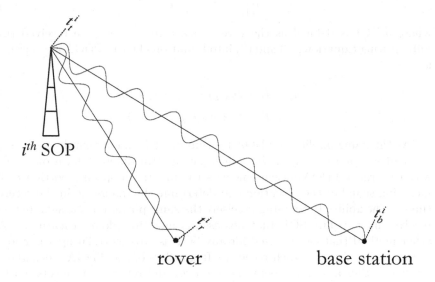

Fig. 1. TDOA Measurement of ith SOP

SOP transmitter locations (such as fixed towers) can be determined a priori through surveying. Moving SOP transmitter locations can be predicted (such as space-based, orbiting transmitters) or determined with additional base stations in a separate, simultaneous algorithm [4]. Each TDOA measurement is formed as the difference in received time of the SOP at the base station compared to the received time of the SOP at the rover. The TDOA measurement of the ith SOP at the base station compared to the rover, δ_{br}^i, can be written as

$$\delta_{br}^i = t_b^i\big|_b - t_r^i\big|_r \tag{1}$$

where $t_b^i\big|_b$ is the received time of the ith SOP at the base station according to the base station clock and $t_r^i\big|_r$ is the received time of the ith SOP at the rover according to the rover clock. The $\big|_b$ and $\big|_r$ denote that the time is measured by the base station and rover, respectively, and allow for imperfect and unsynchronized base station and rover clocks. The received time of the ith signal at the base station according to the base station clock, $t_b^i\big|_b$, is related to the true received time of the ith signal at the base station, t_b^i, by

$$t_b^i\big|_b = t_b^i + \delta_b\left(t_b^i\right) \tag{2}$$

where $\delta_b\left(t_b^i\right)$ is defined as the base station clock error at the true received time. Likewise, the received time of the ith signal at the rover according to the rover clock, $t_r^i\big|_r$, is related to the true received time of the ith signal at the rover, t_r^i, by

$$t_r^i\big|_r = t_r^i + \delta_r\left(t_r^i\right) \tag{3}$$

where $\delta_r\left(t_r^i\right)$ is defined as the rover clock error at the true received time. Substituting Equations (2) and (3) into Equation (1), the TDOA measurement is

$$\delta_{br}^i = t_b^i + \delta_b\left(t_b^i\right) - \left[t_r^i + \delta_r\left(t_r^i\right)\right] \tag{4}$$

$$\delta_{br}^i = t_b^i - t_r^i + \delta_b\left(t_b^i\right) - \delta_r\left(t_r^i\right) \tag{5}$$

As the name implies, a TDOA measurement is fundamentally a measurement of a signal's received time difference at two different locations[1]. One way to form a TDOA measurement is to time-tag a specific portion of the incoming signal at each location and determine the difference in the received times. The ability to time-tag precisely the same portion of the signal at each receiver may require SOP that possess distinct time domain features. Another method that can be used for any SOP is cross-correlating a portion of the incoming signal at each receiver. Regardless of the TDOA measurement method, either a datalink between the rover and reference receivers must be in place for near-real time operation, or data must be stored for subsequent post-processing.

There are many advantages in forming TDOA measurements from SOP. One advantage of TDOA measurements is the transmit time does not need to be known (or determined) to obtain a solution. Consequently, there are no timing requirements placed upon the transmitter for navigation purposes. (A clock may be necessary for determining transmitter frequencies governed by the Federal Communications Commission or other system requirements.) Also, the two receivers do not need to keep accurate time nor be time-synchronized. Finally, TDOA measurements can be formed from a wide range of signals. For example, one can obtain TDOA measurements from AM radio stations, FM radio stations, analog television stations, digital television stations, and cellular tower transmissions. All the measurements can be used in a weighted solution algorithm in a straightforward manner. The only difference is in how the TDOA measurement for each signal type is obtained. Once the TDOA measurements are taken, the herein developed positioning algorithm is applied.

3 Using Measurements from Signals of Opportunity

SOP-derived TDOA measurements have been motived as appealing alternatives to GPS; and the TDOA measurement equation has been given in Equation (5). Although it would appear that the equations governing TDOA measurements, and consequently TDOA GDOP, are quite different from their

---

[1] Not used here, TDOA can also be defined as the difference in received times of two different signals at the same location.

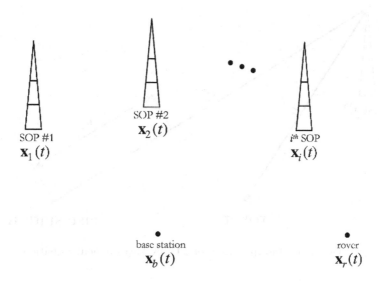

Fig. 2. TDOA Transmitter and Receiver Scenario

GPS counterparts, this section shows that the TDOA measurement equations can be transformed into GPS pseudorange equations. Because SOP are transmitted for purposes other than navigation and are normally controlled by some entity other than the navigation user, in general, the transmit time of a signal is unknown to the navigational user. Furthermore, each SOP transmitter is not synchronized to any of the other transmitters. For example, a local radio station's broadcast is not generally synchronized to other radio stations. A situation may nevertheless occur when transmitters *are* synchronized. The assumption here does not take advantage of this additional constraint in order to encompass a wider range of SOP.

Figure (2) shows the scenario considered. The location of the ith SOP transmitter for $i = 1, 2, \cdots, N$ at time t is assumed to be known as $\mathbf{x}_i(t)$. The equipment needed consists of two receivers: a base station at a known location, $\mathbf{x}_b(t)$, and a rover unit at the location to be determined, $\mathbf{x}_r(t)$. Each receiver may be stationary or moving. The base station location, $\mathbf{x}_b(t)$, may be at a fixed, surveyed site or determined using an available navigation system such as GPS.

The TDOA measurement of the ith signal to the base station relative to the receiver, δ_{br}^i, was given in Equation (5). Converting travel times into distances by multiplying by the speed of propagation of the signal, c, Equation (5) becomes

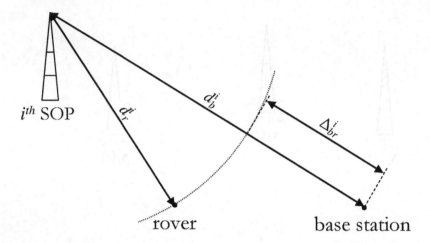

Fig. 3. Geometric Interpretation of TDOA Measurement Equation

$$\Delta_{br}^i \triangleq c\delta_{br}^i = ct_b^i - ct_r^i + c\delta_b\left(t_b^i\right) - c\delta_r\left(t_r^i\right)$$
$$= \left(ct_b^i - ct_t^i\right) - \left(ct_r^i - ct_t^i\right) + \left[c\delta_b\left(t_b^i\right) - c\delta_r\left(t_r^i\right)\right]$$
$$= d_b^i - d_r^i + \left[c\delta_b\left(t_b^i\right) - c\delta_r\left(t_r^i\right)\right] \tag{6}$$

where t_t^i is the unknown transmit time of the ith SOP, $c\delta_{br}^i \triangleq \Delta_{br}^i$, and d_b^i and d_r^i are the distances from the ith SOP transmitter to the base station and rover, repectively. Figure (3) shows the relationship of and d_b^i, d_r^i and Δ_{br}^i. Using this insight, Equation (6) is rearranged as

$$d_r^i - \left[c\delta_b\left(t_b^i\right) - c\delta_r\left(t_r^i\right)\right] = d_b^i - \Delta_{br}^i \tag{7}$$

Equation (7) represents the measurement equation for a single SOP, where Δ_{br}^i is the TDOA measurement (in distance), d_b^i is the known distance from the base station to the ith SOP, $\left[c\delta_b\left(t_b^i\right) - c\delta_r\left(t_r^i\right)\right]$ is the unknown difference in the clock errors of the base station and the rover (expressed in units of distance), and d_r^i is the unknown distance from the rover to the ith SOP.

Using N SOP transmitters, N measurements can be taken. For three dimensional positioning, $N \geq 4$ is required. In general, each of the terms in Equation (7) with the superscript i vary as the SOP transmitter varies. However, the time the ith SOP is received at the rover, t_r^i, can be *selected* by the user by choosing when the SOP is considered "received" at the rover. Using multiple channels, each of the N SOPs can be received simultaneously, or

$$t_r^1 = t_r^2 = \cdots = t_r^N \triangleq t_r$$

It follows that the rover clock error remains the same for each SOP and can be denoted as $\delta_r\left(t_r\right)$.

The time at which the ith SOP is received at the base station, t_b^i, may take on values between $t_r - \frac{B}{c}$ and $t_r + \frac{B}{c}$, or

$$t_b^i \in \left[t_r - \frac{B}{c}, t_r + \frac{B}{c} \right]$$

where B is the distance between the base station and the rover and c is the speed of propagation. If B is constrained such that the base station clock drift is sufficiently small over the possible range of base station received times, then

$$\delta_b(t) \approx k \qquad \forall t \in \left[t_r - \frac{B}{c}, t_r + \frac{B}{c} \right] \tag{8}$$

where k is a constant. Furthermore, Equation (8) holds for each SOP, so that

$$\delta_b\left(t_b^1\right) \approx \delta_b\left(t_b^2\right) \approx \cdots \approx \delta_b\left(t_b^N\right) \triangleq \delta_b(t_b)$$

Finally, replacing $\delta_r\left(t_r^i\right)$ with $\delta_r(t_r)$ and $\delta_b\left(t_b^i\right)$ with $\delta_b(t_b)$, Equation (7) becomes

$$d_r^i - [c\delta_b(t_b) - c\delta_r(t_r)] = d_b^i - \Delta_{br}^i \tag{9}$$

Equation (9) parallels the GPS pseudorange equation, where $d_b^i - \Delta_{br}^i$ is the ith pseudorange measurement, d_r^i is range from the rover to the ith GPS transmitter (or satellite), and $[c\delta_b(t_b) - c\delta_r(t_r)]$ is a bias term constant over all N measurements. Thus, TDOA navigation parallels GPS navigation with a user, a local differential receiver, and the GPS satellites as transmitters. The difference is that in conventional GPS the bias term represents the user clock error, whereas for SOP TDOA measurements the bias term represents the *difference* in the clock errors of the base station and rover. Hence, standard GPS algorithms, such as one presented by Misra and Enge [5], or closed-form solutions such as in [7], can be used to solve for the rover position and the difference in clock errors of the base station clock and rover clock. Note that the algorithm using SOP TDOA measurements cannot be used to estimate true time, since the bias found represents a difference in clock errors and not a single clock error compared to the true time.

Similar to GPS navigation, dilution of precision considerations (a.k.a., observability) also apply and can be addressed in a manner consistent with GPS techniques [5]. In this respect and leveraging GPS knowledge, it may be advantageous to constrain the solution to 2-D positioning while at the same time taking advantage of having an overdetermined system–barring the fortunate circumstance where one of the TDOA receivers is overhead the source. Moreover, it is suggested here to redifine the GDOP metric, namely, consider the condition number of the regressor at the point of convergence of the ILS algorithm. Indeed, both the TDOA and GPS algorithms employ ILS, and for both we propose to transition to the condition number-based characterization of GDOP. Only then is one justified bounding the position error by GDOP $*$ (Measurement Error) [6]. Indeed, when a linear system

$\mathbf{z} = \mathbf{H}\theta$ is solved, the relative error of the estimate is related to the relative measurement error according to:

$$\frac{\|\varDelta\theta\|_2}{\|\theta\|_2} \leq \kappa\{\mathbf{H}\}\frac{\|\varDelta\mathbf{z}\|_2}{\|\mathbf{z}\|_2} \tag{10}$$

where $\kappa\{\cdot\}$ is the condition number of the matrix.

Finally, an example in using SOP for navigation is considered in which an optimal GDOP can be realized. It has been suggested to use TDOA measurements from x-ray pulsars for relative positioning [1, 10, 11]. Without loss of generality, let the reference station be located at the origin; the desired receiver is located at $\mathbf{r} \in \mathbb{R}^3$. The ith emitter is located a distance near ∞ in the known direction, $\mathbf{d}_i$, where $\|\mathbf{d}_i\|_2 = 1$. Finally, assume that ambiguity resolution is acheived. Under these conditions, the measurement equations are

$$z_i \triangleq (\mathbf{d}_i)^T \mathbf{r} + v_i \tag{11}$$

for $i = 1, 2, \cdots, N$ where $N \geq 3$ and v_i is zero-mean, white, Gaussian noise with a covariance σ^2. Now,

$$\mathbf{z} = \mathbf{H}\mathbf{r} + \mathbf{v} \tag{12}$$

where $\mathbf{z} \triangleq [\,z_1\ z_2\ \cdots\ z_N\,]^T$, $\mathbf{v} \triangleq [\,v_1\ v_2\ \cdots\ v_N\,]^T$, and $\mathbf{H}$ is the regressor matrix given as

$$\mathbf{H} = \begin{bmatrix} (\mathbf{d}_1)^T \\ (\mathbf{d}_2)^T \\ \vdots \\ (\mathbf{d}_N)^T \end{bmatrix}_{N\times 3} \tag{13}$$

Notice that

$$\mathbf{H}^T\mathbf{H} = \sum_{i=1}^{N} \mathbf{d}_i(\mathbf{d}_i)^T \tag{14}$$

Thus, if $N = 3$ and the vectors $\mathbf{d}_i$ form an orthonormal triad, then

$$\mathbf{H}^T\mathbf{H} = \mathbf{I}_{3\times 3} \tag{15}$$

$$(\mathbf{H}^T\mathbf{H})^{-1} = \mathbf{I}_{3\times 3} \tag{16}$$

and the condition number is given as

$$\kappa\{(\mathbf{H}^T\mathbf{H})^{-1}\} = 1 \tag{17}$$

If the direction vectors are not orthonormal, without loss of generality, they can be expressed as

$$\mathbf{d}_1 = \begin{bmatrix} 1 \\ 0 \\ 0 \end{bmatrix}, \qquad \mathbf{d}_2 = \begin{bmatrix} \cos\varphi \\ \sin\varphi \\ 0 \end{bmatrix}, \qquad \mathbf{d}_3 = \begin{bmatrix} \cos\theta\cos\psi \\ \cos\theta\sin\psi \\ \sin\theta \end{bmatrix} \tag{18}$$

Now,

$$\mathbf{H}^T\mathbf{H}= \begin{bmatrix} \begin{pmatrix} 1+\cos^2\varphi \\ +\cos^2\theta\cos^2\psi \end{pmatrix} & \begin{pmatrix} \sin\varphi\cos\varphi \\ +\cos^2\theta\sin\psi\cos\psi \end{pmatrix} & \sin\theta\cos\theta\cos\psi \\[2em] \begin{pmatrix} \sin\varphi\cos\varphi \\ +\cos^2\theta\sin\psi\cos\psi \end{pmatrix} & \begin{pmatrix} \sin^2\psi \\ +\sin^2\psi\cos^2\theta \end{pmatrix} & \sin\theta\cos\theta\sin\psi \\[2em] \sin\theta\cos\theta\cos\psi & \sin\theta\cos\theta\sin\psi & \sin^2\theta \end{bmatrix} \quad (19)$$

and

$$\det\left(\mathbf{H}^T\mathbf{H}\right) = \\ = \sin^2\theta\sin\varphi\left(\sin\varphi + 2\sin\varphi\cos^2\varphi + 2\cos\varphi\cos^2\theta\sin\psi\cos\psi\right) \quad (20)$$

$$Trace\left\{\left(\mathbf{H}^T\mathbf{H}\right)^{-1}\right\} = \frac{\begin{pmatrix} \cos^2\theta\left(\cos 2\psi - \cos\left[2\left(\psi-\varphi\right)\right]\right) \\ +2\sin^2\varphi + 4\sin^2\psi\cos^2\theta \end{pmatrix}}{2\sin^2\theta\sin\varphi\begin{pmatrix} \sin\varphi + 2\sin\varphi\cos^2\varphi \\ +2\cos\varphi\cos^2\theta\sin\psi\cos\psi \end{pmatrix}} \quad (21)$$

4 Conclusion

The use of SOP, including x-ray pulsars-based stellar navigation, provides an alternative, and/or augmentation, to current navigation methods such as GPS or INS. A navigation senario using SOP exploitation was described, and the assumptions about SOP were clearly stated. The time difference of arrival (TDOA) measurement process applied to SOP signals was described, and it was shown that SOP TDOA measurements can be processed with existing GPS algorithms to solve for user positions and the rover clock error relative to the base station clock error. Furthermore, existing GDOP techniques (including the "trace" method and the "condition number" method) used for GPS measurements are applicable to TDOA formulation given herein. Furthermore, SOP-based TDOA navigation enables positioning systems where none of the transmitters or receivers require precise clocks, nor do they need to be synchronized.

5 Disclaimer

This is declared work of the U.S. Government and is not subject to copyright protection in the United States. The views expressed in this article are those of the authors and do not reflect the official policy of the United States Air Force, Department of Defense, or the US Government.

References

1. Downs, G. "Interplanetary Navigation Using Pulsating Radio Sources," NASA TR N74-34150, October 1974
2. Hall, T. D., C. C. Counselman, and P. Misra. "Instantaneous Radiolocation Using AM Broadcast Signals," *Proceedings of ION-NTM*, Long Beach, CA, pp. 93-99, January 2001.
3. Hall, T. D. "Radiolocation Using AM Broadcast Signals," Ph.D. Dissertation, Massachusetts Institute of Technology (MIT), September 2002.
4. Mellen, G., M. Pachter, and J. Raquet. "Closed-Form Solution for Determining Emitter Location Using Time Difference of Arrival Measurements," IEEE Transactions on Aerospace and Electronic Systems, Vol. 39, No. 3, pp. 1056-1058, July 2003.
5. Misra, P. and P. Enge. "Global Positioning System: Signals, Measurements, and Performance," Ganga-Jamuna Press, Lincoln, MA, 2001.
6. Pachter, M. and J. B. McKay. "Geometry Optimization of a GPS-Based Navigation Reference System," Journal of the Institute of Navigation, Vol. 44, No. 4, pp. 457-490, Winter 1997-1998.
7. Pachter, M. and T. Nguyen. "An Efficient GPS Position Determination Algorithm," Journal of the Institute of Navigation, Vol. 50, No. 2, pp. 131-141, Summer 2003.
8. Rabinowitz, M. and J. Spilker. "Positioning Using the ATSC Digital Television Signal," Rosum Corporation Whitepaper, August 2001.
9. Rabinowitz, M. and J. Spilker. "The Rosum Television Positioning Technology," *ION 59th Annual Meeting/CIGTF 22nd Guidance Test Symposium*, Albequerque, NM, pp. 528-541, June 2003.
10. Sheikh, S. I., et al. "The Use of x-Ray Pulsars for Spacecraft Navigation," presented at the 14th AAS/AIAA Space Flight Mechanics Conference, Paper #04-109, Maui, Hawaii, February, 2004.
11. Taylor, J. and M. Ryba. "High Precision Timing of Millisecond Pulsars," The Astrophysical Journal, Vol. 371, 1991.
12. Titterton, D. H. and J.L. Weston. "Strapdown Inertial Navigation Technology," IEE Books, Peter Peregrinus Ltd, UK, 1997.

A Comparative Study of Target Localization Methods for Large GDOP

Harold D. Gilbert, Daniel J. Pack and Jeffrey S. McGuirk

Department of Electrical and Computer Engineering
United States Air Force Academy USAF Academy, CO 80840-6236
E-mail: {harold.gilbert,daniel.pack,Jeffery.mcguirk}@usafa.af.mil

Summary. In this chapter, we present a comparative study on two algorithms to localize ground targets using Unmanned Aerial Vehicles (UAVs): an angle-of-arrival (AOA) emitter-location algorithm using triangulation techniques and an angle-rate algorithm. In particular, we focus on the performance of the two algorithms locating targets when a sensor platform is under a large Geographic Dilution of Precision (GDOP) condition. The large GDOP condition occurs when a target is seen by a sensor platform within a small included angle; the total included angle between Line-Of-Bearings (LOBs) is less than five degrees. The comparative study is a part of the United States Air Force Academy's Unmanned Aerial Vehicles (UAVs) research project to develop a group of cooperative UAVs to search, detect, and localize moving ground targets. The GDOP conditions limit the accuracy of the AOA triangulation-emitter-location algorithm's accuracy due to the resulting highly elliptical probable error. In such cases, angle-rate algorithms should be used for better localization accuracy. Usually, a large GDOP condition is encountered during two important operational applications: (1) tasks that use slow-moving-sensor platforms, such as a small UAV, and (2) tasks involving short-up-time emitters that typically are not transmitting signals long enough for any sensor platform to open more than a small total included angle. We investigate the performance of the two algorithms as we vary the included angle for the sensor platform. The performances of angle-rate and triangulation algorithms are compared via MATLAB simulation to determine the preferred regions of operation.

1 Introduction

Ever since aviation became available for military use, developing techniques to accurately locate targets has been the pursuit of numerous researchers. Ideally, localization of a target becomes a straight forward procedure if we can accurately sense the distance of a target from two or three distributed sensor platforms, and if we can compute the positional relationships between the sensor platforms. Typically, the sensed range information is corrupted with noise, making accurate platform position difficult. To further complicate the task, some targets, such as time-varying Radio Frequency (RF) signal emitters, can be seen only within a small time window.

Over the past decade, locating targets cooperatively using multiple sensor platforms generated sizable interests among engineers and scientists in a variety of research communities [1,2,3]. The UAV community, in particular, has been actively advancing the state of this technology [4,5,6]. It is within this context that we are developing a cooperative UAV system at the US Air Force Academy [7]. We have developed a multiple UAV system graphical simulator, and we are in the process of acquiring and constructing hardware for a flying fleet of UAVs.

In our applications, multiple UAVs are working together to detect, locate and exploit radiating electromagnetic emitters. The first UAV to detect a target needs to rapidly estimate an approximate emitter location. This information is passed to other UAVs that will participate in a cooperative localization task. The initial estimation of the target location is important since it helps other UAVs to plan their trajectories to approach the target in an optimal fashion . Once a set of UAVs are within the sensor range from a target, they can collectively and accurately localize that target. The objective of the current work is to compare two types of algorithms based on the principles of angle-rate and AOA triangulation. The results of the comparative study will be used to properly select a technique to increase the target localization accuracy for the first UAV that detects a target.

2 Background

For several decades, passive sensors have been deployed to locate unknown electromagnetic sources (emitters). As shown in Figure 2, three basic techniques allow passive localization of stationary-based emitters from airborne platforms [8]:

1. The azimuth triangulation method where the intersection of successive spatially displaced bearing measurements provides the emitter location;
2. The azimuth/elevation location method that provides a single-pulse instantaneous emitter location from the intersection of the measured azimuth/elevation bearing with the earth's surface; and
3. The time difference of arrival (TDOA) method that computes the unknown emitter location from the measured difference in time of arrival on a single pulse at three spatially remote locations.

Small, inexpensive, UAVs are normally applied to applications that may be solved by the first method described above, because of their limited payload capability. This paper considers applications that employ an azimuth angle-of-arrival (AOA) sensor. Typical sensors are collinear phase interferometers or one-dimensional rotating directional antennas. The measured relative direction-of-arrival (DOA) of the received electromagnetic signal is converted

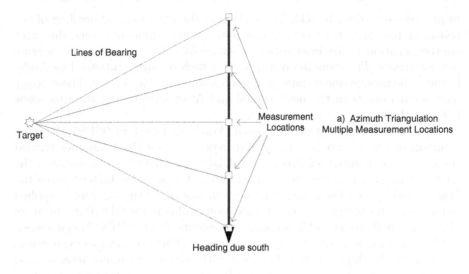

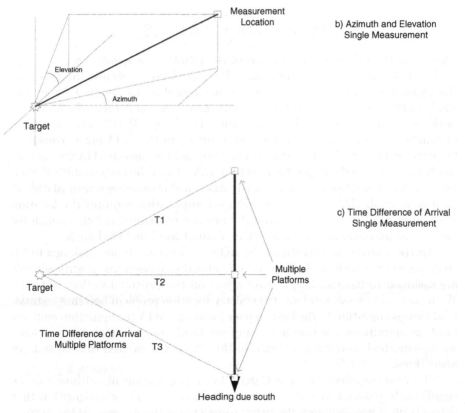

Fig. 1. Passive Location Techniques for Stationary Emitters

to an absolute azimuth AOA by considering the attitude and heading of the collection platform. For the ideal case with no measurement errors, the exact emitter location is the intersection of successive spatially displaced bearing measurements. The more interesting case, which requires statistical methods, is when the measurement data contains random and bias errors. These errors may occur from both the navigational and AOA sensors and the time skew between them.

Since the deployment of the Global Positioning System (GPS), the most common metric to express the satellite geometry is a single number termed the Geometric Dilution of Precision (GDOP) [9]. GDOP is a measure of the spatial distribution of the observed satellites. The smallest GDOP yields the best statistical location accuracy for a given system. If this concept is applied to the azimuth triangulation emitter location problem, the GDOP is a measure of the spatial displacement bearing measurements. A large GDOP occurs when the included angle is small. Hence, for large GDOP the measurement errors yields a highly elliptical probable error with major and minor axes aligned along-range and cross-range, respectively. Since we are considering the two-dimensional azimuth AOA systems, the GDOP is sometimes referred to as HDOP (horizontal DOP) [9]. The generic GDOP notation is used in the paper.

$$GDOP = \frac{\sqrt{\sigma_{\max}^2 + \sigma_{\min}^2}}{\sigma_{\min}} \geq \sqrt{2} \tag{1}$$

where σ is the major and minor axis of the elliptical probable error.

For UAV applications employing GPS navigation systems, the effect on the emitter location accuracy from the positional and time skew errors are small when compared to the typical AOA measurement errors, which include both the heading and AOA sensor errors. For large GDOP cases, the line-of-bearings (LOBs) are almost parallel resulting in the AOA errors causing a large variation in their intersections. For example, the measured LOBs may actually intersect on the opposite side of the UAV from the actual emitter! Such large GDOP conditions normally occur when a slow-moving-sensor platform, such as a small UAV, is used, or when an application requires the location of short-up-time emitters that typically are not transmitting long enough for any sensor platform to open more than a small total included angle.

For the purpose of this study, the AOA measurements are assumed to include zero mean additive Gaussian noise; all other measurement error sources are assumed to have an insignificant affect on the emitter location accuracy. With these above assumptions, the emitter location problem becomes a statistical process to estimate the best emitter location. AOA triangulation emitter-location algorithms are commonly implemented via a Kalman-filter or least squares method. Our work reported in this chapter is limited to least squares algorithms.

The adverse affect of large GDOP for triangulation algorithms can be significantly reduced by considering an alternative *angle-rate* algorithm that depends on a least squares algorithm to calculate the *average* AOA, *average*

AOA angle-rate and *average* included AOA angle that, in turn, are used to calculate the estimated emitter location. This approach is currently used in at least one military application. The remainder of this paper compares the triangulation and angle-rate algorithms for the geometry shown in Figure 2.

A single UAV collection platform, flying due south with a standoff range of 100 Km from the emitter located at the origin, makes multiple symmetrical uniformly spaced measurements as shown in Figure 2. For the specified geometry the emitter is located at point (0,0). The calculated emitter location's x-value is the along-range error, and the emitter location's y-value is the cross-range error; the errors are normalized as a percent of range without a loss of generality. Additionally, we have made the following assumptions:

1. The directional finding system has zero mean Gaussian noise with an error covariance of one;
2. UAVs are equipped with GPS systems that provide them their locations and headings, and the GPS system error is negligible compared to the sensor error; and
3. the sensing equipment on the UAVs is azimuth only.

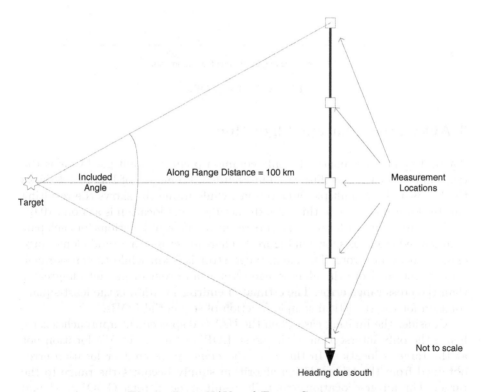

Fig. 2. Basic UAV and Target Geometry

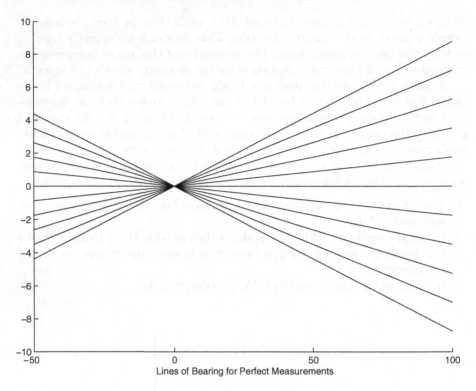

Lines of Bearing for Perfect Measurements

Fig. 3. Perfect LOBs

3 AOA Triangulation Algorithm

For perfect measurements, the only common point for multiple LOBs is the
true emitter location as illustrated in Figure 3. Frame (a) of Figure 1.4 shows
the LOBs and the multiple intersections, while frame (b) shows the resulting
emitter locations. Note in this case, the actual target location is at point (0,0).

Measurements with random errors result in different locations for each pair
of measured LOBs as shown in Figure 4. There are cross-range and along range
emitter location errors. The along-range error is 5 km while the cross-range
error is 100 m. Note the along-range error is an order of magnitude greater
than the cross range error. The estimated emitter location is the least-square
solution for the statistical sample function of sequential LOBs.

Consider the limiting case when the UAV's displacement approaches zero,
hence the only intersection of the noisy LOB's is at the UAV's location not
at the target's location. In this case, the *along-range* emitter location error
obtained from the triangulation algorithm simply becomes the range to the
target. This adverse location error is the result of the extreme GDOP. It should
be obvious the emitter location accuracy of the triangulation algorithm is very
suspect for large GDOP cases. In the next section, an alternative algorithm

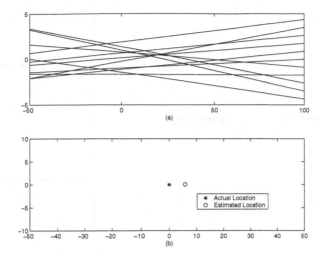

Fig. 4. Measured LOBs and Triangulation Emitter Location Error

is introduced to improve the initial emitter location estimate for large GDOP cases.

4 Angle-Rate Algorithm

Before introducing the angle-rate algorithm, recall for a rotating vector, the tangential velocity is equal to the angular velocity multiplied by the magnitude of the vector. Applying this principle to the emitter location problem, the range to the target is the UAV's velocity divided by the angle-rate of the observed line-of-bearing to the emitter when the UAV is orbiting the emitter with a constant velocity at a constant range. For the geometry described in Figure 5, the emitter's calculated range is one-half of the UAV's y-distance travelled divided by the tangent of one-half of the included angle. The emitter location's calculated y-value, or cross-range, is the UAV's average y-value added to the emitter's range multiplied by the tangent of the average included AOA.

Because of noisy LOB measurements, the *exact* angles in the above figure are not available. The objective of the angle-rate algorithm is to estimate these unknown values. The angle-rate algorithm determines the target's position by using the *average* AOA, *average* AOA angle-rate, and *average* included AOA angle that are obtained from the linear least square regression of the measured AOA data. Note using the *average* AOA obtained via linear regression instead of a tangent regression curve adds a bias to the range estimate. For small included angle where $tan\ \alpha \approx \alpha$, the bias is negligible; as the included angle increases, the bias grows adding error to the algorithm.

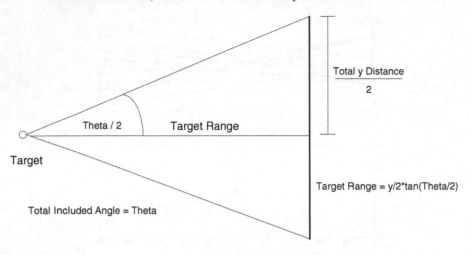

Fig. 5. Angle-Rate Geometry

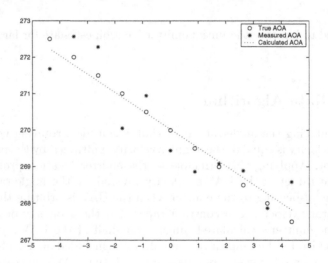

Fig. 6. True, Measured, and Calculated AOAs

The true AOA data from Figure 5 and the measured AOAs and the least squares *average* AOAs are shown in Figure 6. The average AOA angle and average included AOA angle are obtained from the average AOAs regression line.

The estimated emitter location's error using the regression data from Figure 6 is 663 m and 251 m for the along-range and cross-range respectively as shown in Figure 7. The equivalent estimated emitter location for the triangulation algorithm was shown in Figure 4b. Note for this single sample function,

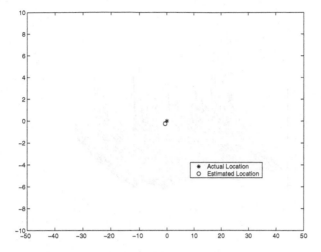

Fig. 7. Angle Rate Location Error

there is a significant error reduction (a factor of 7) by using the angle-rate algorithm instead of the standard triangulation algorithm.

5 Algorithm Comparison

For the case discussed above, it is obvious the angle-rate algorithm is more accurate than AOA triangulation. For this one specific case, there were 11 LOBs and the included angle was five degrees.

To further compare the two algorithms, we varied the parameter values. We kept noise constant with a covariance of one. We collected the accuracy data as we varied the LOBs from 3 to 101, and the included angle from 0.5 to 60 degrees.

As previously mentioned, the along-range error will be the dominant term in the error calculations. This is because the along-range error is along the semi-major axis of the error probable, and for large GDOP, our error is highly elliptical.

When comparing the performance of the two algorithms, we used only the square of the along-range error. The figure of merit (FOM) we used to compare the algorithms' performance is:

$$FOM = \frac{\sigma^2_{along-range-Angle-Rate}}{\sigma^2_{along-range-Triangulation}} \tag{2}$$

In Eq. 2, FOM less than 1 means the angle rate algorithm resulted in a smaller average error, while FOM greater than 1 means the triangulation algorithm resulted in smaller average error.

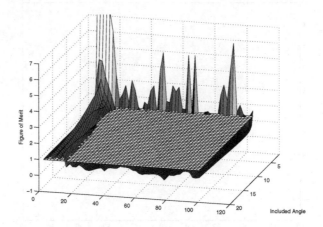

Fig. 8. Figure of Merit

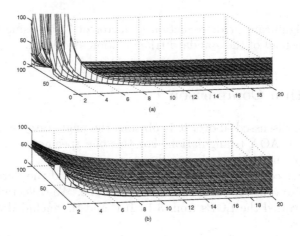

Fig. 9. Error Values for Small Included Angles

Figure 8 shows a three dimensional plot of the FOM vs. included angles ranging from 0 to 20 degrees and LOBs ranging from three to 101. Also shown in the graph is a FOM plane which equals 1. All points on the graph above this plane were for cases in which the triangulation algorithm had a smaller error, while all points below this plane are for cases in which the angle-rate algorithm had better error performance.

Figure 8 shows the angle-rate algorithm performs better than standard triangulation algorithms once the number of LOBs exceeds 10 and the included angle is greater than two degrees.

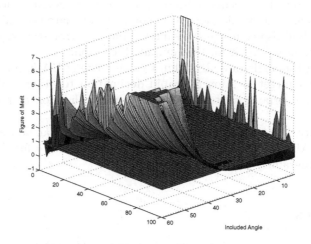

Fig. 10. FOM for Included Angles from 0 to 60 Degrees

It would appear since FOM is so much greater than 1 that triangulation is the algorithm of choice for very small included angles and a small number of LOBs. However, Figure 9 shows the magnitudes of the along-range error for the parameter variations. Figure 9(a) shows the error performance for the angle-rate algorithm while Figure 9(b) shows the error performance for the AOA triangulation algorithm. Note for small included angles and small LOBs, the errors approach 100 km. In many instances, the errors are much larger than 100 km, but in order to keep the scale of the graph meaningful, all error values greater than 100 km were assigned a value of 100 km. Actual error values for the triangulation algorithm at small included angles and small number of LOBs was on the order of 100 km while the error for the same conditions for the Angle Rate algorithm were over 1,000 km. Thus, neither algorithm produces useable results.

Once the included angle passes two degrees and the number of LOBs is more than 10, the angle-rate algorithm produces consistently better results than the AOA triangulation algorithm.

Interestingly enough, when the included angle grows beyond forty degrees, the angle rate error increases. The tangent function results in error in our approximation at these higher angles, as shown in Figure 10.

6 Conclusions and Future Work

The angle-rate emitter location algorithm is a better choice for small-included AOA applications [i.e., large GDOP cases] and provides the better initial target location accuracy for any application. The methods described above may be extended to include recursive linear or non-linear algorithms for either two or three-dimensional applications that not only calculate the unknown emitter location but also compute the sample function error statistics to provide a real-time figure-of-merit for the computed location.

Applying the angle-rate algorithm to slow moving UAVs provides an enhanced initial target location for rapid decimation to other UAVs, thereby enhancing the corporative search, detect, and localization of radiating targets.

The future direction of this work is implementation of angle-rate algorithm into the USAFA's UAV simulator and then into hardware for the USAFA's UAV test program.

References

1. Parker, L., and Emmens, B. (1997) Cooperative Multi-Robot Observation of Multiple Moving Targets, *Proceedings of the 1997 IEEE International Conference on Robotics and Automation,*, pp.2082-2089.
2. Everett, H., Gilbeater, G., Heath-Pastane, T., and Laird, R. (1993), Coordinated Control of Multiple Security Robots, *Proceedings of SPIE Robots VIII*, pp. 292-305.
3. Stone, L., Barlow, C., and Corwin, T., (1999). Bayesian Multiple Target Tracking, *Mathematics in Science and Engineering*.
4. Bourgault, F., Furukawa, T., and Duttant-Whyte, H. (2003), Coordinated Decentralized Search for a Lost Target in a Bayesian World, *Proceedins of the 2003 IEEE/RSJ Intl. Conf. on Intelligent Robts and Systems*.
5. Beard, R., Mclain, T., Goodrich, M., and Anderson, E. (2002). Coordinated Target Assignment of Intercept for Unmanned Air Vehicles, *IEEE Transactions on Robotics and Automation,* vol. 18, no. 6.
6. Vincent, P., and Rubin, I., (2003), A Framework and Analysis for Cooperative Search Using UAV Swarms, *Proceedings of the 2004 ACM Symposium on Applied Computing*, pp. 79-86.
7. Pack, D., and York, G., (2005), Developing a Control Architecture for Multiple Unmanned Aerial Vehicles to Search and Localize RF Time-Varying Mobile Targets: Part I, *Proceedings of the 2005 IEEE International Conference on Robotics and Automation*.
8. Avionics Department of the Naval Air Warfare Center Weapons Division, (1992) *Electronic Warfare and Radar Systems Engineering Handbook*.
9. Yarlagadda, R., Ali, I., Al-Dhahir, N., and Hershey, J. (2000), GPS GDOP Metric, *IEEE Proceedings on Radar, Sonar Navigation*, pp. 259-264.

Leaderless Cooperative Formation Control of Autonomous Mobile Robots Under Limited Communication Range Constraints

Zhihua Qu[1], Jing Wang[1] and Richard A. Hull[2]

[1] Department of Electrical and Computer Engineering,
   University of Central Florida,
   Orlando, FL 32816, USA
   E-mail: qu@mail.ucf.edu; jwang@pegasus.cc.ucf.edu
[2] Lockheed Martin Missiles and Fire Control,
   5600 Sand Lake Rd. MP-450,
   Orlando, FL 32819, USA
   E-mail: Richard.A.Hull@lmco.com

Summary. In this paper, a new leaderless cooperative formation control strategy is proposed for a group of autonomous mobile robots. Through the local state and input transformations, the formation control problem can be recast as the cooperative control design problem for a class of general canonical systems with arbitrary but finite relative degree. A set of less-restrictive sufficient conditions on group communication topology to ensure the success of cooperative control design has been established. The system stability is rigorously proved by studying the convergence of products of row stochastic matrices. The proposed design does not require either that collaborative robots have a fixed communication/control structure (such as leader/follower or nearest neighbor) or that their sensor/communication graph be strongly connected. Detailed simulation results are provided to illustrate the effectiveness of the proposed method.

1 Introduction

Recent years have seen a rapid progress on the study of cooperative and formation control for a group of mobile autonomous robots. The reason is that cooperatively controlled multiple robots have the potentiality to complete the complicated tasks with the advantages of higher efficiency and failure tolerance, such as coordinated navigation to a target, coordinated terrain exploration and search and rescue operations.

To cooperatively control a group of robots with less intervention from the centralized coordinator, a necessary condition is that the robots in the group can exchange information. As a result, the communication topology of the group plays a key role for the success of the coordination tasks. The fundamental problems in the study of cooperative control are thus how to design the decentralized control for the individual robot only using the local information from its neighboring robots while guaranteeing the stability of the overall system; and under what kinds of communication topologies of the group, it is sufficient to design the control to achieve the coordination behavior.

At the early stage, cooperative control studies have been motivated by mimicking the animals' behavior. The basic cohesion, separation and alignment rules were extracted by observing the animal flocking and simulated through the computer animation [26]. The alignment rule was later on modelled mathematically in the study of planar motion of a group of particles [30], and simulation results verified its correctness. In [9], a simple model was given to describe the animal swarm aggregation under the assumption that all the members know the exact positions of all the other members. Using the heuristical behavior method [1], rule based motion schemas were defined to guide and maintain the robot formation while moving to the goal location [2]. The methodology is in essence ad hoc, and no system dynamics are considered. Probabilistic approaches were also employed for multi-robot localization and exploration [7, 21]. As a parallel development, formation control strategies have been sought using graph theory and artificial potentials. In [5], using formation graph, the problem of following a desired trajectory while maintaining a formation was discussed by converting the system dynamics into the domain of relative position and relative angle between robots, and then the feedback linearization technique was applied to stabilize the relative distances of the robots in the formation. Virtual leaders and artificial potential method were used for a group of agents maintaining the group geometry [15, 20], where the closed-loop stability was proved by using the system kinetic energy and the artificial potential energy as a Lyapunov function. In [29], the notion of string stability for line formations was proposed and sufficient conditions for a formation to be string stable was derived. In [12], the action reference concept has been introduced to define the desired trajectory for the individual agent in the group, and the control design reduces to the classical tracking control problem for each agent. The group coordination can be completed by specifying the action reference, such as leader-follower.

More recently, attentions have been particularly paid to the cooperative control problem of making the states of a group of dynamical systems converge to same steady state, the so-called consensus problem, since its solvability has close relation to the solvability of general formation control problem [11, 16, 23, 24, 19, 6, 28, 27, 25]. To seek a rigorous mathematical explanation to the Vicsek model [30], a remarkable work has been done in [11], where it is proved that all the agents' headings converge to a common value provided that the *bidirectional* sensor graphs for all agents are periodically connected. The extensions to the *directed* sensor graph was subsequently done in [16, 19, 25, 27]. For the case of *directed* sensor graph, the condition found is that the sensor graph needs to be *strongly connected* once over a fixed time interval

[16]. In our recent works [23, 24], we posed the consensus problem for a general class of multiple-input-multiple-output dynamical systems in canonical form with arbitrary but finite relative degree. Through the thorough studies on the irreducibility of row stochastic matrix and establishing the new results on convergence of product of a sequence of row stochastic matrix, we presented a general guideline for the cooperative control design.

In this paper, as a continuation of the works in [23, 24], we propose a new leaderless cooperative formation control strategy for a group of autonomous mobile robots. In particular, we consider a group of robotic vehicles that operate individually by themselves most of the time, communicate intermittently among their teammates within their neighboring groups, and vehicles have limited sensing range. Such a setting is typical in many practical applications, such as deploying a group of robots for search or exploration purpose in hazardous environments. The proposed design does not require either that collaborative robots have a fixed communication/control structure (such as leader/follower or nearest neighbor) or that their sensor/communication graph be strongly connected. The convergence of the overall system is rigorously proved by studying the convergence of products of a sequence of row stochastic matrices. Compared to the existing results in the literature, the main contribution of this work lies in two-fold: First, the formation control problem has been solved for a general class of systems in canonical form with higher relative degrees and the agents in the group can be not identical, this makes the proposed method applicable to a broad class of practical systems. To the best of the authors' knowledge, there is no related result reported in the literature. Second, the less-restrictive sufficient conditions on group communication topology to ensure the success of cooperative control design has been established. In the final part of the paper, the illustrative examples are provided to verify the proposed method.

2 Problem Formulation

Throughout the paper, the following notations and definitions are used. Let $\mathbf{1}_p$ be the p-dimensional column vector with all its elements being 1, and $\mathbf{J}_{r_1 \times r_2} \in \Re^{r_1 \times r_2}$ be a matrix whose elements are all 1. A nonnegative matrix has all entries nonnegative. A square real matrix is row stochastic if it is nonnegative and its row sums all equal 1. For a row stochastic matrix E, define $\delta(E) = \max_j \max_{i_1,i_2} |E_{i_1 j} - E_{i_2 j}|$, which measures how different the rows of E are. Also, define $\lambda(E) = 1 - \min_{i_1,i_2} \sum_j \min(E_{i_1 j}, E_{i_2 j})$. Given a sequence of nonnegative matrix $E(k)$, $E(k) \succ 0, k = 0, 1, \cdots$, means that, there is a subsequence $\{l_v, v = 1, \cdots, \infty\}$ of $\{0, 1, 2, \cdots, \infty\}$ such that $\lim_{v \to \infty} l_v = +\infty$ and $E(l_v) \neq 0$, that is, there exists at least one element $E_{ij}(l_v) \geq \epsilon$ for $\epsilon > 0$. A non-negative matrix E is said to be *irreducible* if there does not exist a permutation matrix T such that TGT^T is in the block lower-triangular structure. Otherwise, it is *reducible*.

In this section, we first formulate the cooperative control design problem for a class of dynamical systems in the canonical form, and show that a broad class of practical robotic systems can be converted into the given canonical form through the decentralized state and input transformations. Then, we will show that the formation control problem can be recast to study the solvability of the proposed cooperative control problem for the canonical systems.

2.1 Canonical Form

Consider a group of dynamical systems given by the following canonical form

$$\dot{x}_i = A_i x_i + B_i u_i, \quad y_i = C_i x_i, \quad \dot{\eta}_i = g_i(\eta_i, x_i), \tag{1}$$

where $i = 1, \cdots, q$, $l_i \geq 1$ is an integer, $x_i \in \Re^{l_i m}$, $\eta_i \in \Re^{n_i - l_i m}$, $I_{m \times m}$ is the m-dimensional identity matrix, $\otimes$ denotes the Kronecker product, J_k is the kth order Jordan canonical form given by

$$J_k = \begin{bmatrix} -1 & 1 & 0 & \cdots & 0 & 0 \\ 0 & -1 & 1 & \ddots & 0 & 0 \\ \vdots & \ddots & \ddots & \ddots & \ddots & \vdots \\ 0 & 0 & \cdots & -1 & 1 & 0 \\ 0 & 0 & 0 & \cdots & -1 & 1 \\ 0 & 0 & 0 & \cdots & 0 & -1 \end{bmatrix} \in \Re^{k \times k},$$

$A_i = J_{l_i} \otimes I_{m \times m} \in \Re^{(l_i m) \times (l_i m)}$, $B_i = \begin{bmatrix} 0 \\ I_{m \times m} \end{bmatrix} \in \Re^{(l_i m) \times m}$, $C_i = \begin{bmatrix} I_{m \times m} & 0 \end{bmatrix} \in \Re^{m \times (l_i m)}$, $y_i \in \Re^m$ is the output, $u_i \in \Re^m$ is the cooperative control law to be designed, and subsystem $\dot{\eta}_i = g_i(\eta_i, x_i)$ is input-to-state stable. Without loss of any generality, in this paper we consider the case that $l_1 = l_2 = \cdots = l_q = l$. The objective is to synthesis a general class of cooperative control u_i and establish the less-restrictive conditions on network connectivity requirements such that the all states of the overall system converge to the same steady state.

The following examples illustrate the wider application of the proposed canonical model.

Example 1. Point-mass agent: Given the agent's motion model:

$$\dot{z}_1 = z_2, \quad \dot{z}_2 = v, \tag{2}$$

where $z_1 \in \Re^m$ is the position of the agent, $z_2 \in \Re^m$ is the velocity, and $v \in \Re^m$ is the control. Define the state and input transformations as follows:

$$x_1 = z_1, \quad x_2 = x_1 + z_2, \quad v = -2x_2 + x_1 + u.$$

Then system model can be transformed into (1) with

$$A = \begin{bmatrix} -1 & 1 \\ 0 & -1 \end{bmatrix} \otimes I_{m \times m}, \quad B = \begin{bmatrix} 0 \\ I_{m \times m} \end{bmatrix}, \quad C = \begin{bmatrix} I_{m \times m} & 0 \end{bmatrix}.$$

Example 2. Differential driven mobile robots: Given nonholonomic 4-wheel differential driven mobile robots [14]:

$$\begin{bmatrix} \dot{r}_x \\ \dot{r}_y \\ \dot{r}_\theta \\ \dot{r}_v \\ \dot{r}_\omega \end{bmatrix} = \begin{bmatrix} r_v \cos(r_\theta) \\ r_v \sin(r_\theta) \\ r_\omega \\ 0 \\ 0 \end{bmatrix} + \begin{bmatrix} 0 & 0 \\ 0 & 0 \\ 0 & 0 \\ \frac{1}{m} & 0 \\ 0 & \frac{1}{J} \end{bmatrix} \begin{bmatrix} F \\ \tau \end{bmatrix}, \tag{3}$$

where (r_x, r_y) is the inertial position of the robot, r_θ is the orientation, r_v is the linear speed, r_ω is the angular speed, τ is the applied torque, F is the applied force, m is the mass, and J is the moment of inertia. By taking the robot "hand" position as the guide point (which is a point located a distance L from (r_x, r_y) along the line that is perpendicular to the wheel axis), the robot model in (3) can be feedback linearized to

$$\dot{\phi}_1 = \phi_2, \quad \dot{\phi}_2 = v, \tag{4}$$
$$\dot{r}_\theta = \begin{bmatrix} -\frac{1}{2L} \sin(r_\theta) & \frac{1}{2L} \cos(r_\theta) \end{bmatrix} \phi_2, \tag{5}$$

where $\phi_1 = \begin{bmatrix} r_x + L\cos(r_\theta) \\ r_y + L\sin(r_\theta) \end{bmatrix} \in \Re^2$ is the position of "hand" point, $\phi_2 \in \Re^2$, and

$$\begin{bmatrix} F \\ \tau \end{bmatrix} = \begin{bmatrix} \frac{1}{m}\cos(r_\theta) & -\frac{L}{J}\sin(r_\theta) \\ \frac{1}{m}\sin(r_\theta) & \frac{L}{J}\cos(r_\theta) \end{bmatrix}^{-1} \left(v - \begin{bmatrix} -r_v r_\omega \sin(r_\theta) - L^2 r_\omega^2 \cos(r_\theta) \\ r_v r_\omega \cos(r_\theta) - L^2 r_\omega^2 \sin(r_\theta) \end{bmatrix} \right).$$

As shown by example 1, system (4) can be further put into the form of (1), and (5) is the internal dynamics. Once ϕ_2 and ϕ_1 are controlled to the steady state ϕ_{ss}, we can show that the internal dynamics (5) is also stable. Assume that $\phi_{ss} > 0$. It is easy to see that $\frac{\pi}{4}$ is an equilibrium of system (5). Then by taking Lyapunov function candidate as $V = 1 - \cos(r_\theta - \frac{\pi}{4})$, it follows from (5) that $\dot{V} = -\frac{\sqrt{2}\phi_{ss}}{4L}[\sin(r_\theta) - \cos(r_\theta)]^2 \leq 0$. Thus, we can conclude that the internal dynamics is asymptotically stable.

Example 3. Point-mass aircraft [17]: Given

$$\dot{x} = V \cos\gamma \cos\phi, \tag{6}$$
$$\dot{y} = V \cos\gamma \sin\phi, \tag{7}$$
$$\dot{h} = V \sin\gamma, \tag{8}$$
$$\dot{V} = \frac{T - D}{M} - g\sin\gamma, \tag{9}$$
$$\dot{\gamma} = \frac{g}{V}(n\cos\delta - \cos\gamma), \tag{10}$$
$$\dot{\phi} = \frac{L\sin\delta}{mV\cos\gamma}, \tag{11}$$

where x is the down-range displacement, y is the cross-range displacement, h is the altitude, V is the ground speed and is assumed to be equal to airspeed, γ is the flight path angle, ϕ is the heading angle, T is the aircraft engine thrust, D is the drag, M is the aircraft mass, g is the gravity acceleration, L is the aerodynamic lift, δ is the banking angle. The banking angle δ, the engine thrust T, and the load factor $n = L/gm$ are the control variables.

By differentiating (6), (7) and (8) once with respect to time [17], the kinematic equations describing the aircraft position can be put into (2) (which can be further transformed into the canonical form (1)) with

$$z_1 = [x, y, h]^T, \quad z_2 = [\dot{x}, \dot{y}, \dot{h}]^T, \quad v = [u_1, u_2, u_3]^T,$$

where u_1, u_2 and u_3 are the new control variables. Once cooperative control u_1, u_2, and u_3 are designed, the actual control variables can be computed by

$$\delta = \tan^{-1} \left[\frac{u_2 \cos\phi - u_1 \sin\phi}{\cos\gamma(u_3 + g) - \sin\gamma(u_1 \cos\phi + u_2 \sin\phi)} \right],$$
$$n = \frac{\cos\gamma(u_3 + g) - \sin\gamma(u_1 \cos\phi + u_2 \sin\phi)}{g \cos\delta},$$
$$T = [\sin\gamma(u_3 + g) + \cos\gamma(u_1 \cos\phi + u_2 \sin\phi)]m + D,$$

and the heading angle ϕ and the flight-path angle γ can be computed as $\tan\phi = \dot{y}/\dot{x}$, $\sin\gamma = \dot{h}/V$.

Example 4. One more example is the underwater glider discussed in [4], where by taking the position of shifting mass (r_{p_1}, r_{p_3}) and the buoyancy mass m_b as the outputs, the system can be feedback linearized into

$$\dot{\zeta}_1 = \zeta_2, \quad \dot{\zeta}_2 = \omega_1, \tag{12}$$
$$\dot{m}_b = \omega_2, \tag{13}$$
$$\dot{\eta} = q(\eta, \zeta, \omega), \tag{14}$$

where $\zeta_1 = [r_{p_1}, r_{p_3}]^T$, $\omega_1 \in \Re^2$ and $\omega_2 \in \Re$ are the new control inputs. It is obvious that (12) is in the form of (2) with relative degree 2, (12) is in the form (1) with relative degree 1. It can also be verified that internal dynamics (14) is stable [4].

Remark 1. As shown by above examples, in general, any input-output feedback linearizable dynamical systems with stable internal dynamics can be transformed into the canonical form (1) [22, 13].

2.2 Formation Control Problem

Now, let us consider the formation control problem for a group of q point-mass agents whose dynamics are given by:

$$\dot{z}_{i1} = z_{i2}, \quad \dot{z}_{i2} = v_i, \quad i = 1, \cdots, q, \tag{15}$$

where $z_{i1} = [z_{i11}, z_{i12}] \in \Re^2$ is the planar position of the ith agent, $z_{i2} = [z_{i21}, z_{i22}] \in \Re^2$ is its velocity, and $v_i = [v_{i1}, v_{i2}]^T$ is its acceleration input.

A formation is defined in a coordinate frame, which moves with the desired trajectory. Let $e_1(t) \in \Re^2$ and $e_2(t) \in \Re^2$ be the orthonormal vectors which forms the moving frame $F(t)$. Let $z_d(t) = [z_{d1}(t), z_{d2}(t)] \in \Re^2$ be any desired trajectory of the origin of the moving frame. A formation consists of q points in $F(t)$, denoted by $\{P_1, \cdots, P_q\}$, where

$$P_i = d_{i1}(t)e_1(t) + d_{i2}(t)e_2(t), \quad i = 1, \cdots, q \tag{16}$$

with $d_i(t) = [d_{i1}(t), d_{i2}(t)] \in \Re^2$ being the desired relative position for the ith agent in the formation. It is obvious that $d_i(t)$ being constant refers to the rigid formation. The desired position for the ith agent is then

$$z_i^d(t) = z_d(t) + d_{i1}(t)e_1(t) + d_{i2}(t)e_2(t). \tag{17}$$

Figure 1 illustrates a formation setup for 3 agents.

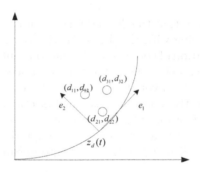

Fig. 1. Illustration of Formation Setup

In this paper, the formation control objective is to design a decentralized control $v_i(t)$ according to the feedback information from the agents within its limited sensor range so as to make the group of agents converge to the desired formation given by (16) while moving along the specified trajectory. More specifically, two problems will be addressed:

(a) Establish the less restrictive connectivity conditions among agents under which the coordination behavior of the group of agents can be achieved.
(b) Explicitly design the cooperative control according to the connectivity among agents and prove the stability of the overall closed-loop system.

To solve the problems stated above, the following assumptions are made throughout the paper:

A1:Each agent is represented by a point.

A2:The agent's motion is instantaneous and there are no communication time-delays within the group.

In what follows, we show that, through state and input transformations, the formation control problem for (15) can be recast as the cooperative control design problem for (1). Let the input transformation be

$$x_{i1} = z_{i1} - z_i^d(t), \quad x_{i2} = z_{i2} + x_{i1} - \dot{z}_i^d(t) \tag{18}$$

and the decentralized control be

$$v_i(t) = -2x_{i2} + x_{i1} + u_i. \tag{19}$$

Substituting (18) and (19) into (15), we obtain the canonical model (1) with $x_i = [x_{i1}^T, x_{i2}^T]^T \in \Re^4$, $u_i \in \Re^2$, and $y_i \in \Re^2$. To this end, if we can design the cooperative control u_i such that states x_{i1} and x_{i2} for all i converge to the same steady state x_{ss}, then it follows from (18) that

$$z_{i1} \rightarrow x_{ss} + z_i^d(t), \quad z_{i2} \rightarrow \dot{z}_i^d(t),$$

from which it can be seen that the desired formation is achieved for the whole group while agents moving with the desired trajectory shape.

In the basic formulation of cooperative formation control problem, collision among agents is not explicitly considered. In practice, special local control strategy can be activated to avoid the possible collision, and then switch back to the proposed cooperative control. In the next section, we will show how to design cooperative control for system (1) according to the local information within the limited sensor range and find the less-restrictive conditions under which the coordination behavior of the whole group can be guaranteed.

3 Leaderless Cooperative Control Design

In this section, we first analyze the connectivity of the overall system by introducing signal transmission matrix. Then, the analytical conditions on connectivity and the design of cooperative control are provided with the rigorous mathematical proof.

3.1 Limited Sensor Range

A typical scenario of a group of agents moving in a plane is depicted in figure 2.

As shown by figure 2, the ith agent has the limited and directed sensor range, and it is preferred to design $u_i(t)$ by only using the local output information coming from its sensor range at the current time instant t. Under

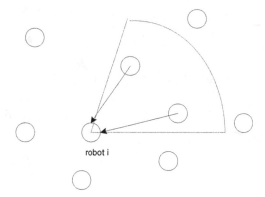

Fig. 2. Scenario

such a preference for control design, it is easy to understand that the overall closed-loop system convergence (in the sense of cooperative control objective being achieved) is determined by the connectivity topology of the group. In general, the connectivity topology of the group will change dynamically with the evolution of the system. The intuition is that the more the system is connected, the higher the possibility of system convergence will be. However, there may be the possible uncertainties with system communication channels, such as the communication dropout. To reduce the effect of such uncertainties, a robust cooperative control will be that requiring the information from other agents as less as possible, that is, we prefer to seek the design of the successful cooperative control under the connectivity topologies that is sparse. In this paper, we analyze that under which kind of less restrictive requirements on connectivity topology, it is sufficient to design the cooperative control to ensure the convergence of the overall system.

The connectivity of the agents can be described by the following signal transmission matrix:

$$S(t) = [S_{ij}(t)] \in \Re^{q \times q}, \tag{20}$$

where $S_{ii} = 1$ which means that agent always has sensor information itself; $S_{ij} = 1$ for $i \neq j$ if the ith agent can sense the jth agent, otherwise $S_{ij} = 0$. Figure 3 gives an example to illustrate the connectivity topologies of a group of 4 agents at two different time instants. In the graph, the line with arrow indicates the directed communication channel between agents. If the ith agent receives information from the jth agent, then there is a line from the jth agent pointing at the ith agent. The corresponding signal transmission matrices are

$$S(t_k^S) = \begin{bmatrix} 1 & 1 & 0 & 1 \\ 0 & 1 & 0 & 0 \\ 1 & 0 & 1 & 0 \\ 0 & 0 & 1 & 1 \end{bmatrix}, \quad S(t_{k1}^S) = \begin{bmatrix} 1 & 1 & 0 & 0 \\ 0 & 1 & 0 & 1 \\ 0 & 0 & 1 & 0 \\ 0 & 0 & 1 & 1 \end{bmatrix}. \tag{21}$$

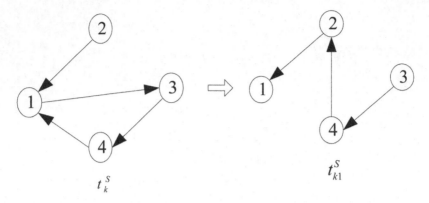

Fig. 3. Connectivity topologies

The property of signal transmission matrix $S(t)$ presents an exact description about the connectivity among the agents. A directed graph represented by $S(t)$ is *strongly connected* if between every pair of distinct nodes n_i, n_j there is a directed path of finite length that begins at n_i and ends at n_j. It is shown in [10] that the fact that a directed graph represented by $S(t)$ is strongly connected is equivalent to that matrix $S(t)$ is irreducible. To this end, the task becomes to study the property of $S(t)$ under which the overall system convergence can be achieved. In general, $S(t)$ will change with the system evolution and/or the change of communication sensor mechanisms installed on each agent (such as the sensor may rotate periodically on each agent to enlarge the ability of receiving more information). The following time sequence is introduced to describe the change of $S(t)$.

Definition 1. *Let* $\{t_k^S : k = 0, 1, \cdots\}$ *with* $t_0^S = t_0$ *be the sequence of time instants at which the topology (that is, $S(t)$) of the multi-agent network changes.*

In general, time sequence $\{t_k^S\}$ and the corresponding changes of matrix $S(t)$ are detectable instantaneously but not known apriori. Nonetheless, it can be assumed without loss of generality that $0 < \underline{c}_t \le t_{k+1}^S - t_k^S \le \bar{c}_t < \infty$.

The property of $S(t)$ plays a key role on the analysis of system convergence. Recently, some excellent works have been done to establish the link between the property of sensor graph (equivalently, the $S(t)$ in this paper) and the coordination behavior [16, 11, 19]. It is proved that if the system sensor graph is periodically *strongly connected*, then systems' coordination behavior can be reached using the nearest neighbor rules for the control design. In particular, the *undirected* sensor graph case is considered in [11], while the *directed* sensor graph case is studied in [16, 19]. In those works, the system model is assumed to be the single integrator. In this paper, two improvements will be made:

(i) We further relax the conditions on $S(t)$;
(ii) The results are built upon the system model (1) which has arbitrary but finite relative degree.

That is, on the one hand, we will show that even $S(t)$ are always reducible, it is still possible to establish the conditions for overall system convergence; on the other hand, the extension to higher relative degree enlarges the applicability to more general agent control problem, such as motion control of UAVs. The extension itself is not trivial as we need to prove a propagation property of the irreducibility, and this will be clarified in subsection 3.3.

In what follows, we will give the cooperative control according to $S(t)$ and explicitly obtain the conditions for system convergence.

3.2 Cooperative Control Design

Let the cooperative control be given by: for $i = 1, \cdots, q$,

$$u_i = G_i(t)y, \qquad (22)$$

where $y = [y_1^T \ \cdots \ y_q^T]^T$, and $G_i = [G_{i1} \ \cdots \ G_{iq}]$ with $G_{ij} = S_{ij}(t)K_{ij}(t)$, where $K_{ij}(t) \in \Re^{m \times m}$ are piecewise constant non-negative matrices and chosen according to the changes of $S(t)$ such that $\sum_{j=1}^{q} S_{ij}(t)K_{ij}(t)\mathbf{1}_m = \mathbf{1}_m$. It follows that $G_{ij}(t) = G_{ij}(t_k^S)$ for $t \in [t_k^S, t_{k+1}^S)$, $G_{ij} \geq 0$, $G_{ii} > 0$ and $G_i(t)\mathbf{1}_{mq} = \mathbf{1}_m$. That is, $G_i(t)$ are piecewise constant for all i, row stochastic, and determined by $S(t)$.

It follows from (1) and (22) that the overall closed-loop system is given by

$$\dot{x} = (A + BGC)x = (-I_{N_q \times N_q} + D(t))x, \qquad (23)$$

where $x = [x_1^T, \ \cdots, \ x_q^T]^T \in \Re^{N_q}$, $N_q = mlq$, $x_i = [x_{i1}^T, x_{i2}^T, \cdots, x_{il}^T]^T \in \Re^{ml}$, $x_{ij} = [x_{ij1}, x_{ij2}, \cdots, x_{ijm}]^T \in \Re^m$ with $i = 1, \cdots, q$, and $j = 1, \cdots, l$, $A = \mathrm{diag}\{A_1, \ \cdots, \ A_q\} \in \Re^{N_q \times N_q}$, $C = \mathrm{diag}\{C_1, \ \cdots, \ C_q\} \in \Re^{(mq) \times N_q}$, $B = \mathrm{diag}\{B_1, \ \cdots, \ B_q\} \in \Re^{N_q \times (mq)}$, $G = [G_1^T \ \cdots \ G_q^T]^T \in \Re^{(mq) \times (mq)}$, and $D(t) = [D_{ij}]$ with $(i = 1, \cdots, q)$

$$D_{ii} = \begin{bmatrix} 0 & I_{(l-1) \times (l-1)} \otimes I_{m \times m} \\ G_{ii} & 0 \end{bmatrix} \in \Re^{lm \times lm}, \qquad (24)$$

and

$$D_{ij} = \begin{bmatrix} 0 & 0 \\ G_{ij} & 0 \end{bmatrix} \in \Re^{lm \times lm}, \quad i, j = 1, \cdots, q, \ i \neq j. \qquad (25)$$

It is obvious that $D(t)$ is piecewise constant and row stochastic.

Since the stability of the overall closed-loop system (23) is dependent on the group connectivity topology $S(t)$, we start the analysis by studying the property of $S(t)$. Given signal transmission matrix $S(t)$, it is shown that there is a permutation matrix $T_1(t) \in \Re^{q \times q}$ such that [3, 18]

$$S_{T_1}(t) = T_1^T(t)S(t)T_1(t) = \begin{bmatrix} S_{T_1,11}(t) & 0 & \cdots & 0 \\ S_{T_1,21}(t) & S_{T_1,22}(t) & \cdots & 0 \\ \vdots & \vdots & \ddots & \vdots \\ S_{T_1,p1}(t) & S_{T_1,p2}(t) & \cdots & S_{T_1,pp}(t) \end{bmatrix}, \qquad (26)$$

where $S_{T_1,ii} \in \Re^{q_i \times q_i}$, $\sum_{i=1}^{p} q_i = q$, and $S_{T_1,ii}(t)$ are irreducible. If $p = 1$, it is obvious that $S(t)$ is irreducible. Otherwise, $S(t)$ is reducible. It is worth mentioning that the case of $q_i \neq 1$ corresponds to the situation that q subsystems are regrouped into n subgroups with the ith subgroup consisting of q_i subsystems, $i = 1, \cdots, p$.

Corresponding to the permutation matrix $T_1(t)$ for $S(t)$, we have augmented permutation matrices $T_2(t) = T_1(t) \otimes I_{m \times m} \in \Re^{mq \times mq}$ such that

$$G_{T_2}(t) = T_2^T(t)G(t)T_2(t) = \begin{bmatrix} G_{T_2,11}(t) & 0 & \cdots & 0 \\ G_{T_2,21}(t) & G_{T_2,22}(t) & \cdots & 0 \\ \vdots & \vdots & \ddots & \vdots \\ G_{T_2,p1}(t) & G_{T_2,p2}(t) & \cdots & G_{T_2,pp}(t) \end{bmatrix}, \quad (27)$$

where $G_{T_2,ii}(t)$ is irreducible. Similarly, we have augmented permutation matrices $T_3(t) = T_1(t) \otimes I_{lm \times lm} \in \Re^{lmq \times lmq}$ such that

$$D_{T_2}(t) = T_3^T(t)D(t)T_3(t) = \begin{bmatrix} D_{T_3,11}(t) & 0 & \cdots & 0 \\ D_{T_3,21}(t) & D_{T_3,22}(t) & \cdots & 0 \\ \vdots & \vdots & \ddots & \vdots \\ D_{T_3,p1}(t) & D_{T_3,p2}(t) & \cdots & D_{T_3,pp}(t) \end{bmatrix}. \quad (28)$$

Example 5. Let us consider the signal transmission matrices illustrated in figure 3. The corresponding $G(t)$ are as follows:

$$G(t_k^S) = \begin{bmatrix} G_{11}(t_k^S) & G_{12}(t_k^S) & 0 & G_{14}(t_k^S) \\ 0 & G_{22}(t_k^S) & 0 & 0 \\ G_{31}(t_k^S) & 0 & G_{33}(t_k^S) & 0 \\ 0 & 0 & G_{43}(t_k^S) & G_{44}(t_k^S) \end{bmatrix},$$

$$G(t_{k1}^S) = \begin{bmatrix} G_{11}(t_{k1}^S) & G_{12}(t_{k1}^S) & 0 & 0 \\ 0 & G_{22}(t_{k1}^S) & 0 & G_{24}(t_{k1}^S) \\ 0 & 0 & G_{33}(t_{k1}^S) & 0 \\ 0 & 0 & G_{43}(t_{k1}^S) & G_{44}(t_{k1}^S) \end{bmatrix}.$$

For $S(t_k^S)$, given permutation matrices

$$T_1(t_k^S) = \begin{bmatrix} 0 & 1 & 0 & 0 \\ 1 & 0 & 0 & 0 \\ 0 & 0 & 1 & 0 \\ 0 & 0 & 0 & 1 \end{bmatrix}, \quad T_2(t_k^S) = T_1(t_k^S) \otimes I_{m \times m}, \quad T_3(t_k^S) = T_1(t_k^S) \otimes I_{lm \times lm},$$

we have

$$S_{T_1}(t_k^S) = \begin{bmatrix} S_{T_1,11}(t_k^S) & \emptyset \\ S_{T_1,21}(t_k^S) & S_{T_1,22}(t_k^S) \end{bmatrix}, \quad G_{T_2}(t_k^S) = \begin{bmatrix} G_{T_2,11}(t_k^S) & \emptyset \\ G_{T_2,21}(t_k^S) & G_{T_2,22}(t_k^S) \end{bmatrix},$$

and

$$D_{T_3}(t_k^S) = \begin{bmatrix} D_{T_3,11}(t_k^S) & \varnothing \\ D_{T_3,21}(t_k^S) & D_{T_3,22}(t_k^S) \end{bmatrix},$$

where

$$S_{T_1,11}(t_k^S) = 1, \quad S_{T_1,21}(t_k^S) = \begin{bmatrix} 1 \\ 0 \\ 0 \end{bmatrix}, \quad S_{T_1,22}(t_k^S) = \begin{bmatrix} 1 & 0 & 1 \\ 1 & 1 & 0 \\ 0 & 1 & 1 \end{bmatrix},$$

with $S_{T_1,11}(t_k^S)$ and $S_{T_1,22}(t_k^S)$ being irreducible and $p = 2$, $G_{T_2,11}(t_k^S) = G_{22}(t_k^S)$ and

$$G_{T_2,21}(t_k^S) = \begin{bmatrix} G_{12}(t_k^S) \\ 0 \\ 0 \end{bmatrix}, \quad G_{T_2,22}(t_k^S) = \begin{bmatrix} G_{11}(t_k^S) & 0 & G_{14}(t_k^S) \\ G_{31}(t_k^S) & G_{33}(t_k^S) & 0 \\ 0 & G_{43}(t_k^S) & G_{44}(t_k^S) \end{bmatrix},$$

with

$$G_{ii} = \begin{bmatrix} 0 & I_{(m-1)\times(m-1)} \\ S_{ii}/\sum_{j=1}^q S_{ij} & 0 \end{bmatrix}, \quad G_{ij} = \begin{bmatrix} 0 & 0 \\ S_{ij}/\sum_{j=1}^q S_{ij} & 0 \end{bmatrix},$$

$D_{T_3,11}(t_k^S) = D_{22}(t_k^S)$ and

$$D_{T_3,21}(t_k^S) = \begin{bmatrix} D_{12}(t_k^S) \\ 0 \\ 0 \end{bmatrix}, \quad D_{T_3,22}(t_k^S) = \begin{bmatrix} D_{11}(t_k^S) & 0 & D_{14}(t_k^S) \\ D_{31}(t_k^S) & D_{33}(t_k^S) & 0 \\ 0 & D_{43}(t_k^S) & D_{44}(t_k^S) \end{bmatrix},$$

with D_{ii} and D_{ij} given by (24) and (25).

Similarly, for $S(t_{k1}^S)$, let

$$T_1(t_{k1}^S) = \begin{bmatrix} 0 & 0 & 1 & 0 \\ 0 & 0 & 0 & 1 \\ 0 & 1 & 0 & 0 \\ 1 & 0 & 0 & 0 \end{bmatrix},$$

it is easy to obtain $S_{T_1(t_{k1}^S)}$, $G_{T_2}(t_{k1}^S)$ and $D_{T_3}(t_{k1}^S)$. $\diamond$

Remark 2. It should be noted that the permutation matrices T_1, T_2 and T_3 are introduced here only for system stability analysis purpose.

3.3 Conditions and Convergence Analysis

In this subsection, we present the main result of the paper, and establish the less-restrictive conditions on connectivity topologies of the overall system. The following lemmas are used in the proof of the main theorem. Lemma 1 states a propagation property of the irreducibility of matrix, which will be used to solve the problem with system of higher relative degree. An easy-to-check conditions for the convergence of lower-triangular row stochastic is given in lemma 2, which lays one of the basic foundations for the proof of the main theorem. Lemma 3 gives a necessary and sufficient condition on the convergence of combination of sequence of lower-triangular row stochastic matrices with another sequence of row stochastic matrices, and this lays another foundation for the proof of the main theorem.

Lemma 1. *[23] Given any non-negative matrix $E \in \Re^{(qm) \times (qm)}$ with sub-blocks $E_{ij} \in \Re^{m \times m}$, let $\overline{E} = [\overline{E}_{ij}] \in \Re^{(Lm) \times (Lm)}$ with $L = l_1 + \cdots + l_q$ be defined by*

$$\overline{E}_{ii} = \begin{bmatrix} 0 & I_{(l_i-1)} \otimes I_m \\ E_{ii} & 0 \end{bmatrix}, \quad \overline{E}_{ij} = \begin{bmatrix} 0 & 0 \\ E_{ij} & 0 \end{bmatrix}$$

where $l_i \geq 1$ are positive integers for $i = 1, \cdots, q$. Then, $\overline{E}$ is irreducible if and only if E is irreducible.

Lemma 2. *[24] Consider a sequence of nonnegative, row stochastic matrices in the lower-triangular structure*

$$P(k) = \begin{bmatrix} P_{11}(k) \\ P_{21}(k) & P_{22}(k) \\ \vdots & \vdots & \ddots \\ P_{p1}(k) & P_{p2}(k) & \cdots & P_{pp}(k) \end{bmatrix} \in \Re^{R \times R},$$

where $R = \sum_{i=1}^{m} r_i$, sub-blocks $P_{ii}(k)$ on the diagonal are square and of dimension $\Re^{r_i \times r_i}$, sub-blocks $P_{ij}(k)$ off diagonal are of appropriate dimensions. Suppose that $P_{ii}(k) \geq \epsilon_i \mathbf{J}_{r_i \times r_i}$ for some constant $\epsilon_i > 0$ and for all $(i = 1, \cdots, p)$, and in the ith row of $P(k)$ $(i > 1)$, there is at least one j $(j < i)$ such that $P_{ij} \succ 0$. Then,

$$\lim_{k \to \infty} \prod_{l=0}^{k-1} P(k-l) = \mathbf{1}_R c,$$

where constant vector $c = [c_1, 0, \cdots, 0] \in \Re^{1 \times R}$ with $c_1 \in \Re^{1 \times r_1}$.

Lemma 3. *[24] Given sequences of row stochastic matrices $P(k) \in \Re^{R \times R}$ and $P'(k) \in \Re^{R \times R}$, where $P(k)$ is in the lower-triangular structure and $P'(k)$ satisfying $P'_{ii}(k) \geq \epsilon_p > 0$. Then,*

$$\lim_{k \to \infty} \prod_{l=0}^{k-1} P(k-l)P'(k-l) = \mathbf{1}_R c_1, \tag{29}$$

if and only if $\lim_{k \to \infty} \prod_{l=0}^{k-1} P(k-l) = \mathbf{1}_R c_2$, where c_1 and c_2 are constant vectors.

The following assumption shows the less-restrictive conditions on the connectivity topologies of the agents.

Assumption 1 *Suppose that there exists a sub-sequence $\{s_v, v = 0, 1, \cdots, \infty\}$ of $\{0, 1, 2, \cdots, \infty\}$, such that $S_{T_1}(t_{s_v}^S)$ is in the same lower-triangular structure (that is, $T_1(t_{s_v}^S) = T_{1c}$ for all v, where T_{1c} is a fixed permutation matrix), and satisfies the conditions that (i) $S_{T_1,ii}(t_{s_v}^S)$ is irreducible and (ii) for every $i > 1$, there is at least one j such that $S_{T_1,ij}(t_{s_v}^S) \succ 0$, $j < i$.*

Theorem 1. *Consider the cooperative control of system (1) under assumption 1. Given control (22) with the corresponding feedback matrix $G(t_k^S)$ is chosen according to $S(t_k^S)$, in particular, $G(t_{s_v}^S)$ is chosen according to*

$$
G_{ii}(t_{s_v}^S) = \begin{bmatrix} 0 & I_{(m-1) \times (m-1)} \\ \dfrac{S_{ii}(t_{s_v}^S)}{\sum_{j=1}^q S_{ij}(t_{s_v}^S)} & 0 \end{bmatrix},
$$
$$
G_{ij}(t_{s_v}^S) = \begin{bmatrix} 0 & 0 \\ \dfrac{S_{ij}(t_{s_v}^S)}{\sum_{j=1}^q S_{ij}(t_{s_v}^S)} & 0 \end{bmatrix}, \ j \neq i. \tag{30}
$$

Then, the stability of the overall closed-loop system can be guaranteed with

$$
\lim_{t \to \infty} x(t) = 1_{N_q} c x(t_0^G), \tag{31}
$$

where constant vector $c \in \Re^{1 \times N_q}$.

Proof. The proof starts with finding the expression of the solution of the overall system, and then the convergence is proved by studying the convergence properties of product of a sequence of row stochastic matrices.

Step 1. Define the state transformation

$$
x = T_3 z, \tag{32}
$$

where $T_3 = T_1 \otimes I_{lm \times lm} \in \Re^{lmq \times lmq}$. Then, the system dynamic (23) becomes

$$
\dot{z} = -(I - T_3^T D T_3)z = -(I - D_{T_3})z. \tag{33}
$$

The solution of (33) is

$$
z(t_{k+1}^S) = P(t_k^S)z(t_k^S), \ k = 0, 1, \cdots, \tag{34}
$$

where

$$
P(t_k^S) = e^{-(I - D_{T_3}(t_k^S))(t_{k+1}^S - t_k^S)}. \tag{35}
$$

For notational convenience, denote $T_3(k) = T_3(t_k^S)$ and $P(k) = P(t_k^S)$. It then follows from (32) and (34) that

$$
x(t_{k+1}^S) = \prod_{l=0}^k T_3(k-l)P(k-l)T_3(k-l)^T x(t_0^S). \tag{36}
$$

To prove (31), it suffices to prove that

$$\lim_{k\to\infty} \prod_{l=0}^{k} T_3(k-l)P(k-l)T_3(k-l)^T = \mathbf{1}_{N_q}c. \tag{37}$$

Step 2. We then show the convergence of (37). It follows from the lower-triangular structure of $D_{T_3}(t_k^S)$ that $P(k)$ are also in the lower-triangular structure. Moreover, $P(k)$ is row-stochastic matrix and its diagonal elements are lower-bounded by a positive value [8]. Define $P'(s_v) = T_3(s_v)^T T_3(s_v - 1)P(s_v - 1)T_3^T(s_v - 1) \cdots T_3(s_{v-1} + 1)P(s_{v-1} + 1)T_3^T(s_{v-1} + 1)T_3(s_{v-1})$, the proof of (37) is equivalent to that of

$$\lim_{v\to\infty} P(s_v)P(s_v)'P(s_{v-1})P(s_{v-1})' \cdots P(s_1)'P(s_0) = \mathbf{1}_{N_q}c. \tag{38}$$

By assumption 1, $S_{T_1,ii}(t_k^S)$ is irreducible, then under the choice of $G(t_{s_v}^S)$ in (30), we know that $G_{T_2,ii}(t_{s_v}^S)$ is irreducible by invoking lemma 1. Invoking lemma 1 again and noting (24) and (25), we have that $D_{T_3,ii}(t_{s_v}^S)$ is irreducible and $P_{ii}(s_v) > 0$. On the other hand, $G_{T_2,ij}(t_{s_v}^S) \succ 0$ leads to $D_{T_3,ij}(t_{s_v}^S) \succ 0$ and $P_{ij}(s_v) \succ 0$. It then follows from lemma 2 that

$$\lim_{v\to\infty} P(s_v)P(s_{v-1}) \cdots P(s_0) = \mathbf{1}_{N_q}c_s, \tag{39}$$

where c_s is a constant vector. To this end, the proof can be done by noting $P'_{ii}(\cdot) > 0$ and invoking lemma 3.

Remark 3. Compared to the recent results in [11, 16], the conditions presented in theorem 1 are less-restrictive. In particular, the fact that the dimension p in (26) can be any value from 1 to q says that the system sensor graph do not need to be strongly connected. Assumption 1 corresponds to that system connectivity topologies can change and fixed leader-follower structure is not required for system convergence.

The following remarks illustrate that some system connectivity topologies belong to the special cases of the general result presented in theorem 1.

Remark 4. The fixed leader-follower structure. In this case, the signal transmission matrix $S(t)$ is in the fixed lower-triangular structure, that is, $T_1(t_k^S) = T_{1c}$ for all k. To this end, under the satisfaction of assumption 1, the system convergence directly follows from theorem 1. Two typical leader-follower connectivity topologies are given by figure 4. The corresponding signal transmission matrices are

$$S_1 = \begin{bmatrix} 1 & 0 & 0 & \cdots & 0 \\ 1 & 1 & 0 & \cdots & 0 \\ \vdots & \vdots & \vdots & \ddots & \vdots \\ 1 & 0 & 0 & \cdots & 1 \end{bmatrix}, \quad S_2 = \begin{bmatrix} 1 & 0 & 0 & \cdots & 0 \\ 1 & 1 & 0 & \cdots & 0 \\ \vdots & \vdots & \ddots & \ddots & \vdots \\ 0 & 0 & \cdots & 1 & 1 \end{bmatrix},$$

with $p = q$. It is obvious that the conditions in theorem 1 are satisfied.

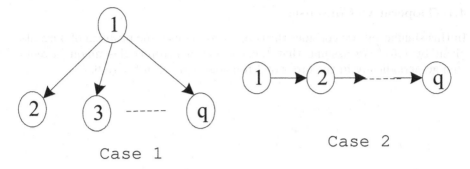

Case 1   Case 2

Fig. 4. Two leader-follower topologies

Remark 5. Circular Pursuit. In this case, the connectivity topology is shown

by figure 5. The signal transmission matrix is $S = \begin{bmatrix} 1 & 0 & 0 & \cdots & 1 \\ 1 & 1 & 0 & \cdots & 0 \\ \vdots & \vdots & \ddots & \ddots & \vdots \\ 0 & 0 & \cdots & 1 & 1 \end{bmatrix}$. In this

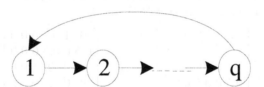

Fig. 5. The topology of circular pursuit

case, S is irreducible, and $p = 1$.

Remark 6. The nearest neighbor rules in [11, 16] correspond to the case that there exists a sub-sequence $\{s_v, v = 0, 1, \cdots, \infty\}$ of $\{0, 1, 2, \cdots, \infty\}$, such that $S(t_{s_v}^S)$ is irreducible, which is stronger than the condition in assumption 1.

4 Simulation

In this section, we provide the simulation results to illustrate the basic results stated in this paper.

4.1 Cooperative Consensus

In this simulation, we consider the cooperative consensus problem of 3 agents given by (15). We assume that the connectivity topologies among agents change periodically in the order of 4 topologies given in figure 6.

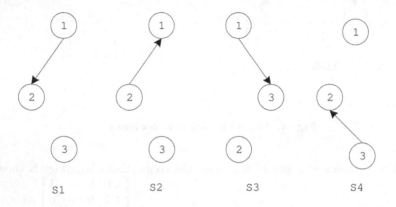

Fig. 6. Connectivity topologies

It is easy to check that signal transmission matrices S_1 and S_3 are in the same structure and assumption 1 is satisfied. According to theorem 1, let the corresponding feedback gain matrix $G(t)$ be

$$G^1 = \begin{bmatrix} 0 & 1 & 0 & 0 & 0 & 0 \\ 1 & 0 & 0 & 0 & 0 & 0 \\ 0 & 0 & 0 & 1 & 0 & 0 \\ 0.5 & 0 & 0.5 & 0 & 0 & 0 \\ 0 & 0 & 0 & 0 & 0 & 1 \\ 0 & 0 & 0 & 0 & 1 & 0 \end{bmatrix}, \quad G^2 = \begin{bmatrix} 0 & 1 & 0 & 0 & 0 & 0 \\ 0.5 & 0 & 0.5 & 0 & 0 & 0 \\ 0 & 0 & 0 & 1 & 0 & 0 \\ 0 & 0 & 1 & 0 & 0 & 0 \\ 0 & 0 & 0 & 0 & 0 & 1 \\ 0 & 0 & 0 & 0 & 1 & 0 \end{bmatrix},$$

$$G^3 = \begin{bmatrix} 0 & 1 & 0 & 0 & 0 & 0 \\ 1 & 0 & 0 & 0 & 0 & 0 \\ 0 & 0 & 0 & 1 & 0 & 0 \\ 0 & 0 & 1 & 0 & 0 & 0 \\ 0 & 0 & 0 & 0 & 0 & 1 \\ 0.5 & 0 & 0 & 0 & 0.5 & 0 \end{bmatrix}, \quad G^4 = \begin{bmatrix} 0 & 1 & 0 & 0 & 0 & 0 \\ 1 & 0 & 0 & 0 & 0 & 0 \\ 0 & 0 & 0 & 1 & 0 & 0 \\ 0 & 0 & 0.5 & 0 & 0.5 & 0 \\ 0 & 0 & 0 & 0 & 0 & 1 \\ 0 & 0 & 0 & 0 & 1 & 0 \end{bmatrix}.$$

The initial positions are $[1,0]^T$, $[0,1]^T$ and $[-1,0]^T$, respectively. Figure 7 shows the convergence of agents' position, which verifies the proposed design in this paper.

4.2 Circling Around a Center

In this example, assume that in the group there are 3 agents given by (15), and two kinds of connectivity topologies appear alternatively during the agent

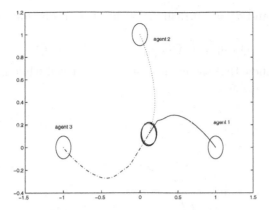

Fig. 7. Consensus of three agents

motion: (i) agent 1 as the leader, agent 2 follows agent 1 and agent 3 follows agent 1; (ii) each agent runs by itself. The corresponding agent sensor matrices are

$$S^1(k) = \begin{bmatrix} 1 & 0 & 0 \\ 1 & 1 & 0 \\ 1 & 0 & 1 \end{bmatrix}, \quad S^2(k) = \begin{bmatrix} 1 & 0 & 0 \\ 0 & 1 & 0 \\ 0 & 0 & 1 \end{bmatrix}.$$

We can choose the corresponding $G(t)$ as

$$G^1(k) = \begin{bmatrix} 0 & 1 & 0 & 0 & 0 & 0 \\ 1 & 0 & 0 & 0 & 0 & 0 \\ 0 & 0 & 0 & 1 & 0 & 0 \\ 0.5 & 0 & 0.5 & 0 & 0 & 0 \\ 0 & 0 & 0 & 0 & 0 & 1 \\ 0.5 & 0 & 0 & 0 & 0.5 & 0 \end{bmatrix}, \quad G^2(k) = \begin{bmatrix} 0 & 1 & 0 & 0 & 0 & 0 \\ 1 & 0 & 0 & 0 & 0 & 0 \\ 0 & 0 & 0 & 1 & 0 & 0 \\ 0 & 0 & 1 & 0 & 0 & 0 \\ 0 & 0 & 0 & 0 & 0 & 1 \\ 0 & 0 & 0 & 0 & 1 & 0 \end{bmatrix}.$$

It is easy to verify that S^1 satisfies assumption 1, and G^1 satisfies the conditions in theorem 1.

The formation control objective is to make the agents circle around a center, and all agents try to keep the same radius and speed, and relative angle offset. Let

$$z^d(t) = [\cos(t), \ \sin(t)]^T.$$

The moving frame $F(t)$ is defined as

$$e_1(t) = \begin{bmatrix} -\sin(t) \\ \cos(t) \end{bmatrix}, \quad e_2(t) = \begin{bmatrix} \cos(t) \\ \sin(t) \end{bmatrix}.$$

The formation is defined by the three points:

$$P_1 = \text{the origin of } F(t), \ P_2 = d_1 e_1 + d_2 e_2, \ P_3 = -d_1 e_1 + d_2 e_2,$$

where $d_1 = 0.5$ and $d_2 = -0.1340$. The initial positions for three agents are given by

$$[0, \ 0]^T, \ \ [-0.5, \ -0.5]^T, \ \ [0.5, \ -0.5]^T$$

Figures 8 to 13 show that the circle motion is achieved while maintaining the formation among agents.

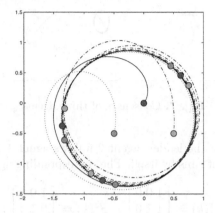

Fig. 8. Circle motion under cooperative control

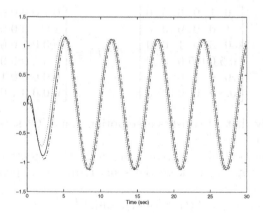

Fig. 9. Longitudinal velocities of 3 agents (z_{121}: solid line, z_{221}: dotted line, z_{321}: dashed line)

4.3 Formation Switching

In this simulation, we consider the formation control problem of 6 agents given by (15). The agents first move into a triangular formation, and switch to the

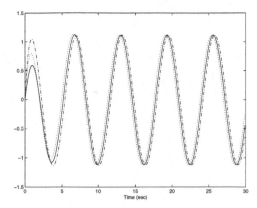

Fig. 10. Lateral velocities of 3 agents (z_{122}: solid line, z_{222}: dotted line, z_{322}: dashed line)

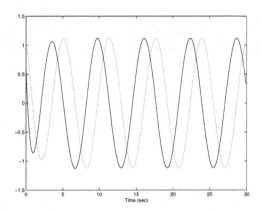

Fig. 11. Cooperative control for agent 1 (v_{11}: solid line; v_{12}: dotted line)

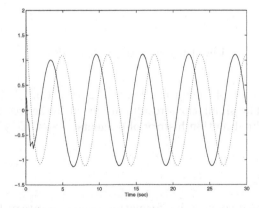

Fig. 12. Cooperative control for agent 2 (v_{21}: solid line; v_{22}: dotted line)

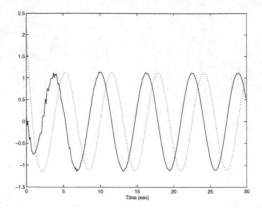

Fig. 13. Cooperative control for agent 3 (v_{31}: solid line; v_{32}: dotted line)

line formation when getting into the narrow space, after that switch back to the triangular formation. During the motion, we assume that two kinds of connectivity topologies appears alternatively, as shown by figure 14.

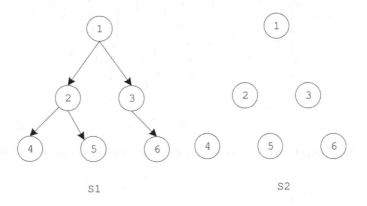

Fig. 14. Connectivity topologies

The initial position of agents are $[2.5, 1]^T$, $[2, 1]^T$, $[1.5, 1]^T$, $[1, 1]^T$, $[0.5, 1]^T$, $[0.5, 1]^T$, $[0, 1]^T$. The simulations results are given by figures 15 to 18, which show the longitudinal motion history of 6 agents.

5 Conclusion

In this paper, a new cooperative formation control strategy has been developed. In summary, the proposed control-design methodology enables us to analyze, understand and achieve cooperative behavior and autonomy for a

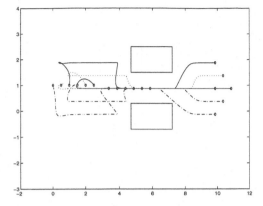

Fig. 15. Formation History ($t \in [0, 100]$)

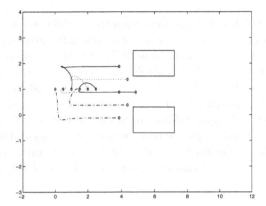

Fig. 16. Formation History Snapshot($t \in [0, 40]$)

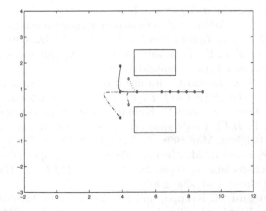

Fig. 17. Formation History Snapshot($t \in [40, 80]$)

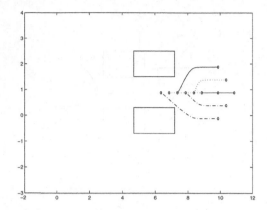

Fig. 18. Formation History Snapshot($t \in [80, \; 100]$)

team of robotic vehicles that are autonomous by themselves to operate, and communicate with and/or sense each other intermittently under the limited communication range constraints. In addition to the development of new and effective cooperative formation controls, the framework can also be used to gain fundamental understandings about and develop analytical analysis tools for the rules and behaviors inspired by nature. Being analytical, the proposed results have wide applicability because its ability of handling such requirements as adaptability, guaranteed performance (asymptotic cooperative behavior), robustness (against uncertainties in communication channels, in sensors and detection, in nonlinear dynamics of the vehicles), and scalability.

References

1. R. Arkin. *Behavior-Based Robotics*. MIT Press, 1998.
2. T. Balch and R. C. Arkin. Behavior-based formation contorl for multirobot teams. *IEEE Trans. on Robotics and Automation*, 14:926–939, 1998.
3. R. B. Bapat and T. E. S. Raghavan. *Nonnegative Matrices and Applications*. Cambridge University Press, Cambridge, 1997.
4. P. Bhatta and N. E. Leonard. Stabilization and coordination of underwater gliders. In *Proc. 41st IEEE Conf. Decision and Control*, 2002.
5. J. P. Desai, J. Ostrowski, and V. Kumar. Controlling formations of multiple mobile robots. In *IEEE Conference on Robotics and Automation*, pages 2864–2869, Leuven, Belgium, May 1998.
6. J. Alexander Fax and R. M. Murray. Finite-horizon optimal control and stabilization of time-scalable systems. In *Proc. 39th IEEE Int. Conf. on Decision and Control*, Sydney, Australia, 2000.
7. D. Fox, W. Burgard, H. Kruppa, and S. Thrun. A probabilistic approach to collaborative multi-robot localization. *Autonomous Robots*, 8(3), 2000.
8. D. Freedman. *Markov Chains*. Springer-Verlag, New York, 1983.
9. V. Gazi and K. M. Passino. Stability analysis of swarms. *IEEE Trans. on Automatic Control*, 48:692–697, 2003.

10. R. A. Horn and C.R. Johnson. *Matrix Analysis*. Cambridge University Press, Cambridge, 1985.
11. A. Jadbabaie, J. Lin, and A. S. Morse. Coordination of groups of mobile autonomous agents using nearest neighbor rules. *IEEE Trans. on Automatic Control*, 48:988–1001, 2003.
12. W. Kang, N. Xi, and A. Sparks. Theory and applications of formation control in a perceptive referenced frame. In *IEEE Conference on Decision and Control*, pages 352–357, Sydney, Australia, Dec. 2000.
13. H. K. Khalil. *Nonlinear Systems, 3rd edition*. Prentice Hall, Upper Saddle River, New Jersey, 2003.
14. J. R. T. Lawton, R. W. Beard, and B. J. Young. A decentralized approach to formation maneuvers. *IEEE Trans. on Robotics and Automation*, to appear, 2003.
15. N. E. Leonard and E. Fiorelli. Virtual leaders, artificial potentials and coordinated control of groups. In *IEEE Conference on Decision and Control*, pages 2968–2973, Orlando, FL, Dec. 2001.
16. Z. Lin, M. Brouchke, and B. Francis. Local control strategies for groups of mobile autonomous agents. *IEEE Trans. on Automatic Control*, 49:622–629, 2004.
17. P. K. Menon and G. D. Sweriduk. Optimal strategies for free-flight air traffic conflict resolution. *J. of Guidance, Control and Dynamics*, 22:202–211, 1999.
18. H. Minc. *Nonnegative Matrices*. John Wiley & Sons, New York, NY, 1988.
19. L. Moreau. Leaderless coordination via bidirectional and unidirectional time-dependent communication. In *Proceedings of the 42nd IEEE Conference on Decision and Control*, Maui, Hawaii, Dec 2003.
20. R. Olfati and R. M. Murray. Distributed cooperative control of multiple vehicle formations using structural potential functions. In *15th Triennial World Congress*, Barcelona, Spain, 2002.
21. L. E. Parker. Alliance: An architecture for fault-tolerant multi-robot cooperation. *IEEE Trans. on Robotics and Automation*, 14:220–240, 1998.
22. Z. Qu. *Robust Control of Nonlinear Uncertain Systems*. Wiley-Interscience, 1998, New York.
23. Z. Qu, J. Wang, and R. A. Hull. Cooperative control of dynamical systems with application to mobile robot formation. In *The 10th IFAC/IFORS/IMACS/IFIP Symposium on Large Scale Systems: Theory and Applications*, Japan, July 2004.
24. Z. Qu, J. Wang, and R. A. Hull. Products of row stochastic matrices and their applications to cooperative control for autonomous mobile robots. In *Proceedings of 2005 American Control Conference*, Portland, Oregon, 2005.
25. W. Ren and R. W. Beard. Consensus of information under dynamically changing interaction topologies. In *Proceedings of the American Control Conference*, Boston, Jun 2004.
26. C. W. Reynolds. Flocks, herds, and schools: a distributed behavioral model. *Computer Graphics (ACM SIGGRAPH 87 Conference Proceedings)*, 21(4):25–34, 1987.
27. R. O. Saber and R. M. Murray. Agreement problems in networks with directed graphs and switching topology. In *Proceedings of the 42nd IEEE Conference on Decision and Control*, Maui, Hawaii, Dec 2003.
28. R. O. Saber and R. M. Murray. Consensus protocols for networks of dynamic agents. In *Proceedings of the American Control Conference*, Denver, CO, Jun 2003.

29. D. Swaroop and J. Hedrick. String stability of interconnected systems. *IEEE Trans. on Automatic Control*, 41:349–357, 1996.
30. T. Vicsek, A. Czirok, E. B. Jacob, I. Cohen, and O. Shochet. Novel type of phase transition in a system of self-driven particles. *Physical Review Letters*, 75:1226–1229, 1995.

Alternative Control Methodologies for Patrolling Assets With Unmanned Air Vehicles

Kendall E. Nygard, Karl Altenburg, Jingpeng Tang, Doug Schesvold, Jonathan Pikalek, Michael Hennebry

Department of Computer Science and Operations Research
North Dakota State University
Fargo, ND 58105, USA
E-mail: {Kendall.Nygard,Karl.Altenburg,Jingpeng.Tang}@ndsu.edu
{Doug.Schesvold,Jonathan.Pikalek,Micheal.Hennebry}@ndsu.edu

Summary. We consider the problem of controlling a system of many Unmanned Air Vehicles (UAVs) whose mission is to patrol and protect a set of important assets on the ground. We present two widely differing methods, employing emergent intelligent swarms and closed-form optimization. The optimization approach assumes complete communication of all newly sensed information among all of the UAVs as it becomes available. The optimization problem is a network flow model that is readily solvable to obtain optimum task allocations to configure patrols for the UAVs in the swarm. Reapplication of the optimization algorithm upon demand yields the benefit of cooperative feedback control. The swarm procedure establishes patrol patterns by utilizing decentralized, reactive, behaviors. Global communication is unnecessary, and control is established only through passive sensors and minimal short-range radio communication. Both models have been implemented and successfully demonstrated in an agent-based, simulated environment. The strengths, weaknesses, and relative performance the two approaches are compared and discussed.

1 Introduction

Unmanned air vehicles (UAVs) have great promise, especially in the area of airborne reconnaissance and surveillance. To date, most UAV missions involve a single UAV for a task. There is great interest in cooperatively controlling multiple UAVs that can work as a team to efficiently and effectively carry out a range of mission tasks, including search, target classification, attack, and battle damage assessment. An example mission is for a team of UAVs to persistently patrol high-value assets in a confined urban area and protect them from mobile threats. The patrolling problem is constrained by limitations in communication, computational power, and incomplete a priori situation awareness.

We describe two widely differing approaches to control for the problem, based on an optimization model and on swarms. In the optimization approach, a time-phased network-based model is solved iteratively to build task assignments for the UAVs. The method works best if there is complete global communication and information knowledge sharing. Running the model simultaneously and independently on all platforms at coordinated points in time ensures that all of the UAVs have knowledge of the same solution without having to explicitly communicate them.

In the swarm approach, we use a highly distributed paradigm based on autonomous UAVs following behaviors that are reactions to information locally available through their sensors. By structuring the available behaviors in prescribed ways, reasonable patrol patterns emerge without centralized control. The appearance of cooperation is a epiphenomenon of their collective interaction.

We demonstrate that the swarm based bottom-up approach is more robust than the centrally optimized one, particularly for operations in uncertain and dynamic environments. This is because the emergent intelligence approach is insensitive to numbers of UAVs, and relies little on situational knowledge, or radio-based inter-agent communication. The solutions are also highly adaptive and flexible. However, the solutions will likely have sub optimal performance measures that are at best feasible and satisficing. The optimization-based approach provides the best possible solutions when complete information about all of the UAVs and the situation is available. The solutions also have the desirable property of not following easily discernible patterns, making it less likely that an enemy could implement countermeasures.

2 Related Works

Work in reactive autonomous agents comes from diverse fields. The swarm approach we developed draws upon graphical multi-agent systems, such as Reeves' particle systems, Resnick's StarLogo, and Reynold's Boids [10,11,12]. The basic control philosophy is inspired by work in autonomous mobile robots, especially the subsumption architecture of Brooks [3] and the behavioral primitives such as aggregate, disperse, and follow, developed by Mataric [6]. The swarm agents are similar to minimalist robotic systems, such as reported in Kube and Zhang and Werger [5,14]. The swarm intelligence approach to patrolling was first suggested by Nygard et. al. [8].

Network flow optimization algorithms have a rich history. The seminal work of Ford and Fulkerson [4] provides an early presentation of network-based algorithms. Classical uses in transportation resource allocation and scheduling can be found in work such as that of Brown and Graves [2] and Nygard, Chandler and Pachter [9]. An application to task assignments using closed form optimization is also presented in Murphey [7].

3 Methods

3.1 Swarm Intelligence

In the swarm solution the patrol patterns are regular flight tracks with different radii and altitudes around the protected asset. Following the tracks while patrolling maintains a persistent presence around the asset for surveillance and possible destruction of hostile intruders. The tracks also provide multiple viewpoints for surveillance as well as multiple layers of protection. The behaviors are structured to favor populating the inner tracks over the outer ones, by using a track switching protocol understood by all of the units. Collision avoidance is a high priority behavior within the protocol. The altitude of a given patrol track is proportional to its radius, so that lower tracks are smaller than upper ones, resulting in an "upside-down wedding cake" configuration. Each patrol track consists of a fixed number of waypoints that form a regular polygon with the asset at its center.

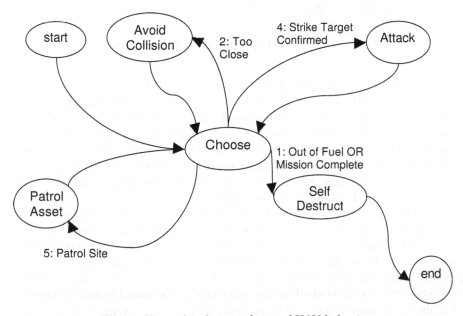

Fig. 1. Hierarchical state charts of UAV behavior.

The high level mission objective emerges from the local UAV behaviors of collision avoidance, patrolling, and attacking. The collision avoidance and attacking behaviors are similar to those used in the sweep search mission described in [13]. We focus here on the patrolling behaviors. The high-level

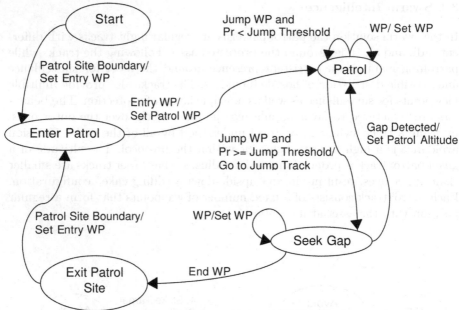

Fig. 2. Detailed state charts of UAV behavior.

control structure is illustrated in the state chart of Figure 1. The control is hierarchical, with the Choose module of Figure 1 being a state diagram module that identifies which lower-level behavioral module is in control at a particular point in time. The agent's environment and internal state is assessed at fixed time intervals by the Choose module. At each cycle, sensory input is processed to determine the best choice of action. Figure 2 illustrates an expansion of the Patrol Asset module into its lower-level state chart consisting of the behaviors enter patrol, patrol, seek gap, and exit patrol.

The enter patrol behavior allows the UAV to enter the outermost patrol track. A UAV maneuvers to orient itself in the predetermined direction of patrol flight. The UAV will enter the outer patrol track unless it encounters another UAV. If another UAV is encountered, it will fly away from the asset for a prescribed distance before repeating the enter patrol behavior, thereby preventing congestion control for the outer track. The patrol behavior consists of orbiting around the asset by flying from waypoint to waypoint. Each UAV in a patrol track maintains a set cruise speed. A probability distribution is sampled to govern whether the UAV will attempt to switch to the next inner track. This decision is made at a pre-specified waypoint, which limits track switching and avoids potential collisions.

A decision for a UAV to switch tracks is based on sensor information that reveals the extent of congestion of the target track. A UAV that is unsuccessful in switching tracks will record that information, and lower its probability of attempting a switch at the next opportunity. Initially the UAVs attempt track switches with 100% probability. As depicted in Figure 3, a track switch attempt requires the following three steps: 1) Move from the patrol track to the jump track, 2) Move from the jump track to the patrol track if an adequate entry gap is detected, and 3) Start over if a gap is not detected. After entering the jump track, the UAV accelerates to a fixed faster speed and begins the seek gap behavior. The UAV seeks a minimum separation distance or gap between UAVs on the lower patrol track. The gap is determined using forward scanning visual sensors and a timer. If the elapsed time between flying over a lower UAV is sufficient and there are no UAVs ahead, the gap seeking UAV enters the new patrol track. The gap calculation is illustrated in Figure 4. The gap calculations are shown in Equations 1 and 2. If the UAV does not find a large enough gap enough it will exit the patrol area immediately prior to completing a full orbit.

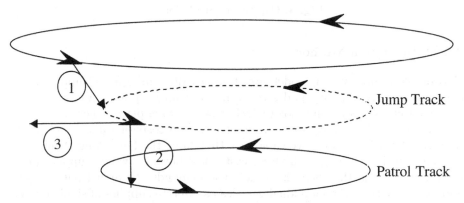

Fig. 3. Track switch protocol.

$$t(v_{fast} - v_{cruise}) \geq \frac{minGapDist}{2} \tag{1}$$

$$t_{fast} = \frac{\Delta l + \frac{minGapDist}{2}}{v_{fast} - v_{cruise}} \tag{2}$$

The purpose of the exit patrol behavior is to leave the patrol area after a failed track switch to avoid collisions with other patrolling UAVs. A UAV flies away from the asset until it is beyond the outermost patrol track, then begins climbing to the altitude of the outermost patrol track. Then the enter patrol behavior is invoked. The exit patrol behavior may also be used when UAVs are low on fuel and must return to base.

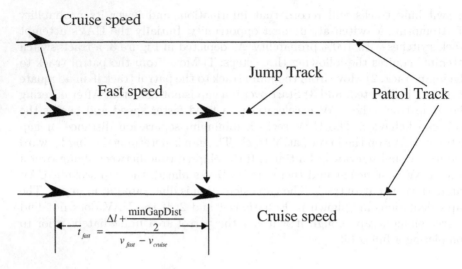

Fig. 4. Gap timer calculation.

3.2 Optimization Method

Network flow optimization models are often described in terms of supplies and demands for a commodity, nodes which model transfer points, and arcs that interconnect the nodes and along which flow can take place. There are typically many feasible choices for flow along arcs, and costs or values associated with the flows. Arcs can have capacities that limit the flow along them. An optimal solution is the globally least cost set of flows for which supplies find their way through the network to meet the demands. To model patrol visit allocation, we treat the individual vehicles as discrete supplies of single units, tasks being carried out as flows on arcs through the network, and ultimate disposition of the vehicles as demands. The numerical values of the flows are 0 or 1. Each assignment of tasks is determined by a solution of the network optimization model. The receipt of new target information is an event that triggers the formulation and solving of a fresh optimization problem that reflects current conditions, thus achieving feedback action. At any point in time, the database onboard each vehicle contains a site set, consisting of indexes, types and locations for vantage points that have been registered for patrolling. Figure 5 illustrates the network model at a particular point in time.

The model is demand driven, with the large rectangular node on the right exerting a demand pull of N units (labeled with a supply of $-N$), so that each of the UAV nodes on the left (with supply of $+1$ unit each) must flow through the network to meet the demand. In the middle layer, the M nodes represent all of the vantage point sites that are registered for patrolling. An arc exists from a specific vehicle node to a site node if and only if it is a feasible UAV / site pair. At a minimum, the feasibility requirement means that there is

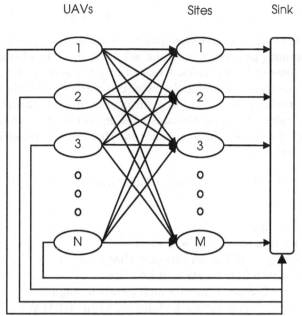

Fig. 5. Network optimization model structure at a specific point in time.

enough fuel remaining to patrol the site if tasked to do so. Other feasibility conditions could also be invoked, if, for example, there were differences in the onboard sensors that precluded certain UAV/site combinations, or if the available observation angles were unsuitable. Finally, each node in the vehicle set on the left has a direct arc to the far right node labeled sink, modeling the option of deferred assignment. The capacities on the arcs from the sites are fixed at 1. Because of the integrality property of network flow problems, the flow values are either 0 or 1 at optimality. This enforces a condition that at most one vehicle can patrol any given site, avoiding the need to model the nonlinear scoring situation that would occur if multiple UAVs could simultaneously patrol a single site. Each unit of flow along an arc has an expected future value (which can be viewed as a negative cost). The optimal solution maximizes total value.

The network optimization model is expressed in closed form as follows:

$$Z = max \sum_{i,j \in I, i \neq j} c(i,j)x(i,j) \tag{3}$$

$$\sum_{j \in I, i \neq j} x(i,j) - \sum_{k \in I, k \neq j} x(k,j) = 0 \quad i \in I \tag{4}$$

$$x(i,j) \leq b(i,j) \quad \{i,j \mid i,j \in I, i \neq j\} \tag{5}$$

$$x(i,j) \geq 0 \qquad \{i,j \mid i,j \in I, i \neq j\} \tag{6}$$

The model is a capacitated transshipment problem (CTP), a special case of a linear programming problem. Constraint set 4 enforces a condition that flow-in must equal flow-out for all nodes. Constraint set 5 mandates that flows on arcs must not exceed specified upper bounds. Basis matrices for the constraints are totally unimodular, which enforces integer solutions. Restricting these capacities to a value of one on the arcs leading to the sink, along with the integrality property, induces binary values for the decision variables x(i,j).

The cost function is:

$$c(i,j) = d(i,j) \times u(i, t(i,j), t_0(i,j), t_1(i,j)) \times n(i,j) \tag{7}$$

where
j = index of a UAV
i = index of a vantage site to be visited
dist_travelable(j) = Maximum distance that j can fly
dist(i,j) = distance from location of j to site i
end_time(j) = scheduled time at which j ends its flight
MIN(end_time,j) = minimum scheduled end time for UAVs other than
    j
MAX(end_time) = maximum scheduled end time over all UAVs
n(j) = (end_time(j) - MIN(end_time,j)/(MAX(end_time) -
   MIN(end_time,j))
d(i,j) = 1 - (dist(i,j)/dist_travelable(j)) × n(j)
t(i,j) = time at which j could reach and visit vantage site i
$t_0(i,j)$ = latest time at which site i would be visited prior to the time at
    which j could visit
$t_1(i,j)$ = earliest time at which sit i would be visited after the time at
    which j could visit
u(i, t(i,j), t_0(i,j), t_1(i,j)) = site_value(i) × (min(t(i,j), (t_i(i,j) - t_0(i,j))-
      t(i,j))) × site_rate(i)

The factor n(j) is designed to favor assigning sites to UAVs with larger amounts of fuel over those with less. The distance function d favors sites that are near over those that are far away, which supports an earlier is better criterion for visiting sites. The site visit urgency function u increases incentive to visit sites as a function of the intrinsic value of visiting the site and elapsed time since the last visit.

The CTP is solved iteratively, with the first iteration solution fixing the first site to be visited by each UAV. For each of these sites assigned to be visited, the data is updated to reflect being visited at the projected time the UAV would arrive. The corresponding UAV is moved in a virtual sense to the location of its first visited site and the appropriate amount of fuel is deducted. The process is repeated iteratively until no further site visits can be accomplished, indicating no further patrolling can be done without violating fuel constraints. The structure of the cost function results in the sequencing of waypoints not following regular patterns. Sufficient fuel is reserved for the UAVs to return to a designated recovery point, at which time replacement UAVs could be launched. Launches of the UAVs can be phased in time to avoid clustering of launches.

3.3 Contrasting the two methods

Both methods have been coded and extensively evaluated with multiple scenarios within simulation environments. Comparisons are given below:

Robustness. The swarm approach relies little on a priori, situational knowledge or high-bandwidth, inter-agent communication, and is indifferent to numbers of, losses, or additions of UAVs. These characteristics make the operation resistant to intrusion by attackers who might try to jam radio frequencies, or impersonate friendly units. The swarm approach is also less vulnerable to loss of global positioning, communication network saturation, lack of battlefield intelligence, and dynamic battlefield conditions. In terms of robustness, the approach is inherently of high performance and intrinsically superior to the optimization-based method.

Scalability. The swarm approach runs on completely separate threads of execution, and is thus highly scalable. The optimization approach has a computational burden for setting parameters and communicating data for the cost function. However, solutions are implicitly shared and the optimization solver is extremely fast. Thus, the optimization approach also scales well.

Performance measures. The swarm approach is at best capable of providing feasible and satisficing solutions that adhere to structured flight patterns. When compared with the optimization approach on measures that pertain to visiting sites at times that maximize surveillance return, the swarm approach is inferior.

Deconfliction. Both the swarm and optimization approaches inherently support deconfliction. However, if many UAVs are employed in a crowded geographical area, the units in the swarms may spend more time carrying out behaviors to avoid collisions than doing useful mission execution. The sequencing accomplished in optimization approach directly avoids collisions. However, if communication is interrupted and the model computes site assignments without full knowledge of all the UAVs tasked in the mission, then the approach may generate collisions. However, it is possible to simply mandate that the UAVs return for recovery if complete information is not available.

4 Conclusions and Future Work

Cooperative control is not possible without communication. In the swarm approach, global communication is assumed to be unavailable, and control is supported through passive sensors and minimal, short-range radio communication that provide simple signals and cues. In the optimization approach, global and perfect communication is assumed, but implicit communication of solutions is accomplished by the solvers running in parallel on all of the UAVs. In both approaches all units share common control programs. The assumption is that all agents will act in a rational and predictable fashion, which reduces the need for communication.

Evaluation carried out through simulation establishes that the swarm approach is effective, robust, and scalable. The approach is particularly well suited for numerous, small, inexpensive, and expendable UAVs. The use of virtual beacons (waypoints), signal-based communication, and simple rules provide a robust and effective method for cooperative control among n UAVs for purposes of patrolling an asset. However, the solutions provided by the swarm approach is not globally optimal.

In the optimization approach, complete communication of data to parameterize the model is called for, but communicating solutions is unnecessary because the model is shared and solved locally and independently. Unlike traditional uses of optimization in weapons system task allocation, the model requires very little preplanning, and is dynamically driven by new information reassigning roles on the fly. Bandwidth requirements are very low. There is a low computational burden on each vehicle, tractable even with modest computers. The potential weaknesses stem from the large burden on accurately specifying cost functions; synchronization that relies on good communication, and acknowledgements of messages.

The optimization approach is preferred if reliable communication is available. The swarm approach is preferred in complex, dynamic, and uncertain environments.

The ultimate goal of the research is to develop robust, intelligent, and high-performance cooperative control strategies for multi-vehicle missions. There is some potential for combing attractive elements of the two approaches. To this end, we are currently evaluating supporting heterogeneous UAV teams and developing hybrid UAV agents. This might lead, for example to an architecture in which optimizing UAVs serve as local leaders and managers of swarms. The managing UAVs could select sites that needed patrolling and communicate that need to the swarming UAVs. The swarming UAVs would follow the manager from site to site and provide a robust, fault tolerant surveillance system at each site. The implementation of such a system would add complexity and new vulnerable points of potential failure, but is likely manageable.

References

1. Altenburg K, Schlecht J, Nygard KE. (2002). An Agent-based Simulation for Modeling Intelligent Munitions", Athens, Greece: Advances in Communications and Software Technologies, pp. 60-65.B. Smith, "An approach to graphs of linear forms (Unpublished work style)," unpublished.
2. Brown, G.G. and Graves, G., "Real-Time Dispatch of Petroleum Tank Trucks," Management Science, 27, 1, pp. 19-32, 1981.

3. Brooks, R.A., "A Layered Control System for a Mobile Robot," IEEE Journal of Robotics and Automation, Vol. 2:1, pp. 14-23, 1986.
4. Ford, L.R., and D.R. Fulkerson, Flows in Networks, Princeton University Press, 1962.
5. Kube, C.R. and Zhang, H., "Collective Robotic Intelligence," Second International Conference on Simulation of Adaptive Behavior, pp. 460-468, December 7-11, 1992.
6. Mataric, M.J., "Designing and Understanding Adaptive Group Behavior," Adaptive Behavior, Vol. 4:1, pp. 51-80, December 1995.
7. Murphy, R.A., "An Approximate Algorithm for a Weapon Target Assignment Stochastic Program," Approximation and Complexity in Numerical Optimization: Continuous and Discrete Problems, editor: P. M. Pardalos, Kluwer Academic, 1999.
8. Nygard, Kendall E., Karl Altenburg, Jingpeng Tang, Doug Schesvold, A Decentralized Swarm Approach to Asset Patrolling with Unmanned Air Vehicles, 4th International Conference on Cooperative Control and Optimization, Destin, FL, USA. November 19 - 21, 2003
9. Nygard, Kendall E., Phillip R. Chandler, and Meir Pachter, Dynamic Network Flow Optimization Models for Air Vehicle Resource Allocation in Proceedings of the American Control Conference, 2001.
10. Reeves, W.T., "Particle Systems - A Technique for Modeling a Class of Fuzzy Objects", Computer Graphics, Vol. 17:3, pp. 359-376, 1983.
11. Resnick, M. Turtles, Termites, and Traffic Jams: Explorations in Massively Parallel Microworlds, Cambridge, MA, MIT Press, 1994.
12. Reynolds, C.,"Flocks, Herds, and Schools: A Distributed Behavioral Model," Computer Graphics, Vol. 21:4, pp. 25-34, 1987.
13. Schlecht J, Altenburg K, Ahmed BM, Nygard KE. (2003). Decentralized Search by Unmanned Air Vehicles using Local Communication", Las Vegas, NV: Proceedings of the International Conference on Artificial Intelligence 2003 (Volume II), pp. 757-762.W.-K. Chen, Linear Networks and Systems (Book style). Belmont, CA: Wadsworth, 1993, pp. 123-135.
14. Werger, B.B., "Cooperation without Deliberation: A Minimal Behavior-based Approach to Multi-robot Teams," Artificial Intelligence, Vol. 110:2, pp. 293-320, June 1999.

6. Healey, A.J.: Layered Control Approach for a Mobile Robotic Field Operator. Robotics and Automation, vol. 30, pp. 18–23, 1999

7. Fenn, J.B., and Lin, Laborory Blaze in Networks of Free Flight, February 1993

8. Khoe, V.L. and Zhang, H.: Collective Robotic Intelligence. Second International Conference on Simulation of Adaptive Behavior, pp. 1–9, December 7–11, 1990

9. Albus, J.S.: "Designing and Understanding Autonomous Robot Behavior," Intelligence, vol. 21, pp. 51–89, December 1998

10. Murphy, R.R.: "An Antonomoos Searching for a Weapons Target Acquisition Feedbacke Program," Approximation and Simulation Series, Springer and Optimization, Computation and Practical Techniques, edited by P.M., Springer and Heidelberg, 1999

11. Arkin, Randall J., Kurt Arkowitz, Ifeachor, Tamar Tian and Sukenick, A Decentralized Action Approach to a Sync Patrolling with Unmanned Air Vehicle, International Conference on Cooperative Control and Optimization, December 11–13, Proceedings 10–23, 2008

12. Shapiro, Randall G., Phillip R. Chandler, and Meir Packhan Durand. Search Path Optimization, to Sell for Air Vehicle Reserved Allocation. Proceedings of the American Control Conference, 2001

13. Reeves, W.R.: Particle Systems, A Technique for Modeling a Class of Fuzzy Objects, ACM Computer Graphics, Vol. 177, pp. 359–376, 1983

14. Resnick, M.: Turtles, Termites, and Traffic Jams, Exploring Vast Massively Parallel Microworlds. Cambridge, MA, MIT Press, 1994

15. Reynolds, C.: Flocks, Herds, and Schools, A Distributed Behavioral Model, Computer Graphics, Vol. 21, pp. 25–34, 1987

16. Schoenwald, D., Stinnett K., Arkin, D.A., Byrne K., (2000) Decentralized Control for a Distributed, Ad-hoc Networks of Mobile Robots, Las Vegas, NV, Proceedings of the International Conference on Artificial Intelligence, 2001, Volume 2, III, pp. 37–54 W. and Chan, Local Networks and Systems, (Book style), Prentice GA, W. Stevens, 1993, pp. 136–145

17. Murphy, R.R., Cooperative Robot Localization, Minimal Behavior-based Approach to Multi-robot Control, Critical Intelligence, Vol. 10.3, pp. 309–326, June 1998

A Grammatical Approach to Cooperative Control

John-Michael McNew and Eric Klavins

Electrical Engineering
University of Washington
Seattle,WA 98195, USA
E-mail: {jmmcnew, klavins}@ee.washington.edu

Summary. In many cooperative control methods, the geometric state of the system is abstracted to the underlying graph or *network topology*. In this paper we present a grammatical approach to modeling and controlling the network topology of cooperative systems based on graph rewriting. By restricting rewrites to small subgraphs, *graph grammars* provide a useful method for programming the concurrent behavior of large decentralized systems of robots. We illustrate the modeling process through an ongoing example and demonstrate mathematical tools for reasoning about the system's behavior. Finally, we briefly describe methods to design continuous controllers that augment the grammar so that geometric requirements may also be satisfied.

1 Introduction

Inexpensive peer-to-peer networking technologies have spurred the investigation of control methods for large-scale networks of complex concurrent systems such as automated highway systems, air-traffic control systems and cooperative systems of robots. Traditional control objectives for individual plants such as stabilization are insufficient to capture the complex behaviors desired from these systems. Furthermore, the scale and complexity of these systems requires that local control of each robot, node, or subsystem must be used exclusively to produce the desired global behavior. In the natural world, members of decentralized systems often *self-organize* in response to environmental stimuli and to each other to produce complex global behaviors. One of the central questions for *engineered self-organization* is: Given a specification of a global behavior, can we synthesize a set of local controllers that produce that global behavior and are robust to uncertainties about the environmental conditions.

In many cooperative control methods, the geometric state of the system is abstracted to the underlying graph or *network topology*. In this paper we focus our efforts on specifying and controlling the evolution of the network topology. In particular, we are interested in controlling the network topology using only local interactions. Some of the control problems that interest us include coordinating multiple vehicles, sequencing tasks in a concurrent environment, and reconfiguring the network topology. Most research in this area assumes a connected network. However, in this paper we are concerned with tasks that often require a partially disconnected network. Our point of departure is the use of *graph grammars* to model how the network topology changes due to local interactions among agents. The graph grammar model is amenable to many standard tools from concurrency theory, which can be used to show that systems meet their specifications.

In this paper, we examine systems that combine exploration and formation forming in response to environmental stimuli. In particular, we consider an example we refer to as "Wandering Scouts." In Section 3 we model this system as a *graph grammar*. In Section 4 we introduce notation to specify behaviors of graph transition systems. In Section 5 we use equivalence classes to partition the set of reachable graphs into *macrostates*. In Section 6.1 we adapt standard concurrency methods to the current setting and prove that for a class of initial systems, eventually it is always the case that the terminal graph of the system meets a desired criterion. In Section 6.2 we show that for a larger class of initial conditions, the grammar proposed has at least one trajectory where deadlock occurs. We augment the system and prove it meets the criterion for the larger class of graphs. Finally, in Section 7 we briefly explain how to design continuous controllers that use the topologies generated by the grammar to guarantee proper formation forming.

2 Related Work

One of the earliest compelling models of self-organization in a continuous state space is the local interaction model proposed by Reynolds to simulate bird flocking behavior. Reynolds motivates motion by three steering behaviors: separation, alignment and cohesion. More recently, Leonard and Fiorelli [9] analyze flocking behavior using potential function theory and Lyapunov methods. Fax and Murray [3] use graph theoretic methods to analyze the stability of such formations, while Tabuada et al [12] show which formation graphs have feasible non-trivial trajectories. In these efforts, the connection topology is fixed and the underlying graph is connected. Jadbabaie, Lin and Morse [4] use ergodic matrix theory to demonstrate that under certain restrictions a discrete time simplification of the Reynolds model is stable for essentially arbitrary switching sequences.

With the occasional exception of a *leader* robot, these results utilize essentially homogeneous controllers on all the robots. We are interested in the concurrent execution of multiple tasks, thus we examine programmed switching between heterogenous controllers. Olfati-Saber [11] describes controlled switching of graph topology using a hybrid automaton for the purpose of squeezing through tight spaces. However, similar to most of the previous results, a fully *connected graph* is assumed. Since we are interested in scenarios where smaller teams of robots complete tasks concurrently, we model systems wherein the overall topology is not necessarily always *connected*.

Klavins [5] describes self-organization of robot formations as a graph process where the discrete states of robots are represented by symbols. Klavins, Ghrist, and Lipsky [7] introduce graph grammars to assemble pre-specified graphs from an initially disconnected graph. By restricting rewrites to small subgraphs, graph grammars provide a useful method to program the concurrent behavior of large decentralized systems of robots. Klavins et al. [5, 6, 7] demonstrate the use of *graph grammars* to define local interaction rules for assembly, replication and other tasks. An application of graph grammars to robotic systems is demonstrated wherein free-floating robots use graph grammars to assemble into larger structures in a predictable and robust manner [1].

3 Systems and Graphs

3.1 A Motivating Example

We informally present an example cooperative control scenario we refer to as "Wandering Scouts." Throughout the paper we use this example to illustrate the process of converting a system to a formal graph grammar model and the process of reasoning about that model.

Suppose a group of robotic scouts with only local communication and sensing capabilities patrols an area to protect against enemy incursions. If three robotic scouts surround an enemy agent, they can capture it, and transport it to a detention center. One possible strategy is to send out the scouts in teams of three. However, we do not know a priori the location or strength of the enemies. We choose rather to send out the scouts to patrol individually, thus covering a greater area. If a scout is in patrol mode and senses an enemy, the scout chases it, thereby disrupting its activities. Once a scout is pursuing an enemy, it may recruit other nearby patrolling scouts to help encircle, capture, and transport the enemy to a detention center. There are four essential subtasks in our problem.

1. Random patrol coverage,
2. Disruption of the enemies' activities,
3. Capture and transport of enemies, and
4. Detention of enemies.

Informally, since we can neither specify the controllers and objectives of the enemy nor their initial density and spatial distribution, we often consider the enemies to be an "environmental stimuli". The graph topology for this system arises from local interactions between the robotic system and the environmental stimuli. Although there is no formal connection between the

Robotic Scouts	Enemies
w: a patrol or *wandering scout*	e: an undisrupted *enemy*
h: a pursuer or *hunter*	d_k: a *disrupted enemy* with degree k
l: a *leader*	c: a *captured enemy*
f: a *follower*	p: a detained enemy or *prisoner*

Fig. 1. Operational modes and the associated labels for the robotic scouts and enemies.

network topology and the spatial distribution, certain initial spatial distributions give rise to characteristic orderings of the local interactions.

3.2 Graph Grammars

A *simple labeled graph* over an alphabet Σ is a triple $G = (V, E, l)$ where V is a set of *vertices*, E is a set of *edges*, and $l : V \rightarrow \Sigma$ is a labeling function. In this paper, a graph is a model of the *network topology* of an interconnected collection of robots, vehicles or particles. A vertex x corresponds to the index of a robot. The presence of an edge xy corresponds to a physical and/or communication link between robots x and y. We use the label $l(x)$ of robot x to keep track of local information and also to indicate the operational mode of the robot.

Example 1. The labels in Figure 1 indicate the operational modes of the robotic scouts and the enemy agents. Additionally we denote by j a detention center or *jail*.

▲

A *graph grammar* consists of a set Φ of rules. Each rule $r = (L, R)$ is a pair of labeled graphs over some small vertex set $V_L = V_R$. Let G be a larger graph representing a possible state of a system and let h be an injective, label and edge preserving map from V_L into G. We call h a witness. The pair (r, h) describes an action on G that produces a new graph $G' = (V, E', l')$ defined by

$$E' = (E - \{h(x)h(y)|xy \in E_L\}) \cup \{h(x)h(y) \mid xy \in E_R\}$$
$$l'(x) = \begin{cases} l(x) \text{ if } x \notin h(V_L) \\ l_R \circ h^{-1}(x) \text{ otherwise.} \end{cases}$$

That is, we replace $h(L)$ (which is a copy of L) with $h(R)$ in the graph G. We write $G \xrightarrow{r,h} G'$ or equivalently $G' = f_{(r,h)}(G)$ to denote that we obtain G' from G by applying action (r, h).

Example 2. In Figure 2 we pose the rule set Φ as a way to model the wandering scouts system.

By convention we refer to the rules in the order they are displayed. So we refer to the rule at the top of Figure 2 as rule one or r_1, the next rule down as

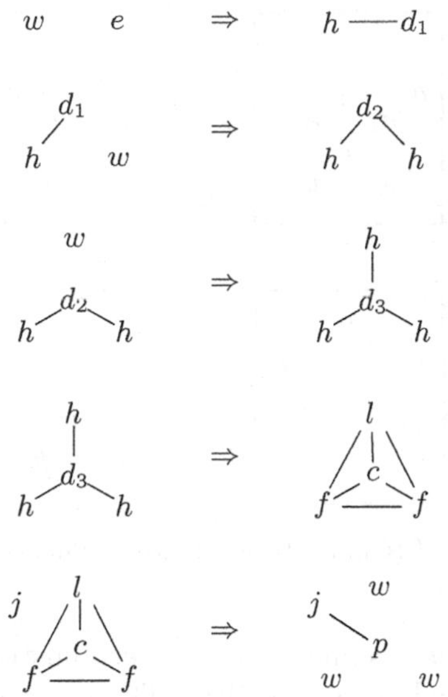

$$w \quad e \quad \Rightarrow \quad h\text{---}d_1$$

$$
\begin{array}{c} d_1 \\ / \\ h \quad w \end{array}
\quad \Rightarrow \quad
\begin{array}{c} d_2 \\ / \ \backslash \\ h \quad h \end{array}
$$

$$
\begin{array}{c} w \\ \nearrow\!\!\!\!\!\!d_2\!\!\!\!\searrow \\ h \qquad h \end{array}
\quad \Rightarrow \quad
\begin{array}{c} h \\ | \\ \nearrow\!\!d_3\!\!\searrow \\ h \qquad h \end{array}
$$

$$
\begin{array}{c} h \\ | \\ \nearrow\!\!d_3\!\!\searrow \\ h \qquad h \end{array}
\quad \Rightarrow \quad
\begin{array}{c} l \\ \diagup | \diagdown \\ c \\ f\text{---}f \end{array}
$$

$$
\begin{array}{c} l \\ j\,\diagup | \diagdown \\ c \\ f\text{---}f \end{array}
\quad \Rightarrow \quad
\begin{array}{c} w \\ j \diagdown \\ p \\ w \quad w \end{array}
$$

Fig. 2. A grammar Φ for the wandering scouts example.

rule two or r_2 and so on. In our system, an edge between two vertices indicates one or both of the agents try to maintain a specified interagent distance. Thus, execution of the first rule, r_1, in Φ indicates when a local interaction occurs between a patrolling scout w and an enemy e, the scout gives chase creating an edge and changing its mode to h. This disrupts the enemy's activities, thus its label changes to d_1. (The subscript "1" indicates that one robot is connected to the enemy). The second and third rules recruit additional robots to chase the enemy.

The fourth rule adds edges between the scouts. In the full system (i.e. including spatial aspects), the edges and labels in the right hand side of rule r_4 will be used by the scouts' continuous controllers to "encircle" the enemy and capture it. Since the system will ultimately use leader-follower formation control, rule r_4 also changes one robot's label to l and the other's to f. Note that we informally refer to the right hand side of r_4 as an *encirclement* component. Finally the fifth rule transfers the captured enemy to the detention center and returns the robotic scouts to patrol mode w. ▲

3.3 Systems and Trajectories

A *system* (G_0, Φ) consists of an initial graph G_0 and a set of rules Φ. A *trajectory* is a (finite or infinite) sequence

$$G_0 \xrightarrow{r_1, h_1} G_1 \xrightarrow{r_2, h_2} G_2 \xrightarrow{r_3, h_3} \dots$$

where $r_i \in \Phi$. If the sequence is finite, then we require that there is no rule in Φ applicable to the terminal graph.

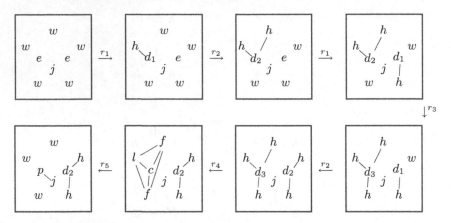

Fig. 3. A trajectory of (G_0, Φ) demonstrating concurrent capturing of two enemies.

A system (G_0, Φ) defines a non-deterministic dynamical system whose states are the labeled graph over V_{G_0}. The system is non-deterministic since, at any step, many rules in Φ may be simultaneously applicable, each possibly via several witnesses. This results in a family of trajectories we denote by $\mathcal{T}(G_0, \Phi)$.

Example 3. Suppose N, M, and K are positive integers such that N is the number of vertices initially labeled by w, M is the number of vertices initially labeled by e and K is the number of vertices labeled by j. For the wandering scouts example, consider initial graphs of the form

$$G_0(N, M, K) = \{\{1, ..., N + M + K\}, \varnothing, l_0\} \tag{1}$$

where initially there are no edges (so that $E_0 = \varnothing$), and the initial labeling l_0 is defined by

$$l_0(i) = \begin{cases} w & i \leq N \\ e & N < i \leq N + M \\ j & N + M < i \leq N + M + K. \end{cases}$$

We can define the class of graphs of interest for the wandering scouts scenario by

$$\mathcal{G}_0 = \{G_0(N, M, K) \mid N, M, K \in \mathbb{N}^+ \wedge N \geq 3\}.$$

To illustrate we choose $G_0 \in \mathcal{G}_0$ with $N = 5, M = 2$, and $K = 1$. Figure 3 shows a partial trajectory of the system (G_0, Φ). Initially there are two enemies that the scouts must capture and transport. In this trajectory all of the scouts concurrently attempt to chase and capture the two enemies. This trajectory models the situation where the enemies and scouts are spatially interspersed.

Figure 4 shows a second possible trajectory of the system. In this trajectory neither of the scouts on the bottom attempts to chase or capture

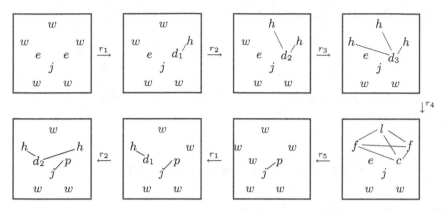

Fig. 4. A partial trajectory of (G_0, Φ) illustrating the sequential capture of two enemies.

the enemy. This trajectory might correspond to a situation where a group of scouts captures and transports an enemy, then returns to patrolling. The next incursion of an enemy occurs in the area they are patrolling. Since we do not know beforehand the spatial distribution of the enemies it is important for our grammar to work for both of these types of trajectories.

▲

The set of all graphs reachable from G_0 via some trajectory is called the *reachable set* $\mathcal{R}(G_0, \Phi)$. The set of all connected components of graphs in $\mathcal{R}(G_0, \Phi)$ up to isomorphism is denoted $\mathcal{C}(G_0, \Phi)$. We suppose that each reachable component type has a single representative in $\mathcal{C}(G_0, \Phi)$. Let $\mathcal{G}$ be the set of all labeled graphs. The components of a grammar $\mathcal{C}(\Phi)$ are given by

$$\mathcal{C}(\Phi) = \bigcup_{G_0 \in \mathcal{G}} \mathcal{C}(G_0, \Phi).$$

If no rules in Φ can alter a reachable component, the component is said to be *stable*. The set of stable components of a grammar is denoted $\mathcal{S}(\Phi)$.

Example 4. Suppose $G_0 = G_0(5, 2, 1)$ is the initial graph defined in Example 3. The components of the system (G_0, Φ) are those pictured in Figure 5. The system only produces these components because there are only two enemies in the initial graph.

We will refer to the components of the grammar as $C_1, C_2, \ldots$ in the order they appear in Figure 5. The ellipsis indicates that while the set of components of the system $\mathcal{C}(G_0, \Phi)$ is finite, the set of components of the grammar, $\mathcal{C}(\Phi)$ is infinite. Specifically, it indicates a sequence of graphs C_k beginning at C_7 where C_k is a star-graph with a vertex labeled j at its center and all other vertices labeled p. Then C_{k+1} is a star graph with one more vertex labeled p.

▲

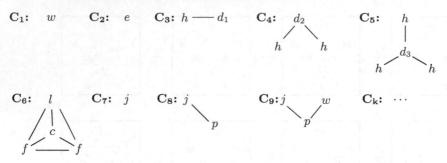

Fig. 5. Components of the grammar Φ, $C(\Phi)$.

4 Propositions About Graphs

Let $\mathcal{G}$ be the set of all labeled, finite graphs. By a *proposition*, we simply mean a subset $P \subseteq \mathcal{G}$ of graphs. By defining propositions in this manner, we avoid having to define a syntax and semantics for logical statements about graphs. Informally, we will describe propositions by logical formula and use double brackets to denote the set of graphs that satisfy the formula. For example,

$$[\![l(1) = b \wedge \exists x \exists y.xy \in E]\!]$$

denotes the set of graphs $G = (V, E, l)$ such that $1 \in V$, $l(1) = b$ and $E \neq \varnothing$. In general, any closed formula about labels and edges using finite quantification over V or E and using constant symbols for elements in V is permitted. If P is a proposition, the we define

$$(P, \Phi) = \{(G_0 \Phi) \mid G_0 \models P\}$$

to be a class of systems.

Example 5. Define the proposition P_0 to be

$$P_0 = [\![G \in \mathcal{G}_0 \mid N > 2M]\!].$$

Then (P_0, Φ) denotes a restricted class of systems to which the system (G_0, Φ) defined in Example 3 belongs. ▲

Definition 4.1 *Let* $\sim \, \subseteq \mathcal{G} \times \mathcal{G}$ *be an equivalence relation on* $\mathcal{G}$. *A proposition* P *is preserved by* $\sim$ *if, for all* $G, G' \in \mathcal{G}$, *if* $G \sim G'$ *then*

$$G \in P \Leftrightarrow G' \in P.$$

If AP *is a set of propositions, then* $AP_\sim$ *is the subset of propositions in* AP *that are preserved.*

We often wish to know what propositions are preserved by a given equivalence relation. For example, suppose $\sim$ is the relation *labeled graph isomorphism*, denoted $\simeq$. Any proposition that can be represented by a formula not using constant symbols to represent vertices in V is preserved by $\simeq$.

This paper makes limited use of Linear Time Logic (LTL) to specify properties. In particular we use temporal logic formulas of the form

$$f = \mathbf{A} \, \mathbf{FG} \, P.$$

Here $\mathbf{A}$ is the path quantifier that denotes "Along all trajectories." The symbol F denotes "eventually" and G denotes "always." Thus for graph grammar systems the above formula reads "Along all trajectories it is eventually always the case that the next graph is in P." We write $(G_0, \Phi) \models f$ if the trajectories of the system are consistent with the formula f. Additionally if P_0 is a proposition we write

$$(P_0, \Phi) \models f$$

when for all $G_0 \in P_0$, $(G_0, \Phi) \models f$.

5 Macrostates

While graph grammars provide a method of programming individual robots, it is often easier to reason about grammars in the abbreviated notation of *macrostates*. Given a temporal logic formula over a set of propositions, $\{P_1, P_2...P_k\}$ it may be possible to find an equivalence relation in which all propositions in the set are preserved. McNew and Klavins [10] use equivalence relations to reduce the size of a graph grammar *model* to make it amenable to *model checking*. Here our goal is to find equivalence relations that preserve the underlying transition system, thus allowing us to reason about issues of progress and safety without reference to the details of rule application and the underlying graph.

The most obvious equivalence relation on graphs is graph isomorphism. This is quite natural given that the grammars we consider regard all vertices as essentially identical. In the context of self-organization, it is often useful to represent an equivalence class generated by the isomorphism relation by listing the number of each component type present in graphs in the class. Thus, suppose that $\mathcal{C}(G_0, \Phi) = \{C_1, C_2, ...\}$. Then $\mathbf{v} : \mathcal{C}(G_0, \Phi) \to \mathbb{N}$ represents all graphs $G \in \mathcal{R}(G_0, \Phi)$ with $\mathbf{v}(1)$ components isomorphic to C_1, $\mathbf{v}(2)$ components isomorphic to C_2 and so on. We write these representatives in vector notation. For the system presented in Example 3 where C_1 is a component of type w, $C_2 = e$ and $C_3 = d_1 - h$, and so on, the vector

$$\mathbf{v} = (\, 4, \, 0, \, 1, \, 0, \, 0, \, 0, \, 0, \, 1, \, 0 \,)^T$$

denotes that $\mathbf{v}(1) = 4$, $\mathbf{v}(3) = 1$, $\mathbf{v}(8) = 1$, and all other entries are zero. If $\mathbf{v}$ represents the equivalence class $[G]_\simeq$ of a graph G, we write $G \models \mathbf{v}$ to denote that G is consistent with $\mathbf{v}$ and we may write $\mathbf{v}_G$ instead of just $\mathbf{v}$. This highlights the fact that $\mathbf{v}$ is a proposition. Note that if $H \models \mathbf{v}_G$, then $H \simeq G$. In keeping with the self-assembly paradigm which is typically addressed in the context of statistical mechanics, we call $\mathbf{v}$ an *isomorphism macrostate*.

Suppose G and G' are reachable graphs where $\mathbf{v}_G$ and $\mathbf{v}_{G'}$ denote the associated isomorphism macrostates. Let (r, h) be the action such that $f_{r,h}(G) = G'$. Let $\mathbf{a} = \mathbf{v}_{G'} - \mathbf{v}_G$. For example, $\mathbf{a}$ may have the form

$$\mathbf{a} = (\,-1,\,-1,\,1,\,0,\,0,\,0,0)$$

indicating that components of type C_1 and type C_2 are combined into a component of type C_3. If $\mathbf{a}(i) = m < 0$, then m components of type C_i are destroyed by applying the action (r, h). If $\mathbf{a}(i) = m > 0$, then m components of type C_i are created. We call the vector $\mathbf{a}$ a *macro-action*.

Definition 5.1 *Fix a rule set Φ. Let (G_0, Φ) be a system. A macro-action $\mathbf{a}$ is in the* action set *of a system, $\mathcal{A}(G_0, \Phi)$, if there exists graphs $G, G' \in \mathcal{R}(G_0, \Phi)$, and an action (r, h) such that $f_{r,h}(G) = G'$ and $\mathbf{a} = \mathbf{v}_{G'} - \mathbf{v}_G$. The action set a grammar, $\mathcal{A}(\Phi)$, is given by*

$$\mathcal{A}(\Phi) = \{\mathbf{a} \mid \mathbf{a} \in \mathcal{A}(G_0, \Phi) \text{ for some initial graph } G_0 \in \mathcal{G}\}.$$

Additionally we call the matrix whose columns are the actions in $\mathcal{A}(\Phi)$ (or $\mathcal{A}(G_0, \Phi)$) the action matrix *denoted by $\mathbf{A}(\Phi)$ (or $\mathbf{A}(G_0, \Phi)$). We write $\mathbf{A}$ to denote the action matrix when its dependence on Φ and possibly G_0 is clear.*

Definition 5.2 *Consider a macro-action $\mathbf{a} \in \mathcal{A}(\Phi)$ and a graph H where for all i, $\mathbf{v}_H(i) = -\mathbf{a}(i)$ if $\mathbf{a}(i) < 0$ and $\mathbf{v}_H(i) = 0$ otherwise. If there exists a rule in Φ and a witness h such that (r, h) is applicable to H and for $H' = f_{(r,h)}(H)$ the new macrostate is given by $\mathbf{v}_{H'}(i) = \mathbf{a}(i)$ if $\mathbf{a}(i) > 0$ and $\mathbf{v}_{H'}(i) = 0$ otherwise, then the macro-action, $\mathbf{a}$, is said to be* transparent.

When a macro-action is transparent, then one may determine its applicability to a macro-state without reference to the rules.

Proposition 5.1 *Let $\mathbf{v}$ be an isomorphism macrostate and $\mathbf{a}$ be a transparent macro-action in $\mathcal{A}(\Phi)$. Then $\mathbf{a}$ is applicable to $\mathbf{v}$ if and only if for every i such that $\mathbf{a}(i) < 0$, $\mathbf{v}(i) + \mathbf{a}(i) \geq 0$. Furthermore, the new macrostate is given by $\mathbf{v}' = \mathbf{v} + \mathbf{a}$.*

Example 6. For the system (G_0, Φ) in Example 3, the action matrix $\mathbf{A}(G_0, \Phi)$ is given by

$$\mathbf{A}(G_0, \Phi) = \begin{pmatrix} -1 & -1 & -1 & 0 & 3 & 3 \\ -1 & 0 & 0 & 0 & 0 & 0 \\ 1 & -1 & 0 & 0 & 0 & 0 \\ 0 & 1 & 1 & 0 & 0 & 0 \\ 0 & 0 & 0 & -1 & 0 & 0 \\ 0 & 0 & 0 & 0 & -1 & -1 \\ 0 & 0 & 0 & 0 & -1 & 0 \\ 0 & 0 & 0 & 0 & 1 & -1 \\ 0 & 0 & 0 & 0 & 0 & 1 \end{pmatrix} . \tag{2}$$

▲

If the number of components in $\mathcal{C}(\Phi)$ is finite, then the macro-actions in $\mathcal{A}(\Phi)$ have finite dimension. Under this condition, it is often easier to reason about the macrostates. If this is not the case we often identify another equivalence relation that does result in macro-action vectors of finite length.

In Example 4 the set of component types of our grammar $\mathcal{C}(\Phi)$ is shown to be infinite. In particular, there exists an infinite sequence of star-shaped graphs with a vertex labeled j at the center and all other vertices labeled by p. In the final state we want all enemies to be labeled p with an edge to some vertex labeled j. The degree of the vertices labeled j is not important. By exploiting the fact that with respect to the desired behavior the star-graphs are essentially the same, we introduce a new equivalence relation, $\sim$ that creates macrostates and macro-action vectors of finite length.

Definition 5.3 *Let* $\mathbf{v}$ *be any isomorphism macrostate. Suppose that* $\mathcal{C}(G_0, \Phi) = \{C_1, C_2, ...\}$ *ordered as in Figure 5. We denote a new truncated macrostate by* $\widetilde{\mathbf{v}}$ *where* $\widetilde{\mathbf{v}}(i) = \mathbf{v}$ *if* $i < 7$ *and* $\widetilde{\mathbf{v}}(7) = \sum_{j=7}^{\infty} \mathbf{v}(j)$.

Thus for any initial graph G_0, any reachable graph in the system (G_0, Φ) may be expressed as a vector of length 7. For G_0 given in Example 3 the reachable isomorphism macrostate

$$\mathbf{v} = (\ 4,\ 0,\ 1,\ 0,\ 0,\ 0,\ 1,\ 0\ )^T$$

becomes the truncated macrostate

$$\widetilde{\mathbf{v}} = (\ 4,\ 0,\ 1,\ 0,\ 0,\ 1)^T$$

Note that no rule changes a vertex labeled j to any other label, and when j appears in the left hand side of a rule it is disconnected from the rest of the vertices in the rule. This implies that Proposition 5.1 is also true for macrostates and macro-actions derived from the $\sim$ equivalence relation.

6 Reasoning About Graph Grammars

For the class of systems (P_0, Φ) in Example 5 we wish to prove that

1. at any given time a scout robot is either patrolling (labeled by w), disrupting (labeled by h and connected to an enemy) or capturing and transporting an enemy (labeled by l or f and connected to a c) and
2. eventually all enemies are detained and remain so.

We may determine whether the rule set Φ in Figure 2 satisfies the first specification by simply examining all possible transition types. Rules r_1, r_2, and r_3 are the only rules whose left hand sides have w and each rule changes the vertex label to h with a connection to a disrupted enemy. Rule r_4 changes h to l or f while rule r_5 relabels vertices with l and f to w and disconnects them from the entire graph. Thus the first specification is met.

Similarly, we may show that the only labels possible for an enemy are $\{e, d_1, d_2, d_3, c, p\}$. And we may show that a vertex labeled p always remains connected to a vertex labeled j since there is no rule that deletes an edge between them. This result implies that the second specification may be written as a temporal logic statement in truncated macrostate notation: That is for any system whose initial graph has N scouts and K detention centers, the second specification can be written as

$$f = \mathbf{A} \, \mathbf{FG} \, \widetilde{\mathbf{z}}$$

where

$$\widetilde{\mathbf{z}} = (\ N,\ 0,\ 0,\ 0,\ 0,\ 0,\ K\ )^T. \tag{3}$$

The formula f states that eventually it is always the case that the system is in a macrostate $\widetilde{\mathbf{z}}$ with N copies of w, and K graphs, some of which have edges to vertices labeled p.

6.1 Lyapunov Functions on Trajectories

In proving that our system meets the second specification, we adapt some standard methods of reasoning about concurrent systems [8] to graph systems represented in macrostate notation.

Definition 6.1 *A discrete Lyapunov function is a function on isomorphism macrostates* $\mathcal{V} : \mathbb{N}^n \to \mathbb{N}$ *such that*

1. *$\mathcal{V}$ is a positive decreasing function over all trajectories,*
2. *$\mathcal{V}(\mathbf{x}) = 0$ implies for all future states $\mathbf{v}$, $\mathcal{V}(\mathbf{v}) = 0$, and*
3. *$\mathcal{V} > 0$ implies that at least one action (r, h) is applicable.*

Note that our definition of a discrete Lyapunov function is related to, but not exactly equivalent to the standard definition found in discrete systems literature.

Proposition 6.1 *Let P be a proposition and $\mathcal{V}$ be a discrete Lyapunov function for a system (G_0, Φ) such that $V(G) = 0$ for some $G \in P$. Then $(G_0, \Phi) \models \mathbf{A} \, \mathbf{FG} \, P$. In other words, along all trajectories it is eventually always the case that the current and next graphs are in P.*

Finding a function V that meets the requirements of Definition 6.1 is highly dependent on the system and the proposition P. We have the following results for the case when the desired proposition is an isomorphism macrostate or a combination of isomorphism macrostates. The results also apply to any type of macrostate for which Proposition 5.1 holds. Although we develop the following results in terms of isomorphism macrostates, they also apply to the truncated macrostates in Definition 5.3.

Proposition 6.2 *Let $\mathcal{A}(\Phi)$ be the set of possible macrostate actions for a system (G_0, Φ). If there exists a vector $\mathbf{w} \in \mathbb{N}^n$ such that for all $\mathbf{a} \in \mathcal{A}(\Phi)$,*

$$\mathbf{w}^T \mathbf{a} < 0$$

then $\mathbf{x} \mapsto \mathbf{w}^T \mathbf{x}$ is a positive decreasing function on all trajectories in (G_0, Φ).

Example 7. For the class of systems (P_0, Φ), we propose the discrete Lyapunov function $V(\widetilde{\mathbf{x}}) = \mathbf{w}^T \widetilde{\mathbf{x}}$ where $\mathbf{w}$ is given by

$$\mathbf{w} = (\, 0\ 5\ 4\ 3\ 2\ 1\ 0\,)^T. \tag{4}$$

The action matrix of Φ in truncated macrostate notation is given by

$$\mathbf{A}(\Phi) = \begin{pmatrix} -1 & -1 & -1 & 0 & 3 \\ -1 & 0 & 0 & 0 & 0 \\ 1 & -1 & 0 & 0 & 0 \\ 0 & 1 & -1 & 0 & 0 \\ 0 & 0 & 1 & -1 & 0 \\ 0 & 0 & 0 & 1 & -1 \\ 0 & 0 & 0 & 0 & 0 \end{pmatrix}. \tag{5}$$

We will often refer to the actions individually, so we note that $\mathbf{A}(\Phi) = (\mathbf{a}_1\ \mathbf{a}_2\ \mathbf{a}_3\ \mathbf{a}_4\ \mathbf{a}_5)$ Then

$$\mathbf{w}^T \mathbf{A} = -(\, 1,\ 1,\ 1,\ 1,\ 1\,).$$

Thus Proposition 6.2 holds for our grammar, which implies that $\widetilde{\mathbf{x}} \mapsto \mathbf{w}^T \widetilde{\mathbf{x}}$ satisfies the first condition of the Lyapunov function definition.   ▲

Proposition 6.3 *Let $\mathbf{w}$ be a vector as described in Proposition 6.2. If for all $\mathbf{a} \in \mathcal{A}(\Phi)$ there exists an element i such that $\mathbf{a}(i) < 0$ and $\mathbf{w}(i) > 0$, then*

$$\mathbf{w}^T \mathbf{x} = 0 \implies \mathbf{G}\, \mathbf{w}^T \mathbf{x} = 0.$$

Example 8. Let $\widetilde{\mathbf{z}}$ in Equation 3 be the desired final macrostate. Then for $\mathbf{w} = (\, 0\ 5\ 4\ 3\ 2\ 1\ 0\,)^T$, $\mathbf{w}^T \widetilde{\mathbf{z}} = 0$. A review of the action set for our system (i.e. the columns of the action matrix in Equation 5) demonstrates that for every action $\mathbf{a}$ there exists an element i such that $\mathbf{a}(i) < 0$ and $\mathbf{w}(i) > 0$. Thus the function $\widetilde{\mathbf{x}} \mapsto \mathbf{w}^T \widetilde{\mathbf{x}}$ satisfies the second condition of Definition 6.1.
▲

Example 9. We wish to show that whenever $\mathbf{w}^T \widetilde{\mathbf{x}} > 0$, then at least one action is applicable. Suppose there exists a vector $\widetilde{\mathbf{x}}$ where $\mathbf{w}^T \widetilde{\mathbf{x}} > 0$ but no action is applicable. Action $\mathbf{a}_4$ is applicable for any macrostate $\widetilde{\mathbf{x}}$ such that $\widetilde{\mathbf{x}}(5) > 0$. Action $\mathbf{a}_5$ is not transparent because a component of type 7 must be present for the macro-action to be applicable. However, we only consider systems where $\widetilde{\mathbf{x}}(7) > 0$, thus $\mathbf{a}_5$ is applicable if $\widetilde{\mathbf{x}}(6) > 0$. Thus, we must show that for all $\widetilde{\mathbf{x}}$ with

1. $\widetilde{\mathbf{x}}(5) = \widetilde{\mathbf{x}}(6) = 0$ and
2. $\mathbf{w}^T \widetilde{\mathbf{x}} > 0$,

actions $\mathbf{a}_1, \mathbf{a}_2$, or $\mathbf{a}_3$ are not applicable. Under these conditions, either $\widetilde{\mathbf{x}}(2), \widetilde{\mathbf{x}}(3)$, or $\widetilde{\mathbf{x}}(4)$ must be non-zero. Action $\mathbf{a}_1$ is applicable if $\widetilde{\mathbf{x}}(2) > 0$ and $\widetilde{\mathbf{x}}(1) > 0$. Action $\mathbf{a}_2$ is applicable if $\widetilde{\mathbf{x}}(3) > 0$ and $\widetilde{\mathbf{x}}(1) > 0$. Action $\mathbf{a}_3$ is applicable if $\widetilde{\mathbf{x}}(4) > 0$ and $\widetilde{\mathbf{x}}(1) > 0$. Thus it must be the case that $\widetilde{\mathbf{x}}(1) = 0$. Since there are M vertices initially marked e and since there is exactly one of these vertices in each component of type C_2, C_3, and C_4, we have the additional constraint that $\widetilde{\mathbf{x}}(2) + \widetilde{\mathbf{x}}(3) + \widetilde{\mathbf{x}}(4) \leq M$. Note that there is one robotic scout in component C_1, zero scouts in C_2, one scout in C_3, etc. Thus the number of robotic scouts in each component type is given by the vector

$$\mathbf{b} = (1\ 0\ 1\ 2\ 3\ 3\ 0).$$

Because the number of robotic scouts remains constant, for our class of initial graphs P_0 and for any macrostate $\widetilde{\mathbf{x}}$, we require that $\mathbf{b}^T \widetilde{\mathbf{x}} = N > 2M$. However,

$$\mathbf{b}^T \widetilde{\mathbf{x}} > 2M$$

subject to the constraints

$$\widetilde{\mathbf{x}}(2) + \widetilde{\mathbf{x}}(3) + \widetilde{\mathbf{x}}(4) \leq M$$
$$\widetilde{\mathbf{x}}(5) = 0$$
$$\widetilde{\mathbf{x}}(6) = 0$$

can only be satisfied if $\widetilde{\mathbf{x}}(1) > 0$, which is a contradiction of our supposition that $\widetilde{\mathbf{x}}(1) = 0$. Thus for all $\widetilde{\mathbf{x}}$, whenever $\mathbf{w}^T \widetilde{\mathbf{x}} > 0$, then at least one action is applicable.

For all systems (G_0, Φ) where $G_0 \in P_0$, the function $\widetilde{\mathbf{x}} \mapsto \mathbf{w}^T \widetilde{\mathbf{x}}$ meets all three conditions in Definition 6.1. We conclude that

$$(P_0, \Phi) \models \mathbf{A}\ \mathbf{FG}(N,\ 0,\ 0,\ ,0,\ 0,\ 0,\ K\ )^T.$$

▲

6.2 Designing Grammars to Avoid Deadlock

For many initial graphs, simple grammars like the one we describe in the previous section generate deadlock conditions on some of the system's trajectories.

Often more complicated grammars are required to guarantee no deadlock occurs. Lyapunov functions are difficult to find for these grammars and we often use *weak Lyapunov* functions to prove such systems satisfy a specification on terminal behavior.

Example 10. In Section 6.1, we prove that

$$(P_0, \Phi) \models f.$$

where $f = \mathbf{A} \, \mathbf{FG} \, \tilde{\mathbf{z}}$ and $\tilde{\mathbf{z}} = (N, \ 0, \ 0, \ , 0, \ 0, \ 0, \ K \ )^T$. Since the size of the enemy force is unknown we would like to expand the class of initial graphs for which the grammar models f. Specifically, we would like to show that

$$(\mathcal{G}_0, \Phi) \models f.$$

However, we may show by counter example that this is not the case. Consider the initial graph $G_0(N, M, K)$ where $N = 3, M = 2$, and $K = 1$. Then the trajectory

$$\tilde{\mathbf{v}}_0 \xrightarrow{\mathbf{a}_1} \tilde{\mathbf{v}}_1 \xrightarrow{\mathbf{a}_2} \tilde{\mathbf{v}}_2 \xrightarrow{\mathbf{a}_3} \tilde{\mathbf{v}}_3 \xrightarrow{\mathbf{a}_4} \tilde{\mathbf{v}}_4 \xrightarrow{\mathbf{a}_5} \tilde{\mathbf{v}}_5 \xrightarrow{\mathbf{a}_1} \tilde{\mathbf{v}}_6 \xrightarrow{\mathbf{a}_2} \tilde{\mathbf{v}}_7 \xrightarrow{\mathbf{a}_3} \tilde{\mathbf{v}}_8 \xrightarrow{\mathbf{a}_4} \tilde{\mathbf{v}}_9 \xrightarrow{\mathbf{a}_5} \tilde{\mathbf{v}}_{10}$$

where $\tilde{\mathbf{v}}_{10} = (3, \ 0, \ 0, \ , 0, \ 0, \ 0, \ 1 \ )^T$ satisfies f. Consider however the trajectory

$$\tilde{\mathbf{v}}_0 \xrightarrow{\mathbf{a}_1} \tilde{\mathbf{v}}_1 \xrightarrow{\mathbf{a}_2} \tilde{\mathbf{v}}_2 \xrightarrow{\mathbf{a}_1} \tilde{\mathbf{u}}_3$$

where $\tilde{\mathbf{u}}_3 = (0, \ 0, \ 1, \ 1, \ 0, \ 0, \ 1 \ )^T$. No progress can be made from $\tilde{\mathbf{u}}_3$ since no macrostate action applies to $\tilde{\mathbf{u}}_3$. Thus the trajectory does not satisfy f.

▲

Example 11. We wish to define a new grammar Υ such that

$$(\mathcal{G}_0, \Upsilon) \models f.$$

We create Υ by adding the following rules to Φ.

$$\Phi' = \begin{cases} \begin{array}{ccc} \overset{h}{\diagup} & & \overset{h}{\diagdown} \\ d_1 \qquad d_1 & \Rightarrow & e \qquad d_2 \end{array} \\[2em] \begin{array}{ccc} \overset{h}{\diagup} & & \overset{h}{\diagdown} \\ d_1 \qquad d_2 & \Rightarrow & e \qquad d_3 \end{array} \\[2em] \begin{array}{ccc} \overset{h}{\diagup} & & \overset{h}{\diagdown} \\ d_2 \qquad d_2 & \Rightarrow & d_1 . \qquad d_3 \end{array} \end{cases}$$

The new grammar is $\Upsilon = \Phi \cup \Phi'$. The components of our new grammar are the same as the components of Φ so that $\mathcal{C}(\Upsilon) = \mathcal{C}(\Phi)$. Because the new rules do not involve labels p or j, we may still use the truncated macrostate notation developed in Definition 5.3. The action matrix of the new grammar is

$$\mathbf{A}(\Upsilon) = \left(\begin{array}{c|ccc} & 0 & 0 & 0 \\ & 1 & 1 & 0 \\ & -2 & -1 & 1 \\ \mathbf{A}(\Phi) & 1 & -1 & -2 \\ & 0 & 1 & 1 \\ & 0 & 0 & 0 \\ & 0 & 0 & 0 \end{array} \right). \tag{6}$$

▲

Systems with the grammar Υ require slightly different proof machinery leading to the following definitions and results.

Definition 6.2 *Let* $\mathbf{v}$ *be a vector in* $\mathbb{N}^m$. *We call* $\mathbf{v}$ *an* application vector *since* $\mathbf{v}(i)$ *indicates that action* $\mathbf{a}_i \in \mathcal{A}(\Phi)$ *is applied* $\mathbf{v}(i)$ *times. If we denote* $\mathbf{A}(\Phi)$ *as* $\mathbf{A}$, *then the vector* $\mathbf{A}\mathbf{v} \in \mathbb{N}^n$ *is the net change in components after applying the actions indicated by the application vector* $\mathbf{v}$.

Proposition 6.4 *There exists a* G_0 *such that there is a cycle in the transition system of* (G_0, Φ) *if and only if the nullspace of the action matrix* $\mathbf{A}$, *Null*$(\mathbf{A})$, *contains a vector* n *where the entries in the vector are all non-negative.*

Definition 6.3 *Let* $\mathbf{z}$ *be a desired final isomorphism macrostate. Let* $\mathbf{w}$ *be a vector in* $\mathbb{N}^n$ *such that* $\mathbf{w}^T \mathbf{z} = 0$ *and for all actions* $\mathbf{a}_i \in \mathcal{A}(\Phi)$, $\mathbf{w}^T \mathbf{a}_i \leq 0$. *We call the function* $\mathbf{x} \mapsto \mathbf{w}^T \mathbf{x}$ *a weak Lyapunov function.*

Proposition 6.5 *Fix a grammar* Φ, *and let* $\mathbf{x} \mapsto \mathbf{w}^T \mathbf{x}$ *be a weak Lyapunov function for desired macrostate* $\mathbf{z}$. *If*

1. *The set of actions of the grammar* $\mathcal{A}(\Phi)$ *cannot generate a cycle,*
2. *For all* $\mathbf{a} \in \mathcal{A}(\Phi)$ *there exists an element* i *such that* $\mathbf{a}(i) < 0$ *and* $\mathbf{w}(i) > 0$, *and*
3. *Whenever* $\mathbf{w}^T \mathbf{x} > 0$, *then at least one macrostate action is applicable,*

then

$$(G_0, \Phi) \models \mathbf{A}\ \mathbf{FG}\ \mathbf{z}.$$

Example 12. Let $\mathbf{w} = (\ 0\ 5\ 4\ 3\ 2\ 1\ 0\ )^T$ as before. Let $\mathbf{A} = \mathbf{A}(\Upsilon)$ given in Example 6. Then $\mathbf{w}^T \mathbf{A} = -(\ 1,\ 1,\ 1,\ 1,\ 1,\ 0,\ 0,\ 0\ )$. Thus $\widetilde{\mathbf{x}} \mapsto \mathbf{w}^T \widetilde{\mathbf{x}}$ is a weak Lyapunov function. A basis for the nullspace of $\mathbf{A}$ is given by the vectors

$$\left\{ \begin{pmatrix} 1 \\ -1 \\ 0 \\ 0 \\ 0 \\ 1 \\ 0 \\ 0 \end{pmatrix}, \begin{pmatrix} 1 \\ 0 \\ -1 \\ 0 \\ 0 \\ 0 \\ 1 \\ 0 \end{pmatrix}, \begin{pmatrix} 0 \\ 1 \\ -1 \\ 0 \\ 0 \\ 0 \\ 0 \\ 1 \end{pmatrix} \right\}.$$

Clearly there are no elements of the nullspace with all non-negative entries. Thus no cycle can be generated from the actions in $\mathcal{A}(\Upsilon)$. In Example 8 the second condition of Proposition 6.5 is shown to be true for the actions $\{\mathbf{a}_1, ..., \mathbf{a}_5\}$. For every new action in Equation 6,there exists an element i such that $\mathbf{a}(i) < 0$ and $\mathbf{w}(i) > 0$. Thus the second condition is met for the function $\widetilde{\mathbf{x}} \mapsto \mathbf{w}^T \widetilde{\mathbf{x}}$.

Finally, we must show that whenever $\mathbf{w}^T \widetilde{\mathbf{x}} > 0$, then at least one macro-action is applicable. Assume no macro-action is applicable to $\widetilde{\mathbf{x}}$. From the previous analysis in Example 9 we know if no macro-action applies to $\widetilde{\mathbf{x}}$ then $\widetilde{\mathbf{x}}(1) = 0, \widetilde{\mathbf{x}}(5) = 0$ and $\widetilde{\mathbf{x}}(6) = 0$. The three new actions imply that if no action is applicable either $\widetilde{\mathbf{x}}(3) = 1$ and $\widetilde{\mathbf{x}}(4) = 0$ or $\widetilde{\mathbf{x}}(4) = 1$ and $\widetilde{\mathbf{x}}(3) = 0$. Since component C_2 contains zero robotic scouts, C_3 contains one robotic scout, and C_4 contains two robotic scouts, if the number of robotic scouts N in the initial graph is greater than two, then clearly there is a contradiction. Thus $\mathbf{w}^T \widetilde{\mathbf{x}} > 0$ must imply at least one macrostate action is applicable.

The grammar Υ, the class of graphs $\mathcal{G}_0$ and the proposed weak Lyapunov function $\widetilde{\mathbf{x}} \mapsto \mathbf{w}^T \widetilde{\mathbf{x}}$, satisfy all three conditions in Proposition 6.5. Thus we may conclude that

$$(\mathcal{G}_0, \Upsilon) \models f.$$

That is, for any initial condition with at least three robotic scouts, eventually all the enemy will be detained.   ▲

7 Simulation Results

In the wandering scouts example, the graph grammar describes the possible evolution of the network topology of a group of robots. A graph grammar does not, however, describe geometry. To incorporate geometry, we must also design continuous controllers, a process we refer to as *embedding* the graph grammar. In the embedding a continuous state $x \in \mathbb{R}^2$ is associated with each vertex. In this section we use "robots" to discuss spatial information and simply vertices if discussing purely topological concepts. For the system in the wandering scouts example, we suppose the presence of an edge between two robots i and j indicates i and j have a communication link and can detect one another's labels. Without an edge, a robot cannot know the operational modes of its neighbors except during the isolated moments in which rules are being locally checked.

We are interested in scenarios where the robots have limited communication and sensing ranges. The primary issues that must be addressed when designing the continuous controllers are:

1. Designing controllers that use the network topology to enforce geometric conditions such as the *encirclement* condition necessary to capture an enemy.

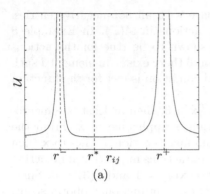 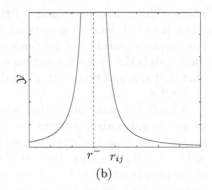

(a)   (b)

Fig. 6. (a) Attractive-repulsive potential function $\mathcal{U}$ for the Wandering Scouts Scenario, (b) Repulsive potential function $\mathcal{Y}$.

2. Designing controllers that guarantee that if progress is possible from a graph G in the graph grammar, it is eventually possible from any spatial state with the same underlying graph G.

We briefly describe a simulation of the wandering scouts example and the continuous controllers used to embed the grammar Υ. In future papers we will present a more formal description of the embedding process in terms of hybrid systems. But these issues are beyond the scope of the current paper.

Suppose r_{ij} denotes the Euclidean distance between two robots i and j. Denote by r_c the communication and sensing radius of the robots. Let $\mathcal{U}$ be an attractive-repulsive potential function.

$$\mathcal{U} : V \times \Sigma \times \Sigma \times E \times \mathbb{R}^{2n} \to \mathbb{R}.$$

Here n is the total number of vertices. Suppose i is the vertex of a robot and j is any other vertex. If the pair of vertices ij is not in the edge set E or if $r_{ij} > r_c$, then $\mathcal{U}(i, l(i), l(j), ij, r_{ij}) = 0$. Otherwise, $\mathcal{U}$ has the form pictured in Figure 6(a).

In the figure, r^* denotes a desired interagent distance between i and j. The maximum separation distance is denoted by r^+. The discontinuity at r^+ is intended to enforce the condition that once an edge is formed it is never broken by moving outside the communication range. The discontinuity at r^- is intended to enforce collision avoidance. For any edge, the parameters r^*, r^+, and r^- may be different for different label pairs.

We also define a purely repulsive potential function

$$\mathcal{Y} : V \times \Sigma \times E \times \mathbb{R}^{2n} \to \mathbb{R}.$$

If $ij \in E$, then $\mathcal{Y} = 0$, otherwise it has the form shown in Figure 6(b). Here r^- is only a function of a robot's own label.

The dynamics of the ith robot are given by

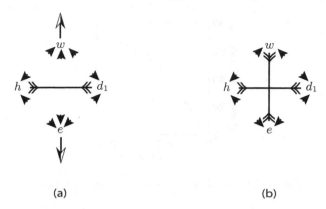

(a)   (b)

Fig. 7. (a)Virtual forces on robots before the application of rule r_1 (b) Undesirable stable equilibrium after the application of rule r_1.

$$\dot{x}_i = - \sum_{\{j|ij\in E\}} \nabla\mathcal{U}(i,l(i),l(j),ij,r_{ij}) - \sum_{\{j|ij\notin E\}} \nabla\mathcal{Y}(i,l(i),ij,r_{ij}) + W.$$

W is continuous random vector process of bounded size that helps guarantee the motion of any two components is only correlated for short periods of time. If this is true and assuming a bounded spatial domain, then with high probability, every macrostate action that is possible in the grammar is also possible in the embedded system.

The function $\mathcal{U}$ has a local minimum at r^*. However the region of attraction is limited to $r^- < r_{ij} < r^+$. In fact outside of this region, $-\nabla\mathcal{U}$ may drive r_{ij} away from r^*. Since $\mathcal{U}$ is only non-zero when there is an edge, and edges are created or destroyed via the application of rules, we require that a rule can only be applied when $r^- < r_{ij} < r^+$. We accomplish this by placing guards on the rules that are boolean functions of the geometry.

Additionally, note that we assume for progress to be guaranteed the motion of the components may only be correlated for small periods of time. Because the potential fields are both attractive and repulsive, if certain geometric conditions are not met, it is possible for components to become permanently interlinked. In particular for any edges ij and kl, we must prevent the embedding of those edges from crossing.

Figure 7 demonstrates the simplest case involving an edge existing between two vertices marked h and d_1. The other vertices marked w and e can apply rule r_1 of the rule set Υ. In panel (a) of Figure 7 the double arrowheads along the edges indicate the direction of the virtual forces generated by $-\nabla\mathcal{U}$. The solid arrowheads show the repulsive virtual forces generated by $-\nabla\mathcal{Y}$. The half-filled arrows indicate the net virtual forces applied to each robot. As these arrows show, the net repulsive force on the robots labeled w and e will eventually drive those robots out of communication and sensing range, thus this configuration is not stable. Note that there are not half-filled arrows

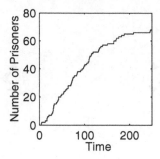

Fig. 8. The number of captured enemy over time.

associated with the vertices marked h and d_1 because in this configuration
the sum of the attractive and repulsive forces is zero. Panel (b) shows how
the forces change if rule r_1 is applied. For each of the four robots, the virtual
forces sum to zero. Thus this configuration represents an undesirable stable
equilibrium in which the motion of the two components will remain correlated.
Suppose h is the label preserving injective mapping of L_{r_1} into G. To ensure
an application of the first rule in Υ creates the proper geometry and that
progress is guaranteed, we place the following guard, gd_1, on rule r_1.

$$gd_1 \iff r^- < r_{h(1)h(2)} < r^+ \bigwedge_{k,m \in sense(h(L))} \neg cross(h(1), h(2), k, m).$$

A robot with vertex k is in the set $sense(h(L))$ if k is not in $h(L)$ and
the distance between i and k, r_{ik} is less than the sensing distance, r_c. The
boolean function $cross$ determines if possible embedded edges between two
pairs of robots cross.

By carefully defining the guards on the rule set Υ and by carefully designing the potential functions $\mathcal{U}$ and $\mathcal{Y}$ we can guarantee that: The only way of
changing the network topology is through the application of rules, the net-
work topology and controllers result in the desired geometries, and with high
probability progress always occurs.

We created a MATLAB simulation of the wandering scouts scenario utiliz-
ing the potential function controllers $\mathcal{U}$ and $\mathcal{Y}$ and the grammar Υ enhanced
by guards. We ran simulations of systems with initial graphs in $\mathcal{G}_0$ ranging
in size from 20 to 500 vertices and various distributions of scouts, enemies
and jails. Figure 8 shows the number of captured enemies over time for a
representative run of a system with 200 robotic scouts, 80 enemies and 5
detention centers. We chose purely random motion in the wandering scouts
mode w as the patrol strategy. The decreasing rate of prisoner detention occurs because the likelihood of randomly encountering an enemy decreases as
the number of enemy not detained decreases. A more structured patrol strategy [2] might result in faster convergence to the point where all enemy are

captured. However, all scenarios regardless of distribution demonstrated converging behavior. Therefore we conclude that the simulation of the embedding of the grammar is consistent with the behavior of the grammar.

8 Discussion

In Section 6.1 we created a one-of-a-kind proof for the simple wandering scouts scenario using Lyapunov methods. These methods are often useful for the design of small subsystems, but as the specifications and systems become more complex, one-of-a-kind methods are more difficult to apply and we expect to eventually use formal verification methods such as *Model Checking*. One of the challenges in model checking graph grammars is the enormous state space generated by graph isomorphism. We demonstrated methods to drastically reduce the size of the model for a limited class of grammars [10]. Model size is a major hindrance in model checking large-scale concurrent networked systems and in future work, we plan to broaden the class of grammars for which we may efficiently compute reduced models.

In section 7 we created continuous controllers and guards on the rule set so that these controllers worked in tandem with the graph grammar to achieve the desired geometries and topologies. This essentially created a locally defined hybrid system we refer to as an *embedded graph grammar*. In future work, we plan to present a formal model of the embedded graph grammar.

We believe the wandering scouts example belongs to a class of problems for which we may be able to automatically synthesize the grammar, controllers, and guards of a solution embedded graph grammar. As the size and complexity of networked systems grows, we expect automatic controller synthesis to become necessary. A key requirement to defining and programming solutions to this class of problems appears to be a method of specification that directly relates the continuous and discrete aspects of the problem and includes a formal notion of "locality."

References

1. J. Bishop, S. Burden, E. Klavins, R. Kreisberg, W. Malone, N. Napp, and T. Nguyen. Self-organizing programmable parts. In *International Conference on Intelligent Robots and Systems*. IEEE/RSJ Robotics and Automation Society, 2005.
2. J. Cortés, S. Martínez, T. Karatas, and F. Bullo. Coverage control for mobile sensing networks. *IEEE Transactions on Robotics and Automation*, 20(2):243–255, 2004.
3. J. Alexander Fax and Richard Murray. Graph laplacians and stabilization of vehicle formations. In *15th IFAC Congress*, 2002.
4. A. Jadbabaie, J. Lin, and A. Morse. Coordination of groups of mobile autonomous agents using nearest neighbor rules. *IEEE Transactions on Automatic Control*, 48(6), 2003.

5. E. Klavins. Automatic synthesis of controllers for distributed assembly and formation forming. In *Proceedings of the IEEE Conference on Robotics and Automation*, Washington DC, May 2002.
6. Eric Klavins. Universal self-replication using graph grammars. In *The 2004 International Conference on MEMs, NANO and Smart Systems*, Banff, Canada, 2004.
7. Eric Klavins, Robert Ghrist, and David Lipsky. A grammatical approach to self-organizing robotic systems. *IEEE Transactions on Automatic Control*, 2005. To Appear.
8. L. Lamport. The temporal logic of actions. *ACM Transactions on Programming Languages and Systems*, 16(3):872–923, May 1994.
9. N.E. Leonard and E. Fiorelli. Virtual leaders, artificial potentials and coordinated control of groups. *Proceedings of the 40th IEEE Conference on Decision and Control (Cat. No.01CH37228)*, vol.3:2968 – 73, 2001.
10. John-Michael McNew and Eric Klavins. Model-checking and control of self-assembly. In *American Control Conference*, 2006. Submitted.
11. Reza Olfati-Saber and Richard M. Murray. Distributed structural stabilization and tracking formations of dynamic multi-agents. In *IEEE Conference on Decision and Control*, 2002.
12. Pedro Lima Paulo Tabuada, George J. Pappas. Motion feasibility of multi-agent formations. *IEEE Transactions on Robotics*, Vol. 21 (3):387–392, 2005.

A Distributed System for Collaboration and Control of UAV Groups: Experiments and Analysis*

Mark F. Godwin [†], Stephen C. Spry, and J. Karl Hedrick

Center for the Collaborative Control of Unmanned Vehicles (C3UV)
University of California, Berkeley, CA, USA

Summary. This chapter describes a distributed system for collaboration and control of a group of unmanned aerial vehicles (UAVs). The system allows a group of vehicles to work together to accomplish a mission via an allocation mechanism that works with a limited communication range and is tolerant to agent failure. This system could be used in a number of applications including mapping, surveillance, search and rescue operations.

The user provides a mission plan containing a set of tasks and an obstacle map of the operating environment. An estimated mission state, described in a high level language, is maintained on each agent and shared between agents whenever possible. This language represents each task as a set of subtasks. Each subtask maintains a state with information on the subtask status, an agent ID, a timestamp, and the cost to complete the subtask. The estimated mission states are based on each agent's current knowledge of the mission and are updated whenever new information becomes available. In this chapter, each subtask is associated with a point in space, although the system methodology can be expanded to more general subtask types.

The agents employ a three-layer hierarchical decision and control process. The upper layer contains transition logic and a communication process. The transition logic manages transitions between tasks and between subtasks, which determine the behavior of the agent at any given time. The communication process manages the exchange of mission state information between agents. Among other capabilities, the subtask transition rules provide time-based fault management; if an agent is disabled or stops communicating, others will assume its subtask after a mission-dependent timeout period. The middle layer contains a trajectory planner that uses a modified potential field method to generate a safe trajectory for a UAV based on the obstacle map and the current subtask objective. The lower layer contains a trajectory-tracking controller that produces heading and airspeed commands for the UAV. Properties of the system are analyzed and the methodology is illustrated through an example mission simulation.

* This work was supported in part by the Office of Naval Research under contract N00014-03-C-0187.
† Corresponding author, markfg@berkeley.edu

1 Introduction

The use of unmanned aerial vehicles (UAVs) to accomplish both military and civilian missions is an active area of research. A significant portion of work to date has focused on control of single UAVs, including trajectory tracking [1] and trajectory planning [2,3]. In addition, coordinated motion of multiple UAVs has been studied and demonstrated with static and dynamic formations of unmanned aircraft [4, 5, 6].

In order to expand the range of missions that UAVs can effectively perform, it is necessary to develop ways for multiple autonomous vehicles to work together in collaborative groups [7,8,14]. This involves applying ideas from the areas of multiagent systems and distributed problem solving to networked multi-vehicle systems, such as UAV groups.

Approaches to team organization and task allocation can be categorized by degree of centralization. On one end of the spectrum, in behavioral [9] or emergent [12] approaches, groups of autonomous agents are designed with individual behaviors that are intended to produce desired group actions and behaviors. On the other end of the spectrum, a group of agents may simply execute commands issued by a central planner. A number of approaches, including auction-based allocation [10,11] and hierarchical dispatching [13], fall between these extremes. The viability of these approaches for a given application is closely tied to the communications topology.

While the more centralized approaches can generally promise more optimal performance, they are also the least scalable and most sensitive to failures of agents or communication links. A centralized planner might work well in ideal cases but may not be feasible in the presence of real-world constraints on communication and computation. Furthermore, any central planner or allocation node represents a single point of failure for the system.

In our system, we consider a group of UAVs with limited communication. The communication topology varies with time as the aircraft move about, and does not generally form a connected graph. This rules out a centralized solution; instead, we seek a distributed solution that does not rely on any fixed communication topology but exploits whatever communication links are present at a given time to coordinate activities and disseminate information throughout the group.

The distributed artificial intelligence community has studied distributed problem solving and task allocation problems for some time now, although mostly in the context of systems of software agents. Work with physical agents includes [9,11,13]. In the ALLIANCE architecture [9], distributed allocation of tasks between a group of robots emerges as a result of agent behavior parameters that describe the agents' tendency to seize tasks from or relinquish tasks to other agents. In this scheme, there is no specific commitment of an agent to complete a task. In the Contract Net Protocol [10,11], tasks are allocated between agents through the use of auctions. The agent with the winning bid is awarded the task and commits to completing the task. The award may be subject to periodic renewal based on task progress.

In this chapter, we describe a distributed task allocation technique based on opportunistic collaboration and exchange of information. Whenever two agents are within communication range, they exchange estimates of the mission 'state.' Following the exchange, each agent merges its current mission state with a mission state received from the other agent.

This approach is similar to the ALLIANCE approach in that tasks are allocated and possibly reallocated between agents without the presence of any third party such as an auctioneer. It is similar to the Contract Net approach in that when allocation or reallocation occurs, it is based on qualification.

The chapter is organized as follows: In section 2, the high level language that is used to describe the state of a mission is discussed. Section 3 describes the mission state estimates that are maintained by each agent in the system and describes the distinction between the local perception of mission state and a global mission limit state. In section 4, we discuss the internal components of an agent that execute a mission. In section 5, we explain the simulation environment used to test the system, and in section 6, we present and discuss an example mission scenario. Finally, section 7 draws conclusions and looks ahead to future work.

2 Mission Plan

Given a group of agents, we want to describe a mission to be accomplished collaboratively. The mission is defined by a mission plan, which consists of a finite set of a distinct tasks:

$$M = \{T_1, T_2, \ldots T_a\}$$

The tasks T_i may describe a wide variety of objectives, such as searching a specified area, patrolling a boundary, or tracking a convoy.

Each task T_i consists of a set of subtasks S_i and a set of task transition rules R_i:

$$Ti = \{S_i, R_i\}$$

where S_i contains one or more subtasks which are defined as one distinct objective that can be accomplished by a single agent. The transition rules, R_i, are Boolean tests used to determine if and when an agent will switch out of one task and into another.

The set S_i consists of the b_i subtasks with in task T_i:

$$Si = \{S_{i1}, S_{i2}, S_{i3}, S_{i4}, \ldots S_{ib_i}\}$$

with each subtask S_{ij} consisting of a set of planner parameters P_{ij} and a set of subtask transition rules R_{ij}:

$$S_{ij} = \{P_{ij}, R_{ij}\}$$

The planner parameter set P_{ij} specifies what type of planner to apply to subtask S_{ij} and contains the necessary parameters for that planner. If subtask S_{ij} consisted of visiting a known location and taking a photograph of the ground, then P_{ij} would contain the coordinates of the location and criteria for path planning. The planner would determine the path to the location and any other actions required to take the picture. The subtask transition rules, R_{ij}, are a set of Boolean tests used to define subtask transitions such as *completion* or *fault*.

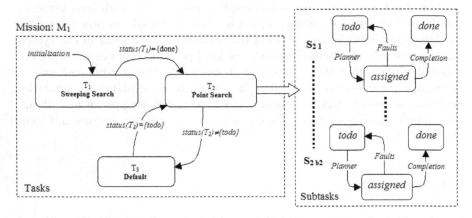

Fig. 1. Diagram of an example mission plan

A diagram of an example mission, M_1 can be seen in Figure 1. As can be seen in the dashed box on the left hand side $M_1 = \{T_1, T_2, T_3\}$. The transition out of T_1 into T_2 would be contained in R_1, the transition out of T_2 into T_3 would be found in R_2 and so on. As described before, a task may contain any number of subtasks. These subtasks are represented on the right hand side. Each subtask can be in three general states, todo, assigned or done. For example, the rules applied to subtask S_{2j} of T_2 are found in R_{2j}. The transition rules R_{2j} define the transitions *assigned→done*, *assigned→todo* and other desired transitions.

All transitions of subtask S_{ij} are governed by R_{ij}, except for the transition between *todo→assigned* which is governed by the planner which will be described in section 4.3.

2.1 Subtasks as Points

Although a subtask could represent many abstract goals, a large set of subtasks can be generalized as a "point" with rules to govern it and a planner to interpret it. The rules specify how to treat the point in space and the planner plans how to accomplish the subtask. The point in space could designate a friend/foe to track or a location to survey. In addition to a stationary or dynamic point a larger number of these points can specify a line, an area or a volume.

Many tasks/subtasks can be represented as traveling to a point and performing some action upon arrival. For example, basic surveillance is just flying between points of interest and gathering information at those points. Tracking a friendly or unfriendly convoy is just trying to reach a point that is continually being updated with the known or perceived position of the convoy.

In these and other scenarios it is a point or series of points that is important and it is up to a trajectory planner to take these points and generate the proper path/plan. For example, in [14] a sinusoidal like trajectory is generated to burn off the greater relative velocity of the UAV to the ground vehicle or convey. This entire trajectory is generated knowing only the position of a ground vehicle and initially specified constant parameters, both of which would be represented in subtask S_{ij} as part of P_{ij}. If it were desired that three unmanned vehicles follow a convoy then there would be three subtasks, all created with similar planner parameters.

3 Mission State Estimates

During operation, the actions of each agent are based on its mission state estimate. The mission state estimate contains the information from the mission plan and is supplemented with additional information on the estimated status of the tasks and subtasks as known by a specific agent k. We will denote the mission state estimate of agent k by $\hat{M}^k$.

As will be explained below, each agent updates its mission state estimate whenever new information becomes available. Due to limited communication between agents, a agent will usually receive some subset of the information updates issued by other agents. Therefore, at any given time, the mission state estimates of two different agents A and B, $\hat{M}^A$ and $\hat{M}^B$, will likely be different.

The mission state estimate of agent k consists of a set of task state estimates, one for each task in the mission plan:

$$\hat{M}^k = \{\hat{T}_1, \hat{T}_2, \ldots, \hat{T}_a\}$$

where we will drop the k superscript on the components of $\hat{M}^k$ in order to simplify notation.

For each task T_i in the mission, the task state estimate $\hat{T}_i$ contains the information:

$$\hat{T}_i = \{\hat{S}_i, \tau_i, R_i\}$$

where $\hat{S}_i$ is a set of subtask state estimates, R_i is the transition rule set, and τ_i is a timestamp that contains the start or end time of task T_i. The set $\hat{S}_i$ consists of the b_i subtask state estimates of task T_i:

$$\hat{S}_i = \{\hat{S}_{i1}, \hat{S}_{i2}, \hat{S}_{i3}, \ldots, \hat{S}_{ib_i}\}$$

For each subtask S_{ij}, the subtask state estimate $\hat{S}_{ij}$ contains the information:

$$\hat{S}_{ij} = \{U_{ij}, A_{ij}, C_{ij}, \tau_{ij}, R_{ij}, P_{ij}\}$$

where U_{ij} is the status of S_{ij}, A_{ij} is an agent ID number, C_{ij} is the cost to accomplish S_{ij}, τ_{ij} is a timestamp for S_{ij}, R_{ij} are the transition rules and P_{ij} are the planner parameters defined in the mission plan. The ^ has been dropped on the elements of $\hat{S}_{ij}$ to simplify notation. The information content is described in more detail below.

U_{ij}: The status of subtask S_{ij}. The value of U_{ij} can be either *todo*, *assigned*, or *done*, and is determined as follows:

1. If S_{ij} is believed to be in progress by any agent, then $U_{ij} = $ *assigned*
2. If S_{ij} is believed to have been completed by any agent, then $U_{ij} = $ *done*
3. Otherwise, then $U_{ij} = $ *todo*

It is important to note that in our system, all subtasks are to be completed exactly one time; once a subtask is *done* it will always be *done* and cannot be restarted. If a cyclic or reoccurring subtask is desired, either a new subtask can be generated periodically, or the subtask transition rules can be defined such that the subtask would never be considered *done*.

A_{ij}: Specifies the identifier of the agent who last set the status of subtask, S_{ij}. If, for example, agent 3 was assigned to subtask S_{24}, then $A_{24}=3$.

C_{ij}: If the status U_{ij} of subtask S_{ij} is *assigned*, then the variable C_{ij} contains the reported cost for agent A_{ij} to accomplish S_{ij}. If, for example, subtask S_{ij} consisted of visiting a point in space then C_{ij} might be an estimate of the time, distance, or energy required to reach that point. In general, C_{ij} may be computed using any desired cost function. However, all subtasks within a task must have comparable cost estimates.

τ_{ij}: A timestamp variable which depends on the value of the subtask status U_{ij}:

1. If $U_{ij} = $ *todo*, then τ_{ij} contains the time U_{ij} was set to *todo*.
2. If $U_{ij} = $ *assigned*, then τ_{ij} contains the time that S_{ij} was assigned to agent A_{ij}.
3. If $U_{ij} = $ *done*, then the variable τ_{ij} contains the time that S_{ij} was completed by agent A_{ij}.

Note that the status, participating agents, and cost to complete a task are not specifically stored in the task state estimate $\hat{T}_i$, as they can be determined directly from the set of subtask state estimates $\hat{S}_i$. On the other hand, the timestamp τ_i that gives the start or end time of a task $\hat{T}_i$ cannot be determined from $\hat{S}_i$ and is therefore stored explicitly in the task state estimate, $\hat{T}_i$.

3.1 Global vs. Local Information

As discussed above, $\hat{M}^k$ is the state of the mission as known by a specific agent k. It is interesting to consider the existence of a limiting mission state estimate. Suppose that time is frozen at time t, communication between agents is unlimited, and the communication algorithm between agents (which will be described in section 4.2) is allowed to run for some finite number of iterations n. Denote the resulting mission state estimate of agent k as $\hat{M}_n^k$. If given a set of K agents there exists an N such that for all $n \geq N$,

$$\hat{M}_n^1 = \hat{M}_n^2 = \cdots = \hat{M}_n^{K-1} = \hat{M}_n^K := \overline{M}$$

then we call $\overline{M}$ the limit state of the mission at time t.

Depending on the level of access to information, at any given time t, $\hat{M}^k$ may be very different from the limit state of the mission $\overline{M}$. The determination of the existence of a limit state $\overline{M}$ as well as a number or upper bound on the number of iterations N required for the information to converge to $\overline{M}$ is part of current and future work.

4 Agents

With an understanding of the information that is contained in a mission plan and a mission state estimate, the contents and functionality of an individual agent can be discussed. For each agent k, we define an internal state $\underline{X}^k$ as

$$\underline{X}^k = \left\{ \begin{array}{l} AgentID \\ Position[x,y] \\ Velocty[v_x, v_y] \\ \hat{M}^k \end{array} \right\}$$

where $\hat{M}^k$ is the mission state estimate as described previously. We also define the message format for the kth agent as

$$\underline{Y}^k = \left\{ \begin{array}{l} AgentID \\ \hat{M}^k \end{array} \right\}$$

As shown in Figure 2, an agent in this system contains four major functional components: communication, transition logic, planning, and low-level control. The communication, transition logic and planner all interact via the internal state $\underline{X}^k$. This interaction occurs primarily through reading and modification of the mission state estimate $\hat{M}^k$. The planner generates a plan based on $\underline{X}^k$ and passes appropriate commands to the low-level controller.

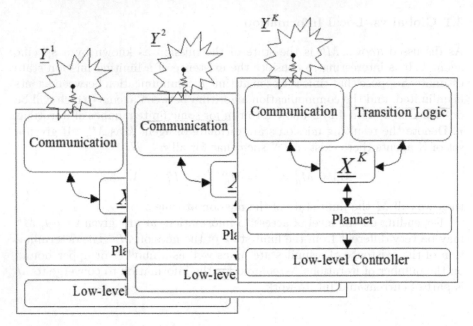

Fig. 2. Internal components of an agent, where K is the total number of agents

4.1 Transition Logic

The transition logic block manages transitions between tasks and between subtasks according to the transition rules specified in R_i and R_{ij} respectively.

The task transition rules R_i are a set of explicit rules that specify under what conditions "big picture" actions should be executed. For example these rules can specify when a convoy should be followed, when an area should be searched or when a group of UAVs should return to base. The subtask transition rules R_{ij} are applied to each subtask. These rules and the specified planner parameters P_{ij} result in more specific behaviors such as mapping, convoy protection or searching. More specifically, the rules contained in R_{ij} define what it means for subtask S_{ij} to be complete as well as decide if a fault has occurred.

Referring back to Figure 1, the rules in R_i specify the transitions found between tasks in the left hand box, and the rules in R_{ij} determine the transitions found in the right hand box, from *assigned→todo* and *assigned→done*. The transition from *todo→assigned* requires a cost estimate, so that responsibility is left to the planner.

From the perspective of an agent there are effectively two different types of *assigned* subtasks: those that assigned to the agent itself, which will be referred to as *current*; and those that are assigned to another agent, which we will refer to as *other*. Because a planner on one agent cannot assign a subtask to another agent, it may be more precise to say the planner of agent

k changes the state of a subtask from $todo \rightarrow current$ in the context of the mission state estimate $\hat{M}^k$. An agent can classify a subtask as $current$ or $other$ by comparing its own agent identifier with the value of A_{ij}^k.

4.2 Communication

The communication block sends and receives messages and integrates received messages into an agent's mission state estimate.

The integration process employs only one operation: overwriting of a subtask state estimate $\hat{S}_{ij}$. The decision to overwrite or not depends on four pieces of information found in each subtask estimate $\hat{S}_{ij}$: status (U_{ij}), time (τ_{ij}), agent ID (A_{ij}), and projected cost to completion (C_{ij}).

Any numbers of subtasks N on two different agents A and B are compared with their equivalent subtask. Only one question is asked: *Will S_{ij} on agent B completely overwrite S_{ij} on agent A?* This takes place on agent A with information sent from agent B. The ij subscripts have been dropped to simplify notation.

For each subtask S: Determine the status of S^A and S^B then check the appropriate condition in Table 1. If the condition is true then overwrite S^A with S^B.

Table 1. Overwrite conditions for a subtask on agent A, from agent B.

Subtask, S	Agent A			
	todo	other	current	done
Agent B todo	$\tau^B \geq \tau^A$	$\tau^B > \tau^A$	$\tau^B > \tau^A$	FALSE
other	$\tau^B \geq \tau^A$	$(A^B = A^A)\&(\tau^B > \tau^A)$	FALSE	FALSE
current	$\tau^B \geq \tau^A$	$\tau^B \geq \tau^A$	$C^B < C^A$	FALSE
done	TRUE	TRUE	TRUE	$\tau^B > \tau^A$

4.3 Planner

The planner block has three principal functions, all of which require the calculation of subtask cost estimates:

1. Choose a subtask when necessary, using current information
2. Calculate an estimate of the cost to complete the current subtask
3. Provide a plan to accomplish the current subtask

When the agent does not have a current subtask, either as a result of completing its previous subtask, or relinquishing it to a better qualified agent, the planner will generate cost estimates for all the subtasks with a *todo* status. The subtask with the lowest estimated cost will then be chosen as the current task for the agent. The planner will then update the state of that subtask, changing its status to *assigned*.

When the agent does have a *assigned* subtask, the planner provides a plan to accomplish that subtask, as well as providing ongoing estimates of the cost to finish the subtask.

In our system, the planner is a path planner and each subtask is character- ized as a point in space. The cost is calculated as the estimated time to reach the target, accounting for the presence of obstacles. The resulting plan is a sequence of waypoints that lead to the point in space. To execute the plan, the low-level controller follows the waypoint sequence to the subtask point.

In general, however, there is no reason the planner couldn't generate differ- ent types of path. For example, the planner could produce a path to a target that minimizes a UAV's radar cross-section. If the subtask point was attached to a moving convoy, then the planner could generate a periodic orbital trajec- tory to sweep out a safe area around the convey, as in [14].

4.4 Low-level Controller

The final layer is the low-level controller that executes a plan. However, as the details of the low-level control are dependent on the physical platform being used this will be discussed, along with the vehicle model, in the simulation section.

5 Simulation

To test and validate our distributed system a MATLAB ® simulation and vi- sualization environment was developed. The simulation environment includes a kinematic aircraft model, a grid-based obstacle map, limited communication and all the internal processes of the agent. All parts of the simulation have been designed in a modular way so that different models, controllers and/or processes can be easily tested, replaced and/or upgraded.

The simulation parameters were chosen to approximate the capability of our experimental platform at UC Berkeley.

5.1 Low Level Controller & Kinematic Model

If an aircraft operates in a flat plane, at low speed and is considered to be small relative to the operating environment then a constrained 2D kinematic model is a reasonable assumption. The governing kinematic equations are:

$$\begin{aligned}
\dot{x} &= V_{aircraft} * \cos(\psi) + V_{windx} \\
\dot{y} &= V_{aircraft} * \sin(\psi) + V_{windy} \\
\dot{\psi} &= u_1
\end{aligned} \qquad (1)$$

where $V_{aircarft}$ is the constant velocity of the aircraft, ψ is the yaw angle and the control action u_1 is equal to the yaw rate. For the simulations the aircraft has a fixed velocity of 20 m/s, wind velocity is set to zero, and $\dot{\psi} \in [-12, 12]^{\circ}/\sec$.

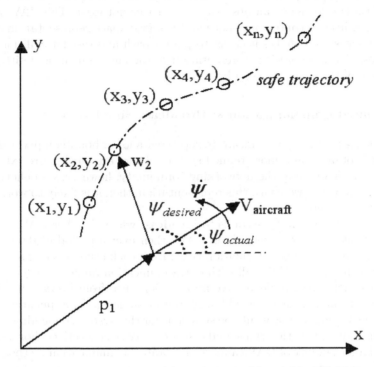

Fig. 3. Waypoint tracker and kinematic model variable definitions

The trajectory-tracking controller tracks waypoints given by the planner. The waypoint tracker controls toward a point, such as (x_2, y_2) in Figure 3, using proportional feedback:

$$u_1 = K_p * (\psi_{desired} - \psi_{actual}) \tag{2}$$

where $\psi_{desired}$ and ψ_{actual} are the angles as specified in Figure 3, K_p is the gain of the P-controller.

When the aircraft comes within some specified radius of the current waypoint the waypoint tracker switches to the next waypoint, in this case (x_3, y_3), until it has reached its final waypoint.

This trajectory-tracking controller should be improved for a real-world system; however in simulation with no physical disturbances it is more than sufficient.

5.2 Grid-Based Obstacle Map

The obstacles are represented in a grid-space where some length in meters is defined as one grid unit. For example, a grid-space of 100 x 100 units where one unit was equal to 50m would represent an area of 5000 m x 5000 m. In each of these grid-spaces an obstacle can exist or not exist. The UAV on the other hand lives in continuous space and the grid-space representation of the environment simply makes it easier to plan a path and test for collision with obstacles. If a more precise representation of the environment is desired, the resolution of the grid-space can be increased.

5.3 Limited Communication & Broadcast Simulation

Our broadcast simulation is meant to represent a lower bound on performance we could obtain from a more refined system. There are much more extensive solutions available from the networking community; however, we believe that if our system works well with this representation, then it is likely to work with many other systems.

Each UAV is given 0.1 seconds to broadcast while all other UAVs are listening. That is, if there were three UAVs then over a period of 0.6 seconds each UAV would receive information twice from each other UAV. This would mean if there were 20 UAVs all within communication range of each other, in 2 seconds each UAV would receive information once from every other UAV. This rate of communication would likely not be sufficient for applications such as collision avoidance but would be sufficient for this system, depending on the maximum communication range and distance between objectives. For a significantly larger number of UAVs a more elaborate communication methodology would be required.

Even if a UAV broadcasts, it is not guaranteed that other UAVs will receive that broadcast. From an inspection of the capabilities of our experimental platform we have developed the distribution below which is used to generate the plot seen in Figure 4. This is not meant to be a completely realistic simulation of a data network and is instead meant to limit information, vary the rate of information exchange, and randomly change the order of communication.

The probability distribution is given by

$$d = \left\{ \begin{array}{ll} 1 & r < p * r_{\max} \\ \frac{1}{(4r/r_{\max})^2} & r \geq p * r_{\max} \end{array} \right\} \tag{3}$$

where r is the Euclidian distance between UAVs, r_{max} is the maximum communication radius and p is the percentage of the max communication radius that will result in guaranteed communication. In the case of Figure 4 $p = 0.10$, $r_{max} = 2000$ meters and $r \in [0, 2000]$ meters.

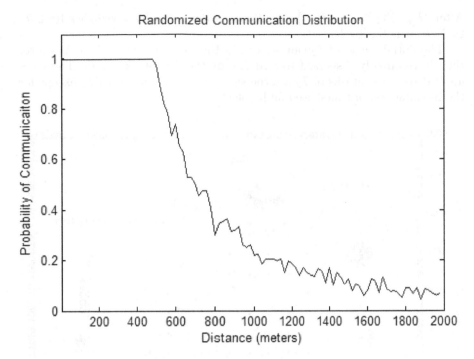

Fig. 4. Normalized simulated communication data of 100 samples every 20 meters with a maximum communication radius of 2000 meters

6 Example Mission Simulation

We will use an example simulation of mission M_1 to highlight some of the capabilities of our system. This is the same mission that is used as a mission example in Figure 1. In Figures 5-7 a curvilinear line represents the path of a UAV. A small circle at the end of a curvy line represents the actual UAV. The subtasks of different tasks are represented by dots and clusters of small circles represent obstacles. Also, when two UAVs communicate, a straight line is drawn between them.

A 2000 second simulation of M_1 was run with no simulated UAV faults. A maximum communication radius of 2000 meters and operation area of 5000 meters x 5000 meters are specified for the simulation. The aircraft and communication model are as described in section 5. *T1* of M_1 can be seen in Figure 5. The UAVs begin at their start positions and then disperse over the grid of subtasks represented by S_1 to complete T_1. T_1 results in a sweeping search of an area defined by $\{P_{11}, P_{12}, P_{13}, \ldots P_{1b_1}\}$ and a set of rules defined by $\{R_{11}, R_{12}, R_{13}, \ldots R_{1b_1}\}$. In this case the greater task T_1 is a result of relatively simple identical rules applied to the set of subtasks S_1. In a sweeping search, one UAV is required to visit each point of a subtask at least once.

After $\{U_{11}, U_{12}, U_{13}, \ldots U_{1b_1}\}$ are all set as *done*, the UAV switches from T_1 to T_2 as denoted by R_1.

The initial stages of T_2 can be seen in Figure 6 and show the UAVs after they have already dispersed toward one of the 7 subtasks of T_2. The rules applied to the subtasks of T_2 are the same as those applied to T_1 except for the introduction of time-based fault tolerance.

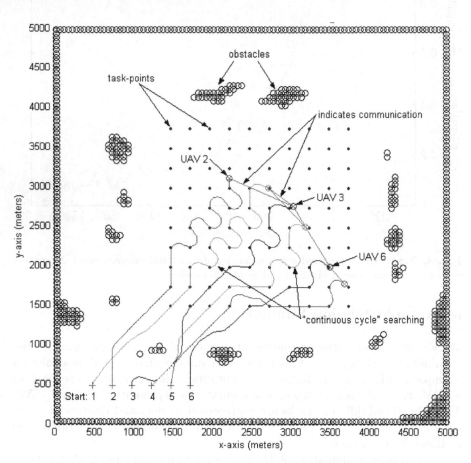

Fig. 5. Six UAVs during ex. M_1, executing T_1 @ t=250 seconds

With time-based fault tolerance, if a failure were to occur out of range of a functioning UAV, the subtask would still be completed. Because a UAV, through the mission state, knows the time at which a subtask was started and also knows what needs to be accomplished, it can estimate when in the future some other UAV should announce completion of that subtask. If this estimated time passes without confirmation that the subtask has been accomplished then the status of the subtask is switched from *assigned→todo* by a UAV. After this is done, an available UAV would proceed to accomplish the subtask in question.

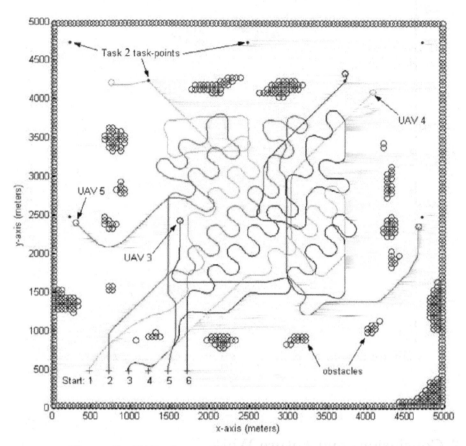

Fig. 6. Six UAVs during M_1, executing T_2 @ t =620 seconds

In addition, communication-based fault tolerance is applied to all subtasks of each task. If failure of a UAV were to occur within range of at least one other UAV, and this could be communicated to the other UAVs, then this knowledge would be integrated back into the mission state. That is, any subtasks currently associated with the faulted UAV would be set from *assigned→todo*.

Finally, Figure 7 displays all six UAVs circling at subtask points of the final task, T_3, in default mode. The visualization no longer displays the trail of each UAV, but it does display all the subtasks of T_1, T_2 and T_3 as their associate points. The default mode is actually the natural behavior of the system and is the result of no rules applied to the subtasks of T_3, that is $\{R_{31}, R_{32}, R_{33}, \ldots R_{3b_3}\}$=NULL.

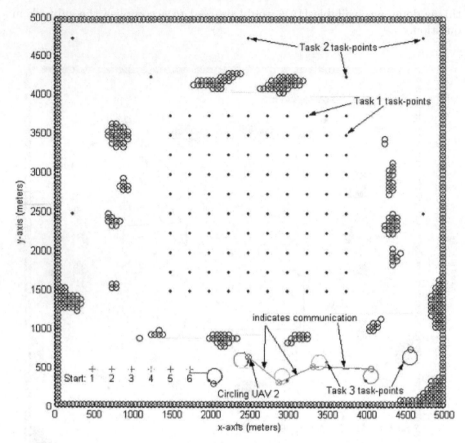

Fig. 7. Six UAVs during ex. M_1, and executing T_3 @ t =1199 seconds

7 Conclusions and Future Work

In this chapter, we have presented a distributed approach to multi-agent collaboration. We first established a description of a mission plan in terms of tasks and subtasks, and identified static and dynamic points as a useful means of defining a large set of possible missions, such as searching or convoy protection. As a superset of the mission plan, we established a high level language in which to describe the estimated state of a mission at any time. This involved augmenting the information contained in the mission plan with information on status, agent identifier, time, and cost. In addition the operation of an agents internal functions were described. A functioning multi-agent system was implemented in a simulation and results of an example mission were discussed.

In simulation, our system produces robust and consistent behavior despite uncertain communication links. It can execute different types of missions/tasks in a fault-tolerant way, and the trajectory planner is able to accommodate the constraints of our aircraft model. In simulation, the number of collaborating UAVs in the group is limited only by computer memory; in a real implementation, communication resources would likely be the limiting factor. However, with reasonable communication bandwidth we believe that this system can be successfully implemented in real-time on a group of real aircraft.

In a distributed multi-agent system information is key. Clearly, an agent's performance is limited by the availability of information. If all agents had complete information, then a group of UAVs could theoretically achieve a globally optimal solution. However this is difficult to achieve in any real distributed multi-agent system. Therefore, the important information must be identified, described in a concise manner, and widely disseminated as efficiently as possible. With this information available, good decisions can be made.

Our current and future work is largely committed to further refinement and formalization of the mission description concepts and the development of formal proofs. We wish to find a minimum set of rules/conditions that are required to guarantee that a mission will be completed if the resources exist. Also, with respect to the communication algorithm, we believe there is a simplified set of rules that govern the information exchange between agents.

We also wish to prove the existence of the limiting mission state described in section 3.1. We would also like to determine an upper bound on the number of required communications required for convergence as it relates to the number of agents in a group. For example, we can conclude from inspection that for a group of two agents only one iteration is required to reach the limiting mission state.

Finally, while each agent only plans one step ahead in the current system, we hope to introduce an algorithm that would allow each agent to plan multiple steps ahead.

References

1. W. Ren, R. W. Beard, "Trajectory Tracking for Unmanned Air Vehicles with Velocity and Heading Rate Constraints," IEEE Transactions on Control Systems Technology, In Press.
2. Y. Kuwata, Real-time Trajectory Design for Unmanned Aerial Vehicles using Receding Horizon Control, Masters Thesis, MIT, June 2003.
3. I. M. Mitchell and S. Sastry, "Continuous Path Planning with Multiple Constraints," IEEE Conference on Decision and Control, Hawaii, USA, December 2003
4. S. Spry and J. K. Hedrick, "Formation Control Using Generalized Coordinates," IEEE Conference on Decision and Control, Bahamas, 2004.
5. R. O. Saber and R. M. Murray, "Flocking with Obstacle Avoidance: Cooperation with Limited Communication in Mobile Networks," IEEE Conference on Decision and Control, Hawaii, USA, December 2003.

6. H. Tanner, A. Jadbabaie, and G. J. Pappas., "Coordination of multiple autonomous vehicles," IEEE Mediterranean Conference on Control and Automation, Rhodes, Greece, June 2003.
7. S. G. Breheny, R. D'Andrea and J. C. Miller, "Using Airborne Vehicle-Based Antenna Arrays to Improve Communications with UAV Clusters," IEEE Conference on Decision and Control, Hawaii USA, December 2003.
8. A. R. Girard, A. S. Howell, and J. K. Hedrick, "Border Patrol and Surveillance Missions using Multiple Unmanned Air Vehicles", Submitted to IEEE Control Systems Technology, 2004.
9. L.E. Parker, "ALLIANCE: An architecture for fault-tolerant multi-robot cooperation," IEEE Transactions on Robotics and Automation, 14(2), pp. 220-240, April 1998.
10. R. Davis and R.G. Smith, "Negotiation as a metaphor for distributed problem solving," Artificial Intelligence, Vol. 20, pp.63-109, 1983.
11. B.P. Gerkey and M.J. Mataric, "Sold! Auction Methods for Multirobot Coordination," IEEE Transactions on Robotics and Automation, 18(??), pp. 758-768, Oct. 2002.
12. J. Kennedy and R. C. Eberhart, Swarm Intelligence, Academic Press, 2001.
13. K. Konolige, D. Fox, C. Ortiz, et al., "Centibots: Very large scale distributed robotic teams," Proc. of the Intl. Symposium on Experimental Robotics, ISER 2004.
14. S.C. Spry, A.R. Girard, and J.K. Hedrick, "Convoy Protection using Multiple Unmanned Aerial Vehicles: Organization and Coordination," Proc. of the 24th American Control Conference, Portland, OR., June 2005.

Consensus Variable Approach to Decentralized Adaptive Scheduling

Kevin L. Moore[1] and Dennis Lucarelli[2]

[1] Division of Engineering
Colorado School of Mines
1610 Illinois Street, Golden CO 80401
E-mail: kmoore@mines.edu
[2] Johns Hopkins University Applied Physics Laboratory
11100 Johns Hopkins Road Laurel, MD 20723-6099
E-mail: dennis.lucarelli@jhuapl.edu

Summary. We present a new approach to solving adaptive scheduling problems in decentralized systems, based on the concept of nearest-neighbor negotiations and the idea of a consensus variable. Exploiting some recent extensions to existing results for single consensus variables, the adaptive scheduling problem is solved by choosing task timings as the consensus variables in the system. This application is illustrated via the example of a synchronized strike mission. The chapter concludes with a discussion of future research directions on this topic.

1 Introduction

In many applications it is possible to decompose a mission or system operation involving multiple agents, interacting through a sequence of inter-dependent tasks, into a set of sequences of dependent tasks that must be executed according to a prescribed schedule with a prescribed allocation of tasks to resources. Typically such problems are initially solved though some type of planning and scheduling algorithm. However, when change occurs that upsets these plans during execution, mission plans must be adapted. In many cases the luxury to replan is not available and in the "heat of battle," new schedules and contingencies for a team must often be determined "on-the-spot," through team-to-team communications, usually without the benefit of advanced planning tools and global domain knowledge. The result is that coordination efforts can distract team members from the task at hand and that mission success can be compromised. Of course, human agents in a system often know how to act on information presented to them. Humans appear to have an almost innate ability to coordinate complex behaviors in near real-time with heterogeneous assets in dynamic, uncertain, and adversarial environments. Moreover, it is often the case that humans can cooperate effectively without full knowledge of the entire coordination plan. There are, however, limits to the complexity and scale of operations that can be effectively coordinated in real-time with human cognition aided by standard technologies. This is especially true when considering cooperative planning and adaptation of time-critical missions amidst pervasive uncertainty and sophisticated adversaries. These observations lead us to consider the use of a computational assistant that can help human agents when change dictates the need for rescheduling mission plans. Such an assistant will need to embody suitable algorithms for reasoning.

In this chapter we propose a fundamentally new approach to distributed adaptation and reasoning based on sophisticated yet scalable mathematical algorithms that are motivated by the idea of a *consensus variable*. The central tenet of our approach is that to coordinate complex behaviors *some* information must be shared by agents in the network. We assume the minimal amount of information required is encapsulated in a time-varying vector, called the *coordination or consensus variable*. Each agent[3] carries "their own" local value of the coordination or consensus variable and updates that value based on the value held by the other agents with whom the agent is able to communicate. Through proper definition of the consensus variable and specification of rules for updating the value of this variable, it is possible to prove convergence of the consensus variable between the communicating agents. To apply the consensus variable approach to the problem of adaptive scheduling, we use recent results that extend the ideas of consensus variables to include both forced consensus and the case of multiple consensus variables separated by hard constraints. We begin by describing the basics of the consensus variable approach and its extensions. We then illustrate the application of consensus variables to the adaptive scheduling problem. This is done via the example of a synchronized strike mission. The chapter concludes with a discussion of future research directions on this topic.

2 Consensus Variables

Suppose we have N agents with a shared consensus variable ξ. Each agent has a local value of the variable given as ξ_i. Each agent updates their value based on the values of the agents that they can communicate with. For continuous-time systems we have the following update rule:

$$\dot{\xi}_i(t) = -\sum_{j=1}^{N} k_{ij}(t)G_{ij}(t)(\xi_i(t) - \xi_j(t)),$$ (1)

---

[3] In this chapter we use the word "agent" loosely. We simply mean an actor or entity or unit in a system and do not assume any type of formal AI definition of the term. Practically speaking, for us each team or unit in a mission, or in a team of teams, is an agent. Further, such agents could be UGVs, UAVs, other autonomous or automatic systems, or even humans or human teams.

where ξ_i is the coordination variable instantiated on the ith agent, $k_{ij}(t) > 0$ is a weighting factor or gain, and $G_{ij}(t)$ equals one if information flows from agent j to agent i at time t and zero otherwise. $G_{ij}(t)$ defines the (possibly time-varying) communication topology between the agents and can describe both unidirectional and bi-directional communications topologies. We say the communication graph is static if $G_{ij}(t) = G_{ij}$. We say a communication graph has spanning tree if there is at least one node from which there is a path that can reach every other node. We say that a node is a spanning node if there is a path from that node that can reach every other node in the graph. Let A be a set of N agents, negotiating about a consensus variable ξ with a communication topology G_{ij}. We say that A is in (global asymptotic) consensus if $\xi_i(t) \to \xi^*$, $\forall i$ as $t \to \infty$. It is well-known that (1) if the communication topology forms a spanning tree then convergence can be assured [1]; and (2) the final value of the convergence variable is a function of both the initial conditions $\xi_i(0)$ and the value of the gains k_{ij}.

The remainder of this section summarizes two extensions to the single consensus variable ideas that appear in [2]. First, it is often the case that we may want to force the consensus negotiation to follow a hard constraint. Suppose each agent updates their value of the consensus variable assuming an input u_i and has an output y_i as follows:

$$\dot{\xi}_i(t) = -\sum_{j=1}^{n} k_{ij}(t)G_{ij}(t)(\xi_i(t) - \xi_j(t)) + b_i u_i$$

$$y_i(t) = \xi_i(t) \tag{2}$$

where $b = [0, \cdots, 0, 1, 0, \cdots, 0]^T$, such that $b_k = 1$ and $b_j = 0, \forall j \neq k$. In our earlier paper we showed that if

$$u_k(t) = k_p(\xi^{sp} - \xi_k)$$

where ξ^{sp} is a constant setpoint and $k_p > 0$ is a constant gain, then the consensus strategy of (2) achieves global asymptotic consensus for A, with

$$\lim_{t \to \infty} \xi_i(t) = \xi^{sp} \quad \forall i$$

if and only if node k is a spanning node for the communication graph G.

In addition to forced consensus, we may also encounter the need for constrained consensus. Specifically, in multi-stage timing missions with heterogeneous assets it is rarely the case that a single coordination variable can capture the information required for the team objective to be achieved. Rather, most problems rely on the complex interplay of *several* coordination variables active in the system at any given time. In general, we can consider multiple sets of agents, A^λ, each with n^λ agents that are negotiating internally about consensus variables ξ^λ over communication graphs G^λ, each defined by a topology $G_{ij}^\lambda(t)$, with local (meaning within the set of agents) evolution of the consensus variable given by (2). Then we assume set-to-set communications taking place via a single node in one agent communicating with a single node in another agent. Further, in many applications one might like to enforce a non-integrable constraint between consensus variables. In our earlier paper we showed a mechanism to achieve this results. Specifically, if we let $u_{k^a}^a$ be the input to node $\xi_{k^a}^a$, where ξ^a is the consensus variable for agent A^a, $u_{k^b}^b$ be the input to node $\xi_{k^b}^b$, where ξ^b is the consensus variable for agent A^b and let Δ_{ab} be the desired constraint between consensus variables ξ^a and ξ^b, then the agent-to-agent constraint rule

$$u_{k^a}^a = -(\Delta_{ab} - (\xi_{k^b}^b - \xi_{k^a}^a))$$
$$u_{k^b}^b = \Delta_{ab} - (\xi_{k^b}^b - \xi_{k^a}^a)$$

leads to global asymptotic global consensus for each set A^a and A^b, with $\xi_i^a \to \xi^{a^*}, \xi_i^b \to \xi^{b^*}$, and $\xi^{b^*} = \xi^{a^*} + \Delta_{ab}$ if and only if nodes k^a and k^b are spanning nodes for the graphs G^a and G^b, respectively.

3 Synchronized Strike Application

In this section we illustrate our ideas about how to use the consensus variable approach for mission timing adaptation by using a Synchronized Strike example. Figure 1 shows the example mission. In the scenario the MH-J (Unit 1) drops the SF Team (Unit 2) and then returns to base to pick up supplies needed for the strike. Meanwhile, the Seal Team moves in on the MK-V boats while the SF Team moves to a forward observation point from which they can identify the supply drop location. This information, and the return of the MH-J to base, is needed before the supplies can be dropped. Following arrival at the proper location, the Seal Team proceeds in its CRRC boat to go to the target. When the supplies have been dropped, the SF Team is at the target, and the Seal Team is at the target, it is possible to engage. Note that we have partitioned the activities of each unit into tasks that separate points of synchronization. Also, we have shown some tasks with two rows, indicating the possibility of separate contingency plans for accomplishing individual tasks. For example, in Task 32 it is possible for the Seal Team to simply speed to the target in their MK-V rather than take the (slower) CRRC boats. For the purposes of this example, however, we assume that the contingency plans to be chosen for a given task are characterized by the task's length. Figure 2 shows the initial plan for the mission. Synchronization times, or consensus variables, are given by ξ^a, ξ^b, ξ^c, and ξ^d. The figure shows nominal durations times for each task and nominal target times for each consensus variable[4].

---

[4] It should also be noted that there are several notions of *time* referred to in this discussion: there is the actual clock time along which the mission proceeds, there are *timings* that units must agree on and take actions to accommodate, and

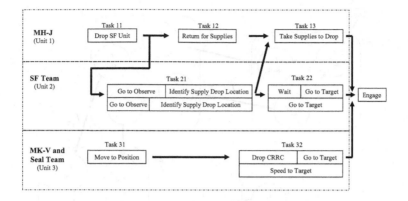

Fig. 1. Synchronized Strike example timeline.

Figure 3 shows the communication topology of the mission. Nodes associated with each consensus variable are mapped to the starting and the ending times of dependent tasks. It is important to note that we show "inputs" into some of the consensus nodes (e.g., T_{11} Start) as well as directed offsets between consensus variables (e.g., T_{21}, representing the duration of Task 21). These represent novel features of our ideas from the previous sections.

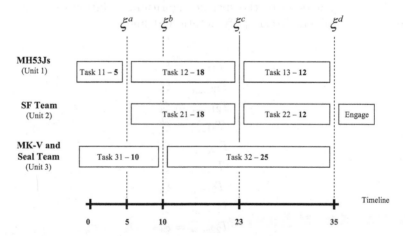

Fig. 2. Synchronized Strike timeline and consensus variables.

finally there is *consensus time* reflected in the time derivative of Equation (1). It is assumed that consensus time evolves on a faster scale relative to the global clock time. The actual "times" of interest are the steady-state values of the consensus time. These reflect the timing of the synchronization between units.

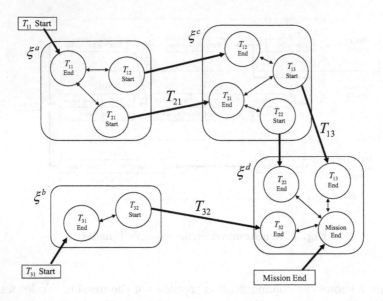

Fig. 3. Communication topology for the example.

Using Fig. 3, following the work on consensus variables from [1], and adding our new results on providing offsets and inputs into the consensus negotiation process, we can now describe the specific algorithmic solution we propose for this particular problem. To simplify notation, define

$$T_{11_{\text{End}}} = \xi_1^a$$
$$T_{12_{\text{Start}}} = \xi_2^a$$
$$T_{21_{\text{Start}}} = \xi_3^a$$
$$T_{31_{\text{End}}} = \xi_1^b$$
$$T_{32_{\text{Start}}} = \xi_2^b$$
$$T_{12_{\text{End}}} = \xi_1^c$$
$$T_{21_{\text{End}}} = \xi_2^c$$
$$T_{13_{\text{Start}}} = \xi_3^c$$
$$T_{22_{\text{Start}}} = \xi_4^c$$
$$T_{32_{\text{End}}} = \xi_1^d$$
$$T_{22_{\text{End}}} = \xi_2^d$$
$$T_{13_{\text{End}}} = \xi_3^d$$
$$\text{MissionEnd} = \xi_4^d$$

Then the consensus negotiation process is defined by the following equations:

$$\dot\xi_1^a = -k_{12}^a(\xi_1^a - \xi_2^a) - k_{13}^a(\xi_1^a - \xi_3^a) + (T_{11} - \xi_1^a)$$
$$\dot\xi_2^a = -k_{21}^a(\xi_2^a - \xi_1^a)$$
$$\dot\xi_3^a = -k_{31}^a(\xi_3^a - \xi_1^a) - k_{32}^{ac}(T_{21} + \xi_3^a - \xi_2^c)$$
$$\dot\xi_1^b = -k_{12}^b(\xi_1^b - \xi_2^b) + (T_{31} - \xi_1^b)$$
$$\dot\xi_2^b = -k_{21}^b(\xi_2^b - \xi_1^b) - k_{21}^{bd}(T_{32} + \xi_2^b - \xi_1^d)$$
$$\dot\xi_1^c = -k_{13}^c(\xi_1^c - \xi_3^c)$$
$$\dot\xi_2^c = -k_{23}^c(\xi_2^c - \xi_3^c) - k_{24}^c(\xi_2^c - \xi_4^c) + k_{23}^{ac}(T_{21} + \xi_3^a - \xi_2^c)$$
$$\dot\xi_3^c = -k_{31}^c(\xi_3^c - \xi_1^c) - k_{32}^c(\xi_3^c - \xi_2^c) - k_{33}^{cd}(T_{13} + \xi_3^c - \xi_3^d)$$
$$\dot\xi_4^c = -k_{42}^c(\xi_4^c - \xi_2^c)$$
$$\dot\xi_1^d = -k_{14}^d(\xi_1^d - \xi_4^d) + k_{12}^{bd}(T_{32} + \xi_2^b - \xi_1^d)$$
$$\dot\xi_2^d = -k_{24}^d(\xi_2^d - \xi_4^d)$$
$$\dot\xi_3^d = -k_{34}^d(\xi_3^d - \xi_4^d) + k_{33}^{cd}(T_{13} + \xi_3^c - \xi_3^d)$$
$$\dot\xi_4^d = -k_{41}^d(\xi_4^d - \xi_1^d) - k_{42}^d(\xi_4^d - \xi_2^d) - k_{43}^d(\xi_4^d - \xi_3^d) + \mathrm{PID}(\mathrm{SP} - \xi_4^d)$$
$$T_{13} = T_{13_\mathrm{nominal}} + \mathrm{PID}(\xi_4^d - \xi_3^d)$$
$$T_{21} = T_{21_\mathrm{nominal}} + \mathrm{PID}(\xi_3^c - \xi_2^c)$$
$$T_{32} = T_{32_\mathrm{nominal}} + \mathrm{PID}(\xi_4^d - \xi_1^d)$$

In these equations "PID" denotes a PID controller acting on its argument and "SP" denotes the setpoint or target objective for the final mission time. When these equations are solved they converge so that

$$\xi_i^a \to \xi^a \to T_{11}$$
$$\xi_i^b \to \xi^b \to T_{31}$$
$$\xi_i^c \to \xi^c$$
$$\xi_i^d \to \xi^d \to \mathrm{SP}$$

with $\xi^c = \xi^a + T_{21}$ and $\xi^d = \mathrm{SP} = \xi^c + T_{13} = \xi^b + T_{32}$.

To see how the approach works we present simulation results for two situations. First, suppose there is a need to adjust the engagement time from 35 time units to 32 time units. Then assume that weather reports indicate that there will be a delay of three time units in Task 13 (dropping the supplies). The first scenario is modeled by a "setpoint" change to the final mission time, ξ_4^d, while the second scenario is modeled as an additive disturbance to T_{13}. We further add the restriction that the controller of Task 13 cannot act to offset more than two units of change over the entire mission (this could reflect, for example, a fuel limit). Figure 4 shows the outcome of the consensus negotiation for the scenarios we have described. The simulation was initialized with the nominal plan. The transients observed in Fig. 4 reflect the negotiation process. The steady-state values in Fig. 4(a) reflect the final negotiated values of the consensus variables. Figure 4(b) shows the final duration of each key task. It can be seen that after the disturbance to Task 13, it was necessary for

Task 21 to become much shorter so the overall goal could still be achieved. In terms of the example mission, we interpret this as the SF Team moving faster to the observation location so as to identify the drop location sooner, so that the weather delay can be accommodated by allowing the MH-J to leave sooner on its second mission. Figure 5 shows the timeline resulting from the negotiations following the change in engagement time and from the task delay.

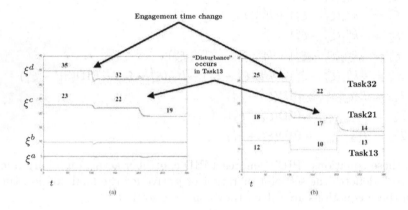

Fig. 4. Negotiation outcomes: (a) consensus variables; (b) task durations.

4 System Architecture

Note that although in the previous section we showed a group of seemingly centralized differential equations, this is perhaps misleading. In fact, the actual computations are decentralized. For example, not all the variables associated with ξ^a, for example, "live" in Unit 1. Only ξ_1^a and ξ_1^a evolve in Unit 1. ξ_3^a evolves in Unit 2. The architecture we envision for each unit system is shown in Fig. 6 for the specific example of the previous section. Each individual unit shows two key components. The Local Consensus Module of a unit negotiates with the Local Consensus Module of the other units to come to consensus on the values of the synchronization points that they share. The Global Consensus Negotiator is an internal controller that requests changes in task length

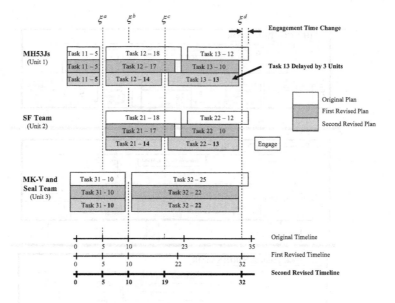

Fig. 5. Timeline for setpoint change and disturbance accommodation.

(i.e., contingency selection) based on mismatches between consensus variables. This module evaluates requests for changes in task length and determines the best possible contingency plan. In our example such a function is implied by the summation points and by the saturation limit we placed on Task 13. However, we expect this module to be implemented as a discrete-event dynamic system-type, formal language-based supervisor that evolves according to an underlying finite state automata. The development of this module is a significant part of our future research plans on this problem.

Not shown in Fig. 6 are several other architectural elements that would be included in a complete system. First, implicit in our architecture is an element to detect and identify change. Second, in later phases of the research we will consider the role of a module to act as a source of constraint definition and enforcement. This is where the organizational rules of military command and control, for example, would be embodied. Finally, a module is needed to embody logic for: (1) reasoning about change and uncertainty; and (2) learning to adapt our actions and timings based on such reasoning.

5 Concluding Comments

In this chapter we have presented a new approach to solving adaptive scheduling problems in decentralized systems. Our approach uses the concept of nearest-neighbor negotiations and the idea of a consensus variable. We be-

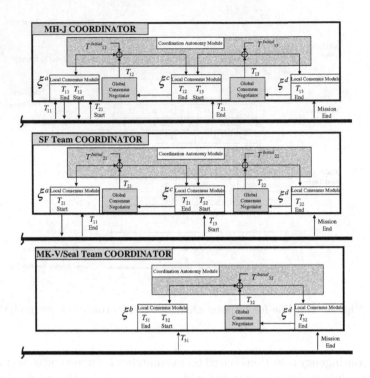

Fig. 6. Architecture for the example.

gan by summarizing the consensus variable approach and describing some recent extensions to those results. We then showed how to solve the adaptive scheduling problem by choosing task timings as the consensus variables in the system. This application was illustrated via the example of a synchronized strike mission. The notion of a consensus variable is a powerful concept. However, a number of enhancements are needed to make it useful. We describe here some key issues that must be addressed:

1. The implementation of our message passing algorithm is essentially a distributed integration process mediated by a wireless network. Understanding the numerical stability of our algorithm with respect to encoding, quantization, and time delay in a real system is an important problem to be addressed. In addition to proper software engineering, theoretical results must be obtained to ensure stability and to avoid unwanted behavior in the system when dealing with the realities of communication time delays. Fortunately, there is a line of research addressing *dynamical systems over networks* as it pertains to flocks of UAVs and other distributed dynamical systems (see [3, 4, 5] and references therein). We are confident

that our algorithms can be effectively addressed in this regard with ideas drawn from this body of work.

2. Operationally, mission adaptation today is largely a manual activity whereby interacting teams communicate by radio and negotiate a consensus on changes in timing and actions. Missions performed by special operations forces (SOF) represent an extreme example of both the benefits and constraints of mission coordination. By adhering to well-established principles such as stealth, speed, and surprise, SOF forces are generally able to achieve tactical control over significantly larger and more capable adversaries for limited time intervals. This efficiency comes at a price, however. Such missions are carefully choreographed and rehearsed, with little flexibility for real-time adaptation. Current doctrine involves execution checklists that specify well constrained communications events derived from mission plan phase diagrams. There are two problems with execution checklists from the coordination perspective. First, each communication event requires a patrol to divert attention from the tactical situation and initiate coordination tasks, resulting in a general increase in vulnerability. Secondly, the fixed format of execution checklists greatly limits contingency options for in-situ patrols. In many cases, mission termination is the only option. Our proposed technology will completely offload basic coordination communications from the patrol and eventually allow in-stride mission replanning to accommodate changes in resource availability or mission objectives. As SOF assets become increasingly sophisticated (e.g. organic UAVs), it will not be feasible to expect the human warfighters to manage this cognitive load in the same manner they do today.

3. There are important characteristics of the problem we are considering that may preclude a classical approach to planning and scheduling for online adaptation of plans and actions resulting from changing battle conditions or mission objectives and the communication required for that adaptation. An operations research (or similar) approach may be amenable to the *generation* of an optimal plan, however, it is unclear how one would resolve the associated global optimization problem in near real-time in a distributed manner when a change is required. Constraint satisfaction problems provide a framework for scheduling and planning and have been extended to handle problems with temporal constraints [6]. Again, a search based approach may be suitable for the generation of a mission plan, but is it unclear how these techniques could be used effectively for real-time adaptation. Researchers in the "agents" community have recognized that multi-agent coordination can only be facilitated by establishing protocols for reactive planning and providing rules for adaptation (see for example [7]). However, it is novel to take a dynamical system point of view for planning and scheduling problems. Our approach is differentiated from classical approaches in that it possesses the following features:
   a) It is provable.
   b) It is scalable.

c) No global communications are required.

d) Uncertainties in constraints and communications can be handled explicitly and algorithmically.

e) It provides an explicit way to handle structural changes, such as node loss.

f) The approach can be extended to handle consensus problems about other variables, such as resources.

g) It can be formulated to include probabilistic considerations, making it possible to place confidence intervals on contingency options.

4. In operational settings things always go wrong. It is often the case that communications fail and that information is uncertain. For the system to be effective the consensus process must properly account for the real-world "noise" it will encounter. This can be done in two ways. First, the noise can be handled *a priori* through proper coordination algorithm design. In particular, the authors of [1] have developed methods for designing the gains k_{ij} so as to minimize the effects of uncertainty [8]. Second, though the process of system identification, structural changes can be detected and identified and new gains can be design that will be optimal for the new situation. The development of learning mechanisms for detecting and identifying the need for adaptation will be an important part of our research effort. We will also explore the related concepts of situational observability based on consensus variable trajectories and the idea of predictive/preemptive coordination based on situational observability.

5. Though not necessarily an explicit part of consensus negotiation, it is the case that the constraints between consensus variables are often a result of the actions to be carried out by an agent either before or after a synchronization point. In future research we will attempt to understand how local constraints affect the consensus negotiation process and its convergence.

References

1. W. Ren, R. W. Beard, and T. W. McLain, "Coordination Variables and Consensus Building in Multiple Vehicle Systems," *Proceedings of the Block Island Workshop on Cooperative Control, Springer-Verlag Series: Lecture Notes in Control and Information Sciences*, vol. 309, 2005.
2. K. L. Moore and D. Lucarelli, "Force and constrained consensus among cooperating agents," in *2005 IEEE International Conference on Networking, Sensing, and Control*, Tuscon, AZ, March 2005.
3. R. Brockett and D. Liberzon, "Quantized feedback stabilization of linear systems," *IEEE Transactions on Automatic Control*, vol. 48, pp. 1279–1289, 2000.
4. N. Elia and S. Mitter, "Stabilization of linear systems with limited information," *IEEE Transactions on Automatic Control*, vol. 46, pp. 1384–1400, 2001.
5. S. Tatikonda and S. Mitter, "Control under communication constraints," *IEEE Transactions on Automatic Control*, vol. 49, pp. 1056–1068, July 2004.

6. R. Dechter, I. Meiri, and J. Pearl, "Temporal constraint networks," *Artificial Intelligence*, vol. 49, pp. 61–95, 1991.
7. M. Tambe, "Towards flexible teamwork," *Journal of Artificial Intelligence Research*, vol. 7, pp. 83–124, 1997.
8. W. Ren, R. W. Beard, and D. Kingston, "Multi-agent Kalman Consensus with Relative Uncertainty," in *2005 American Control Conference*, Portland, OR, June 2005.

A Markov Chain Approach to Analysis of Cooperation in Multi-Agent Search Missions

David E. Jeffcoat[1], Pavlo A. Krokhmal[2], and Olesya I. Zhupanska[3]

[1] Air Force Research Lab, Munitions Directorate
   Eglin AFB, FL 32542, USA
   david.jeffcoat@eglin.af.mil
[2] Department of Mechanical and Industrial Engineering
   University of Iowa
   2403 Seamans Center, Iowa City, IA 52242, USA
   krokhmal@engineering.uiowa.edu
[3] University of Florida, REEF
   1350 Poquito Road
   Shalimar, FL 32579, USA
   zhupanska@gerc.eng.ufl.edu

Summary. We consider the effects of cueing in a cooperative search mission that involves several autonomous agents. Two scenarios are discussed: one in which the search is conducted by a number of identical search-and-engage vehicles, and one where these vehicles are assisted by a search-only (reconnaissance) asset. The cooperation between the autonomous agents is facilitated via cueing, i.e. the information transmitted to the agents by a searcher that has just detected a target. The effect of cueing on the target detection probability is derived from first principles using a Markov chain analysis. Exact solutions to Kolmogorov-type differential equations are presented, and existence of an upper bound on the benefit of cueing is demonstrated.

1 Introduction

In any system-of-systems analysis, consideration of dependencies between systems is imperative. In this chapter, we consider a particular type of system interaction, called cueing. The interaction could be between similar systems, such as two or more wide area search munitions, or between dissimilar systems, such as a reconnaissance asset and a munition. In this chapter, we consider two scenarios: one in which the search is conducted by a number of identical search-and-engage vehicles, and one where these vehicles are assisted by a search-only vehicle. The autonomous agents forming the system cooperatively interact via cueing.

In Shakespeare's day, the word "cue" meant a signal (a word, phrase, or bit of stage business) to a performer to begin a specific speech or action [6]. The word is now used more generally for anything serving a comparable purpose. In this chapter, we mean any information that provides focus to a search; e.g., information that limits the search area or provides a search heading.

Search theory is one of the oldest areas of operations research [8], with a solid foundation in mathematics, probability and experimental physics. Yet, search theory is clearly of more than academic interest. At times, a search can become an international priority, as in the 1966 search for the hydrogen bomb lost in the Mediterranean near Palomares, Spain.

That search was an immense operation involving 34 ships, 2,200 sailors, 130 frogmen and four mini-subs. The search took 75 days, but might have concluded much earlier if cueing had been utilized from the start. A Spanish fisherman had come forward quickly to say he'd seen something fall that looked like a bomb, but experts ignored him.

Instead, they focused on four possible trajectories calculated by a computer, but for weeks found only airplane pieces. Finally, the fisherman, Francisco Simo, was summoned back. He sent searchers in the right direction, and a two-man sub, the Alvin, located the 10-foot-long bomb under 2,162 feet of water [11].

Cueing is a current topic in vision research. For example, Arrington et al. [2] study the role of objects in guiding spatial attention through a cluttered visual environment. Magnetic resonance imaging is used to measure brain activity during cued discrimination tasks requiring subjects to orient attention either to a region bounded by an object or to an unbounded region of space in anticipation of an upcoming target. Comparison between the two tasks revealed greater brain activity when an object cues the subjects attention.

Bernard Koopman [8] pioneered the application of mathematical process to military search problems during World War II. Koopman [4] discusses the case in which a searcher inadvertently provides information to the target, perhaps allowing the target to employ evasive action. The use of receivers on German U-boats to detect search radar signals in World War II is a classic example. Koopman referred to this type of cueing as target alerting.

This chapter uses a detection rate approach to examine the effect of cueing on probability of target detection. Koopman [5] used a similar approach in his discussion of target detection. In Koopman's terminology, a quantity γ was called the "instantaneous probability of detection." From this starting point, Koopman derived the probability of detection as a function of time. It is very clear that Koopman's instantaneous probability of detection is precisely the individual searcher detection rate used here. The main difference is that Koopman considered a single searcher, while we consider the case of multiple interdependent searchers.

Wasburn [10] examines the case of a single searcher attempting to detect a randomly moving target at a discrete time. Given an effort distribution, bounded at each discrete time t, Washburn establishes an upper bound on the probability of target detection. It is noteworthy that Washburn mentions that the detection rate approach to computation of detection probabilities has proved to be more robust than approaches relying on geometric models.

Alpern and Gal [1] discuss the problem of searching for a submarine with a known initial location. Thomas and Washburn [9] considered dynamic search games in which the hider starts moving at time zero from a location known to both a searcher and a hider, while the searcher starts with a time delay known to both players; for example, a helicopter attempts to detect a submarine that reveals its position by torpedoing a ship.

In this chapter, we use a Markov chain analysis to examine cueing as a coupling mechanism among several searchers. A Markov chain approach to target detection can be found in Stone [8], which deals with the optimal allocation of effort to detect a target. A prior distribution of the target's location is assumed known to the searcher. Stone uses a Markov chain analysis to deal with the search for targets whose motion is Markovian. In Stone's formulation, the states correspond to cells that contain a target at a discrete time with a specified probability. In this research, the states correspond to detection states for individual search vehicles.

The rest of the chapter is organized as follows. The next section discusses the effect of cueing on the performance of a cooperative system of several identical search agents. Section 3 presents analysis of a search system that involves a search-only vehicle that provides cues to a number of search-and-engage vehicles.

2 Cooperative Search

Consider a system of N agents engaged in a cooperative search mission, where the objective of every agent is to find (*detect*) an object of interest (a *target*). The search capabilities of any agent are characterized by the detection rate θ, i.e. the probability of detecting a target within time interval Δt:

$$\text{P[agent } i \text{ detects a target during time } \Delta t] = \theta \Delta t + o(\Delta t). \qquad (1)$$

Upon detecting a target, an agent discontinues its search by *engaging* the detected target; we also say that such an agent becomes *inactive*. For example, in a search-and-rescue mission for passengers of a sinking ship the searchers will try to rescue the passenger(s) they find, instead of continuing the search; on the battlefield, an autonomous wide-area search munition will attack the detected target, etc. Moreover, it is assumed that upon engaging a target, the searcher immediately cues the remaining *active* agents, thereby potentially increasing their detection capabilities. Within the presented framework the informational content of the cueing signal is not important; instead, we are interested in the degree by which cueing impacts the search capabilities of individual agents in a cooperative system. In accordance to this, at any time $t \geq 0$ the detection rate of a searcher may change values as

$$\theta_0 \rightarrow \theta_1 \rightarrow \ldots \rightarrow \theta_{N-1}, \tag{2}$$

where θ_k is the detection rate common to $N - k$ active searchers. Clearly, θ_0 is equal to the initial "uncued" detection rate: $\theta_0 = \theta$. Also, it is natural to assume that cueing generally leads to improvement of search capabilities, whence $\theta_k \geq \theta$, $k = 1, \ldots, N - 1$. The search mission terminates when there are no active searchers left, i.e., each searcher has found a target.

The outlined model, complemented by the usual assumptions of independence of target detections on non-overlapping time intervals etc., can be formulated as a continuous-time discrete-state Markov chain, or, furthermore, as a *pure death process* [3, 7]. The states of the system are identified by the number k of inactive searchers, $k = 0, 1, \ldots, N$. The presented formulation, however, differs slightly from the traditional models for birth-death processes in that it defines the transition rate between states by the search rates of individual agents (2), versus the rates defined with respect to the entire population [3]. Let $\mathcal{S}_{N-k,k}$ be *a state* of the system in which there are k inactive searchers, and, correspondingly, $N - k$ active searchers; note that there are $\binom{N}{k}$ different states $\mathcal{S}_{N-k,k}$. Defining probabilities $P_{N-k,k}(t)$ as

$$P_{N-k,k}(t) = \mathsf{P}[\text{system is in one of the states } \mathcal{S}_{N-k,k} \text{ at time } t],$$

one can write the system of Chapman-Kolmogorov ODEs governing these probabilities as

$$\frac{\mathrm{d}}{\mathrm{d}t} P_{N-k,\,k}(t) = -\bar{\delta}_{kN}\,(N-k)\,\theta_k\,P_{N-k,\,k}(t) + \bar{\delta}_{k0}\,k\theta_{k-1}\,P_{N-k+1,\,k-1}(t),$$

$$k = 0, \ldots, N, \tag{3}$$

where $\bar{\delta}_{ij}$ is the negation of the Kroneker symbol δ_{ij}

$$\bar{\delta}_{ij} = 1 - \delta_{ij} = \begin{cases} 0, & \text{if } i = j, \\ 1, & \text{if } i \neq j. \end{cases} \tag{4}$$

Equations (3) admit a simple interpretation: since in a state $\mathcal{S}_{N-k,k}$ there are $N - k$ active searchers with search rate θ_k, the probability of the system being in this state decreases at rate $(N - k)\theta_k P_{N-k,k}(t)$ as each of $N - k$ searchers may detect a target and turn the system into a state $\mathcal{S}_{N-k-1,k+1}$. On the other hand, probability $P_{N-k,k}(t)$ increases at rate $k\theta_{k-1}P_{N-k+1,k-1}(t)$ as there are exactly $\binom{k}{1} = k$ states $\mathcal{S}_{N-k+1,k-1}$ that may lead to a (given) state $\mathcal{S}_{N-k,k}$. Indeed, let $A_k = \{i_1, \ldots, i_k\}$ be any set containing k inactive searchers, $|A_k| = k$. Trivially, A_k can be represented in k different ways as $A_k = A_{k-1}^j \cup \{i_j\}$, where $A_{k-1}^j = A_k \backslash \{i_j\} \subset A_k$, $|A_{k-1}^j| = k - 1$, $j = 1, \ldots, k$. Thus, a state $\mathcal{S}_{N-k,k}$ with k inactive searchers can only be obtained from exactly k states $\mathcal{S}_{N-k+1,k-1}$ with $k - 1$ inactive searchers. Factors $\bar{\delta}_{N,k}$ and $\bar{\delta}_{k,0}$ in (3) have the obvious function of handling the extreme cases of $k = 0$ and $k = N$. Denoting $P_{N-k,k}(t) = \hat{P}_k(t)$, $k = 0, \ldots, N$, equations (3) can be rewritten in the matrix form

$$\frac{\mathrm{d}}{\mathrm{dt}}\hat{\mathbf{P}} = M\hat{\mathbf{P}},\tag{5}$$

where $\hat{\mathbf{P}} = (\hat{P}_0, \ldots, \hat{P}_N)^T$, and matrix $M = \{m_{ij}\} \in \mathbf{R}^{(N+1)\times(N+1)}$ has the following non-zero elements:

$$\begin{aligned}
m_{ii} &= -(N-i)\theta_i, & i = 0, \ldots, N,\\
m_{i,i-1} &= i\theta_{i-1}, & i = 1, \ldots, N.
\end{aligned}\tag{6}$$

Explicitly, the matrix M in (5) can be written as

$$M = \begin{pmatrix}
-N\theta_0 & 0 & 0 & \cdots & 0 & 0 & 0\\
\theta_0 & -(N-1)\theta_1 & 0 & \cdots & 0 & 0 & 0\\
0 & 2\theta_1 & -(N-2)\theta_2 & \cdots & 0 & 0 & 0\\
& & & \ddots & & &\\
& & & \cdots & -2\theta_{N-2} & 0 & 0\\
0 & 0 & 0 & \cdots & (N-1)\theta_{N-2} & -\theta_{N-1} & 0\\
0 & 0 & 0 & \cdots & 0 & N\theta_{N-1} & 0
\end{pmatrix}.\tag{7}$$

The initial conditions for equations (5) reflect the fact that at $t = 0$ the system is in the state $\mathcal{S}_{N,0}$ with probability 1:

$$\hat{P}_0(0) = 1, \hat{P}_1(0) = \hat{P}_2(0) = \ldots = \hat{P}_N(0) = 0.\tag{8}$$

Using (5) and (8) it is straightforward to verify that the probabilities $\hat{P}_k(t)$ satisfy the identity

$$\sum_{k=0}^{N} \binom{N}{k} \hat{P}_k(t) = 1, \quad t \geq 0.\tag{9}$$

Clearly, matrix (7) is lower-triangular, hence system (5) has the characteristic equation of the form

$$\lambda \prod_{j=0}^{N-1} \left(\lambda + (N-j)\theta_j\right) = 0,\tag{10}$$

Assuming that all eigenvalues of M are distinct,

$$(N-i)\theta_i \neq (N-j)\theta_j, \quad 0 \leq i < j \leq N-1,\tag{11}$$

the solution of the Cauchy problem (5), (8) can be written in a simple closed form

$$\hat{P}_i(t) = \sum_{j=0}^{i} a_{ji} e^{m_{jj}t}, \quad i = 0, \ldots, N,\tag{12}$$

with coefficients a_{ji} defined recursively as

$$a_{ji} = \frac{m_{i,i-1}\, a_{j,i-1}}{m_{jj} - m_{ii}}, \quad 0 \le j < i \le N, \tag{13}$$

$$a_{ii} = -\sum_{s=0}^{i-1} a_{si}, \quad i = 1, \dots, N, \quad a_{00} = \hat{P}_0(0) = 1. \tag{14}$$

One of the possible measures of performance (MOE) for a cooperative search system is the time required for all N searchers to find targets. In the scope of the presented approach, this characteristic is embodied by the probability $P_{0,N}(t) = \hat{P}_N(t)$ of the system being in the state $S_{0,N}$ at time t. Noting that dependencies (14)–(13) can be reexpressed for all $0 \le j < i \le N-1$ as

$$a_{ji} = \beta_{ji} a_{ii}, \quad \text{where} \quad \beta_{ji} = \frac{\theta_j}{\theta_i} \frac{\binom{N-1}{j}}{\binom{N-1}{i}} \prod_{s=j+1}^{i} \left[1 - \frac{(N-j)\theta_j}{(N-i)\theta_i}\right]^{-1}, \tag{15}$$

the probability $\hat{P}_N(t)$ can be represented from equality (12) in the form

$$\hat{P}_N(t) = 1 - \sum_{j=0}^{N-1} \binom{N}{j} \prod_{s=j+1}^{N-1} \left[1 - \frac{(N-j)\theta_j}{(N-s)\theta_s}\right]^{-1} a_{jj}\, e^{-(N-j)\theta_j t}, \tag{16}$$

where a_{jj} are calculated due to (14) and (15) as

$$a_{jj} = -\sum_{r=0}^{j-1} \beta_{rj} a_{rr}, \quad j = 1, \dots, N-1, \tag{17}$$

and the usual convention $\prod_{i=i_1}^{i_2} (\cdot)_i = 1$ for $i_1 > i_2$ is adopted.

Expression (16) can now be used for the analysis of the effect of cueing on the cooperation among the searchers. For example, it is easy to see that when cueing has no impact on the detection rates of the searchers, then $\hat{P}_N(t)$ is equal to the probability of all N agents detecting targets independently:

Proposition 1. *If $\theta_0 = \theta_1 = \dots = \theta_{N-1} = \theta$, i.e. the cueing has no effect on the detection capabilities of the searchers, then*

$$\hat{P}_N(t) = \left(1 - e^{-\theta t}\right)^N. \tag{18}$$

Proof: If all the cued detection rates (2) are equal to the uncued rate θ, then the product term in expressions (15) and (16) for β_{ji} and $\hat{P}_N(t)$ reduces to

$$\prod_{s=j+1}^{i} \left[1 - \frac{(N-j)\theta_j}{(N-s)\theta_s}\right]^{-1} = (-1)^{i-j},$$

which immediately yields

$$\beta_{ji} = (-1)^{i-j} \binom{i}{j}, \quad 0 \le j < i \le N - 1, \tag{19}$$

and

$$\hat{P}_N(t) = 1 + \sum_{j=0}^{N-1} a_{jj} \binom{N}{j} (-1)^{N-j} e^{-(N-j)\theta t}.$$

The last expression verifies the statement (18) of the proposition provided that $a_{jj} = 1$, $j = 0, \ldots, N-1$. Using the induction argument, we have that $a_{00} = \hat{P}_0(0) = 1$ from (14), and, assuming that $a_{11} = \ldots = a_{jj} = 1$ for some j, by means of (17) and (19) we obtain

$$a_{j+1,j+1} = -\sum_{s=0}^{j} a_{s,j+1} = -\sum_{s=0}^{j} (-1)^{j+1-s} \binom{j+1}{s} a_{ss} = 1. \tag{20}$$

The last equality in (20) follows from the Newton binom formula $(1-1)^{j+1} = \sum_{j=0}^{j} (-1)^{j+1-s} \binom{j+1}{s} + 1$. ∎

Obviously, cueing has the purpose of improving the system's effectiveness by increasing the cued detection rates $\theta_1, \ldots, \theta_{N-1}$. Indeed, it can be verified that higher values of $\theta_1, \ldots, \theta_{N-1}$ increase the probability of detection $\hat{P}_N(t)$ for a given t. Below we demonstrate that under quite general conditions the effect of cueing on the system's performance is bounded, i.e., there exists an upper bound for the state probability $\hat{P}_N(t)$ when the cued detection rates $\theta_1, \ldots, \theta_{N-1}$ increase indefinitely.

Proposition 2. *Let the cued detection rates approach infinity, $\theta_i \to \infty$, $i = 1, \ldots, N - 1$, in such a way that*

$$\lim_{\theta_i, \theta_j \to \infty} \frac{(N-i)\theta_i}{(N-j)\theta_j} \ne 1, \quad i \ne j.$$

Then the probability $\hat{P}_N(t)$ of all agents having detected a target at time t has the limit

$$\lim_{\theta_1, \ldots, \theta_{N-1} \to \infty} \hat{P}_N(t) = 1 - e^{-N\theta t}. \tag{21}$$

Proof: To establish the statement of the proposition, it suffices to rewrite expression (16) for the probability $\hat{P}_N(t)$ in the form

$$\hat{P}_N(t) = 1 - \prod_{s=1}^{N-1} \left[1 - \frac{N\theta_0}{(N-s)\theta_s} \right]^{-1} a_{00} \, e^{-N\theta_0 t}$$

$$- \sum_{j=1}^{N-1} \binom{N}{j} \prod_{s=j+1}^{N-1} \left[1 - \frac{(N-j)\theta_j}{(N-s)\theta_s} \right]^{-1} a_{jj} \, e^{-(N-j)\theta_j t}. \tag{22}$$

Taking into account that $a_{00} = 1$, and that under the conditions of the proposition $\theta_0/\theta_s \to 0$ for all $s = 1, \ldots, N-1$, and $R(\theta_j, \theta_s) e^{-(N-j)\theta_j t} \to 0$ for all rational functions $R(\cdot)$ and all $1 \leq s, j \leq N-1$, we observe that the second term in the right-hand side of (22) is approaching $\left(-e^{-N\theta t}\right)$, while the third term vanishes. This yields expression (21). ∎

A numerical illustration of the performance of the cooperative search system as defined by (16) is presented in Figures 1, 2, and 3. In these examples, the cueing rates dynamics is assumed to follow

$$\theta_i = \theta \kappa^{\frac{N-i}{N-1}}, \quad i = 1, \ldots, N-1, \quad \text{for some } \kappa > 1. \tag{23}$$

Then in the case of 5 searchers ($N = 5$) the probability of detection $\hat{P}_5(t)$ for various values of κ is displayed in Figure 1. The black line, marked by $\kappa \gg 1$ corresponds to the upper bound (21) for the detection probability $\hat{P}_5(t)$.

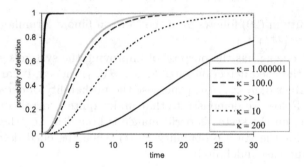

Fig. 1. Probability of detection $\hat{P}_N(t)$ for 5 searchers ($N = 5$).

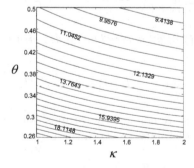

Fig. 2. Isolines of the *time to engage*, $T_{0.95}$, for 2 searchers.

Another measure of effectiveness of the considered cooperative search system may be considered to be the *time to engage*, i.e., such a time $t = T_{0.95}$

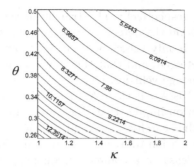

Fig. 3. Isolines of the *time to engage*, $T_{0.95}$, for 10 searchers.

that the probability that all N agents have detected a target by this time is 95 percent:

P[all searchers have detected targets by the time $T_{0.95}$] $= \hat{P}_N(T_{0.95}) = 0.95$.

Figures 2 and 3 display the curves $T_{0.95} =$ const for systems comprising 2 and 10 searchers. From these charts it is evident, for example, that in large systems, where the cueing dynamics obeys relation (23), a more efficient time to engage is achieved by increasing the initial detection rate θ.

3 Bi-level Cooperative Search Model

In this section we consider a cooperative search mission that involves a dedicated search-only vehicle and N search-and-engage vehicles. As follows from the adopted nomenclature, the search-and-engage vehicles are capable of conducting stand-alone search for a target, and they engage the target upon detecting one. Generally, the search capabilities of the search-only vehicle are assumed to be superior to those of search-and-engage vehicles, and its role is to provide cues to them, thereby increasing their detection rates. In contrast to the model considered above, now it is assumed that search-and-engage vehicles do *not* cue each other. The model for the described scenario builds upon the Markov chain approach presented in the previous section.

For example, the search-only vehicle has two possible states, as shown in the left portion of Figure 4. Note that the search-only vehicle transitions from the "search" state to the "detect and cue" state at rate λ, and that the transition back to the "search" state is instantaneous, denoted by an infinite transition rate. That is, once the search vehicle cues a search-and-engage vehicle, the search vehicle immediately resumes its search for additional targets.

The states and transition rates for the search-and-engage vehicles are shown in the right portion of Figure 4. Search-and-engage vehicles have three possible states, "search uncued," "search cued," and "detect and engage," denoted as $\mathcal{U}$, $\mathcal{C}$, and d, respectively. The transition rate from the uncued to the cued state depends on the detection rate λ of the search-only vehicle. We assume that the search-only vehicle will cue only one search-and-engage vehicle for each target detected, and that the cues are equally distributed to the uncued search-and-engage vehicles, i.e., the transition rate from "search uncued" to "search cued" is λ/i if there are i search-and-engage vehicles in the state $\mathcal{U}$. This assumption implies that the search-only vehicle is aware of the current state of all the search-and-engage vehicles, and the search-only vehicle can transmit information to a single search-and-engage vehicle. Even if a transmission is broadcast on a common frequency, we assume that the transmitted data can be "tagged" for use only by an individual search-and-engage vehicle. After a search-and-engage vehicle receives a cue, its detection rate changes to θ_1 (recall that search-and-engage vehicles have the ability to search independently, so that a vehicle could make a direct transition from the "search uncued" to the "detect and engage" state at rate θ_0). In general, it is assumed that $\theta_1 \geq \theta_0$.

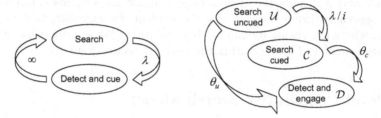

Fig. 4. State diagrams for the search-only vehicle (left) and search-and-engage vehicle (right)

Similarly to the above, let $\mathcal{S}_{ijk}$ be *a state* in which there are i search-and-engage vehicles (also called "searchers") in the state $\mathcal{U}$, j searchers in the state $\mathcal{C}$, and k searchers in the state d, where $i + j + k = N$. It is easy to see that for given i, j, and k there are $\binom{N}{i\ j\ k} = \frac{N!}{i!\,j!\,k!}$ different states $\mathcal{S}_{ijk}$. Further, it is important to note that there are $\binom{N+3-1}{N} = \frac{(N+2)(N+1)}{2}$ different triplets (i, j, k) such that $i + j + k = N$. By defining the probability of the cooperative system occupying a state $\mathcal{S}_{ijk}$ at time t as $P_{ijk}(t)$, one can describe the corresponding Markov model with a finite number of states via the following system of Kolmogorov equations:

$$\frac{\mathrm{d}}{\mathrm{d}t}P_{ijk}(t) = -\bar{\delta}_{kN}\Big[i\theta_0 + j\theta_1 + \bar{\delta}_{i0}\lambda\Big]P_{ijk}(t) + \bar{\delta}_{iN}\bar{\delta}_{j0}\left[\frac{j\lambda}{i+1}\right]P_{i+1,\,j-1,\,k}(t)$$

$$+ \bar{\delta}_{iN}\bar{\delta}_{k0}\big[k\theta_0\big]P_{i+1,\,j,\,k-1}(t) + \bar{\delta}_{jN}\bar{\delta}_{k0}\big[k\theta_1\big]P_{i,\,j+1,\,k-1}(t),$$

$$i + j + k = N. \quad (24)$$

Indeed, in the most general case a state $\mathcal{S}_{i,j,k}$ with i uncued searchers, j cued searchers, and k inactive searchers can be obtained

- from a state $\mathcal{S}_{i+1,j-1,k}$ due to a transition $\mathcal{U} \to \mathcal{C}$, i.e. when one of the $i+1$ uncued searchers receives a cueing signal from the search-only vehicle. Since each of the $i+1$ uncued searchers is being cued at rate $\frac{\lambda}{i+1}$, and there are $j = \binom{j}{1}$ states $\mathcal{S}_{i+1,j-1,k}$ that can result in the given state $\mathcal{S}_{ijk}$, transitions $\mathcal{U} \to \mathcal{C}$ increase the probability $P_{ijk}(t)$ at the rate $\frac{j\lambda}{i+1}P_{i+1,j-1,k}(t)$. This amounts to the second term in equation (24).

- from a state $\mathcal{S}_{i+1,j,k-1}$ due to a transition $\mathcal{U} \to$ d, i.e., when one of the $i+1$ uncued searchers detects a target before receiving a cue by the search-only vehicle. The search rate of an uncued agent is θ_0, and there are $k = \binom{k}{1}$ different states $\mathcal{S}_{i+1,j,k-1}$ that can lead to the given state $\mathcal{S}_{ijk}$. Thus, due to transitions $\mathcal{U} \to$ d the probability $P_{ijk}(t)$ increases at the rate $k\theta_0 P_{i+1,j,k-1}(t)$, which amounts to the third term in (24).

- from a state $\mathcal{S}_{i,j+1,k-1}$ due to a transition $\mathcal{C} \to$ d, when one of the $j+1$ cued searchers detects a target. The search rate of a cued agent is θ_1, and there are $k = \binom{k}{1}$ different states $\mathcal{S}_{i,j+1,k-1}$ that can lead to the given state $\mathcal{S}_{ijk}$. Thus, due to transitions $\mathcal{C} \to$ d the probability $P_{ijk}(t)$ increases at the rate $k\theta_1 P_{i,j+1,k-1}(t)$, which amounts to the fourth term in (24).

- finally, the first term in the right-hand side of (24) accounts for the possibility of transition from the given state $\mathcal{S}_{ijk}$ to states $\mathcal{S}_{i-1,j,k+1}, \mathcal{S}_{i,j-1,k+1}$, and $\mathcal{S}_{i-1,j+1,k}$ correspondingly.

Analogously to (9), probabilities $P_{ijk}(t)$ satisfy

$$\sum_{\substack{j+j+k=N \\ i,\,j,\,k \geq 0}} \binom{N}{i \ \ j \ \ k} P_{ijk}(t) = 1, \quad t \geq 0.$$

Solution of the system of equations (24) is facilitated via representing (24) in a matrix form, with a lower-triangular matrix. A lower-triangular form of equations (24) is obtained by introducing the notation $P_{ijk}(t) = \tilde{P}_\ell(t)$, where the index ℓ runs from 0 to $L = \frac{(N+2)(N+1)}{2} - 1 = \frac{N(N+3)}{2}$ and is determined by the indices i, j, and k as

$$\ell = \sum_{r=0}^{j+k-1}(r+1) + k = \frac{(j+k)(j+k+1)}{2} + k \quad \text{for all } 0 \leq j+k \leq N. \quad (25)$$

Explicitly, the introduced relation between $P_{ijk}(t)$ and $\tilde{P}_\ell(t)$ enumerates as

$$\tilde{P}_0(t) = P_{N00}(t),$$
$$\tilde{P}_1(t) = P_{N-1,1,0}(t),$$
$$\tilde{P}_2(t) = P_{N-1,0,1}(t),$$
$$\tilde{P}_3(t) = P_{N-2,2,0}(t),$$
$$\tilde{P}_4(t) = P_{N-2,1,1}(t),$$
$$\tilde{P}_5(t) = P_{N-2,0,2}(t),$$
$$\vdots$$
$$\tilde{P}_{L-1}(t) = P_{0,1,N-1}(t),$$
$$\tilde{P}_L(t) = P_{00N}(t).$$

It is easy to see that such a correspondence between $P_{ijk}(t)$ and $\tilde{P}_\ell(t)$ allows one to represent equations (24) in a matrix form

$$\frac{\mathrm{d}}{\mathrm{d}t}\tilde{\mathbf{P}} = \tilde{M}\tilde{\mathbf{P}}, \tag{26}$$

where the matrix $\tilde{M} \in \mathbf{R}^{(L+1)\times(L+1)}$ is lower-triangular. The initial conditions for the above system are formulated similarly to (8):

$$\tilde{P}_0(0) = 1, \quad \tilde{P}_\ell(0) = 0, \quad 1 \le \ell \le L. \tag{27}$$

Since $\tilde{M}$ is generally not diagonal, the solution to the Cauchy problem (26)–(27) has the form analogous to (12),

$$\tilde{P}_\ell(t) = \sum_{i=0}^{\ell} a_{i\ell} e^{\tilde{m}_{ii}t}, \tag{28}$$

but the expressions for coefficients $a_{i\ell}$ are more complicated comparing to (13):

$$a_{i\ell} = \sum_{j=i}^{\ell-1} \frac{\tilde{m}_{\ell j}\, a_{ij}}{\tilde{m}_{ii} - \tilde{m}_{\ell\ell}}, \quad i < \ell, \tag{29}$$

$$a_{ii} = -\sum_{j=0}^{i-1} a_{ji}, \quad a_{00} = \tilde{P}_0(0) = 1. \tag{30}$$

Above, it is assumed that the eigenvalues of matrix $\tilde{M}$ are all different:

$$i\theta_0 + j\theta_1 + \bar{\delta}_{i0}\lambda \ne i'\theta_0 + j'\theta_1 + \bar{\delta}_{i'0}\lambda \quad \text{for all } 0 \le i+j \le N \text{ and } 0 \le i'+j' \le N.$$

The developed solution to equations (24) is illustrated on a system comprised by one search-only vehicle and five search-and-engage vehicles ($N = 5$). The value of information transmitted in the cues is determined by parameter κ,

$$\theta_1 = \kappa\theta_0.$$

As before, the system's effectiveness is measured by the probability $\tilde{P}_L(t) = P_{00N}(t)$ that all search-and-engage vehicles have detected targets by time t. Figures 5 and 6 contain graphs of the probability $\tilde{P}_L(t)$ for the case when the initial detection rate θ_0 of the search-and-engage vehicles is equal to 0.1, and the cueing effectiveness κ and the search rate λ of the search-only vehicle vary. In particular, the presented graphs imply that increments in κ have more pronounced effect on increasing the probability $\tilde{P}_L(t)$ than increments in λ. In other words, precise cueing is more valuable than high target detection rate of the search-only vehicle (in the considered case).

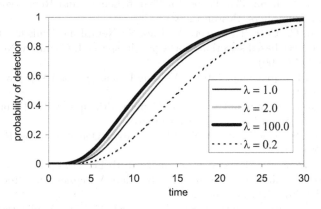

Fig. 5. Probability of detection, $\tilde{P}_L(t)$, for a system of one search-only vehicle and 5 search-and-engage vehicles, where $\theta = 0.1$ and $\kappa = 1.9$.

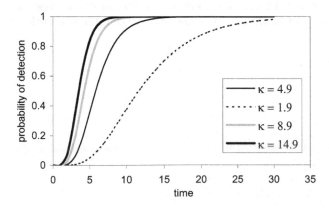

Fig. 6. Probability of detection, $\tilde{P}_L(t)$, for a system of one search-only vehicle and 5 search-and-engage vehicles where $\theta = 0.1$ and $\lambda = 1.5$.

Conclusions

We have proposed a Markov chain approach to quantification of the effect of cueing in cooperative search systems. It has been shown that cueing can dramatically affect the probability of detection over a fixed time interval. We have also shown that there is an upper bound on the benefit of cueing, at least for the problem defined.

References

[1]   Alpern, S., Gal, S.: The Theory of Search Games and Rendezvous. Kluwer Academic Publishers, Boston (2003)

[2]   Arrington, C., Carr, T., Mayer, A., Rao, S.: Neural mechanisms of visual attention: object-based selection of a region in space. J. Cognitive Neuroscience, **12**, 106–117 (2000)

[3]   Kleinrock, L.: Queueing Systems, Volume I: Theory. John Wiley & Sons, New York (1975)

[4]   Koopman, B.: Search and Screening: General Principles with Historical Applications. Pergamon Press, New York (1980)

[5]   Koopman, B.: The Theory of Search II: Target Detection. Ops. Res., **4**, 503–531 (1956)

[6]   Merriam-Webster's Collegiate Dictionary, 10th Ed. (1999)

[7]   Papoulis, A., Pillai, S.U.: Probability, Random Variables, and Stochastic Processes, 4th Ed. McGraw Hill (2002)

[8]   Stone, L.: Theory of Optimal Search, 2nd Ed. Military Applications Section, Operations Research Society of America (1989)

[9]   Thomas, L., Washburn, A.: Dynamic Search Games. Ops. Res., **39**, 415–422 (1991)

[10]   Washburn, A.: The Theory of Search II: Target Detection. In: Haley, B., Stone, L. (eds) Search Theory and Applications. Plenum Press, New York (1980)

[11]   Wools, D.: A Chronicle of Four Lost Nukes. Houston Chronicle (July 2002)

A Markov Analysis of the Cueing Capability/Detection Rate Trade-space in Search and Rescue

Alice M. Alexander and David E. Jeffcoat

Air Force Research Lab, Munitions Directorate
Eglin AFB, FL {alice.alexander,david.jeffcoat}@eglin.af.mil

Summary. This chapter presents a search and rescue scenario modeled as a discrete-state, continuous-time Markov process. In this scenario, there are two vehicle types: search-only vehicles capable of searching for persons in distress, but not engaging or rescuing them, and search-and-engage vehicles with the capability both to search and to rescue. All vehicles have two-way communication with other vehicles. Both vehicle types can search independently, but information provided by other vehicles improves their detection capability. We develop a Markov model and use matrix exponentiation to numerically determine the transient state probabilities for the system. We use as a measure of effectiveness the time required for at least two search-and-engage vehicles to arrive on scene with a threshold probability. We then analyze the trade-space between cueing capability and vehicle detection rates.

1 Introduction

The problem of search and rescue can be formulated as a cooperative search. Baum and Passino [1] investigate the problem of cooperative search for stationary targets using multiple search vehicles. They analyze vehicle search patterns based on a detection function specified for each partitioned cell of the search environment. We consider multiple search vehicles and a single target, but with a focus on cooperation among the vehicles. Curtis and Murphey [3] propose a coordination strategy to simultaneously accomplish area search and task assignment. Slater [10] examines the benefits of cooperation for two vehicles searching for targets in a region containing both real and false targets. We consider cooperation among the vehicles, but not false targets. We analyze a multi-vehicle cooperative search problem in a search and rescue scenario using a Markov chain formulation. Stone [11] uses a Markov chain to examine the optimal allocation of resources to detect a target. Jeffcoat [4] uses a Markov analysis to examine the case of two cooperative searchers and to determine the effect of cueing on the probability of target detection.

This chapter extends [4] by including an additional vehicle type and by using matrix exponentiation to obtain numerical solutions.

1.1 Problem Description

This chapter considers two different types of vehicles. There is a search-only vehicle that performs wide area searches. This vehicle has a detection rate of λ detections per unit time and has the capability to locate persons in distress, but is not capable of engaging, or rescuing, them. There are M search-only vehicles in the scenario, where M is an integer. The second type of vehicle is a search-and-engage vehicle that conducts a limited area search. This vehicle detects persons in distress with a rate of θ detections per unit time and is capable of both searching and engaging. We model the search, but not the engagement. For our purposes, the mission of the search-and-engage vehicle is complete once it has arrived on scene. There are N search-and-engage vehicles in the scenario, where N is an integer. The Coast Guard has two primary search and rescue vehicles, the C-130 Hercules and the HH-60 Jayhawk [8]. In our model, the search-only vehicle is similar in function to the C-130 Hercules, while the search-and-engage vehicle is representative of the HH-60 Jayhawk.

We model cooperation among the vehicles using cueing. In this chapter, a cue is any information that provides focus to a search; e.g., information that provides a search heading or limits a search area. The vehicles have two-way communication: every vehicle has the capability to send and receive cues to and from all other vehicles. In the Coast Guard scenario, a cue could take the form of a radar signal showing the last known location of a boat or a verbal description of the coastline closest to where the boat was last seen. We represent the value of information communicated using the variable k, a real-valued scalar that remains constant throughout a search. The information contained in a cue increases the detection rates of the vehicles still engaged in the search by a factor of k. Once cued, the detection rate of the search-only vehicle increases to $k\lambda$ and the detection rate of the search-and-engage vehicle increases to $k\theta$. Increasing the detection rate by a factor of four is equivalent to decreasing the search area to one-fourth its original size.

The detection rate of a vehicle might represent the sensitivity of a vehicle's sensor set, vehicle speed, or any other factor affecting the capability of the vehicle to conduct a search. The cueing factor, k, might be used to model the latency inherent in a communication network, or to represent the value of information communicated among the search vehicles.

2 Approach

2.1 State Space

We consider a discrete set of possible vehicle states, with transitions between states possible at any real-valued time greater than zero. The vehicles have two modes: search mode and detect mode. The search-only vehicles transition from search mode to detect mode at a rate of λ or $k\lambda$. Similarly, the search-and-engage vehicles transition from search mode to detect mode at a rate of θ or $k\theta$.

A complete state space table is shown in Table 1 for the case of one search-only vehicle ($M = 1$) and four search-and-engage vehicles ($N = 4$). Table 1 contains all the possible combinations of states. The state space changes as the values of M and N change, but throughout the chapter we use the $M = 1$, $N = 4$ case to illustrate the techniques.

Table 1. State Space for $M = 1$, $N = 4$ Case.

State	Number of search-only vehicles	Number of search-and-engage vehicles	Next
1	0	0	2,6
2	0	1	3,7
3	0	2	4,8
4	0	3	5,9
5	0	4	10
6	1	0	7
7	1	1	8
8	1	2	9
9	1	3	10
10	1	4	–

The numbers in the "number of search-only" column indicate how many search-only vehicles have detected the person in distress. The numbers in the "number of search-and-engage" column indicate how many search-and-engage vehicles have detected the person in distress. The "next" column indicates the states to which the system can transition from its current state. For example, from state one (no vehicles found person in distress), the system can transition to state two, indicating that one of the search-and-engage vehicles found the person in distress, or to state six, indicating that the search-only vehicle found the distressed person. We take as an initial condition that all vehicles are searching and thus the system begins in state one. There is no transition from state ten; it is an absorbing state.

2.2 Markov Chain

This stochastic process can be represented as a directed graph (Figure 1). Each node represents a discrete state. The directed arcs represent the possible transition routes, and each arc is labelled with the appropriate transition rate. Defining the system state diagram in this manner allows for ease of scaling, no matter how large or complicated the system. Ross [9] describes a similar system state diagram for large problems involving call centers, internet-access modem banks and wireless networking.

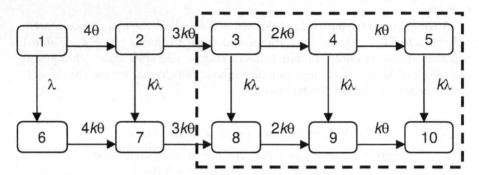

Fig. 1. System state diagram.

The system state diagram in Figure 1 graphically depicts the state space from Table 1. The system begins with all vehicles searching (state one), from which the system could move to the state where one search-and-engage vehicle detects the person in distress (state two) or the state where one search-only vehicle detects the person in distress (state six). The transition from state one to state two occurs at a rate of 4θ. The rate is 4θ because there are four search-and-engage vehicles searching, each having a detection rate of θ. The transition from state one to state six occurs at a rate of λ because this is the detection rate of a single search-only vehicle. From state two, the system can transition to state three at a rate of $3k\theta$ or to state seven at a rate of $k\lambda$. The transition rate from state two to state three is $3k\theta$ because there are now only three search-and-engage vehicles, each searching with a cued detection rate of $k\theta$. Similarly, the transition rate from state two to state seven is $k\lambda$ due to the cue provided by the search-and-engage vehicle in state two. We complete the rest of the system state diagram in a similar manner.

Figure 1 also shows the measure of effectiveness (MOE), which is the time required for at least two search-and-engage vehicles to detect the person in distress (arrive on scene) with a 95% probability. The states within the dashed box are those in which at least two search-and-engage vehicles have arrived on scene. We explain the method for calculating the probabilities in the section on matrix exponentiation.

The system state diagram (Figure 1) represents a Markov chain. In 1907 Andrei A. Markov introduced his ideas on the properties of systems of dependent variables. These systems of dependent variables later became known as Markov processes. A Markov process is a random process whose future state probabilities depend only on its current state [5]. A Markov process with a discrete state space is a Markov chain. We use a continuous time Markov chain for this model, because transitions between states can occur at any time.

2.3 Kolmogorov Equations

A first principles approach to solving a continuous time Markov chain comes from the work of Russian mathematician Andrei Kolmogorov [6], who was the first to derive differential equations for continuous time Markov chains [2]. The Kolmogorov equations are a system of differential equations, one for each discrete state. For example, equation (1) is the differential equation describing the change in probability of state three with respect to time.

$$\frac{d}{dt}P_{0,2}(t) = -k(2\theta + \lambda) \cdot P_{0,2}(t) + 3k\theta \cdot P_{0,1}(t) \tag{1}$$

The differential equation is representative of "probability flow" into and out of a state. In other words, the transition out of state three occurs at a rate of $k(2\theta + l)$, and into state three, from state two, at a rate of $3k\theta$. There is one differential equation for each state, and the solution to this system of equations provides the transient state probabilities. Jeffcoat [4] uses this method for finding the transient state probabilities for a two-vehicle problem. It is obvious that this approach becomes increasingly difficult the larger the system. There are several other methods that can be used to solve Markov chains, one of which is matrix exponentiation.

2.4 Matrix Exponentiation

Antoine Rauzy [7] discusses several methods for computing the transient solutions of large Markov models. One of the methods Rauzy discusses is matrix exponentiation, which is the approach used in this chapter. In order to solve a Markov chain using matrix exponentiation, we first define the transition rate matrix, Q, as shown in Table 2.

We develop the transition rate matrix (Table 2) from the system state diagram (Figure 1). Each off-diagonal element Q_{ij}, $i \neq j$, is the transition rate from state i to state j. If there is no transition from state i to state j, the transition rate is zero, indicated by a blank in Table 2. The diagonal entries (*) are set so that each row sums to zero. For example, $Q(1,1)$ is set to $-(4\theta + \lambda)$.

Equation (2) provides the transient state probabilities,

$$P(t) = P(0)e^{Qt} \tag{2}$$

where $P(0)$ is the vector of initial conditions, Q is the transition rate matrix, and

$$e^{Qt} = I + \sum_{n=1}^{\infty} \frac{Q^n t^n}{n!} \tag{3}$$

In order to solve for $P(t)$, we need to determine the initial condition, $P(0)$. Throughout this chapter, the initial condition is state 1, or $(0,0)$, where no vehicles have detected the person in distress. In other words,

Table 2. Transition Rate Matrix, Q.

State	1	2	3	4	5	6	7	8	9	10
1	*	4θ				λ				
2		*	$3k\theta$				$k\lambda$			
3			*	$2k\theta$				$k\lambda$		
4				*	$k\theta$				$k\lambda$	
5					*					$k\lambda$
6						*	$4k\theta$			
7							*	$3k\theta$		
8								*	$2k\theta$	
9									*	$k\theta$
10										*

$P(0) = [1\ 0\ 0\ 0\ 0\ 0\ 0\ 0\ 0]$. We determine numerical solutions for $P(t)$ using the MATLAB function EXPM. Upon calculation of $P(t)$, we sum $P(t)$ for all states where at least two search-and-engage vehicles have arrived on scene, and then determine the earliest time at which this sum exceeds the 95% probability threshold.

Figure 2 graphically displays the process for determining the MOE. The horizontal axis is time and the vertical axis is the transient state probability. The plot is the sum of the probabilities of the states in which at least two search-and-engage vehicles have arrived on scene (states within the dashed box in Figure 1). The horizontal dashed line is the 95% probability threshold. The time at which the plot crosses the threshold is the earliest time at which at least two search-and-engage vehicles have arrived on scene with a 95% probability.

2.5 Test Setup

A primary purpose of this chapter is to examine the trade space between the cueing capability k, the detection rates λ and θ, and the number of each vehicle type. A design of experiments approach is used to examine the significance of these variables, with each parameter set at one of two levels, "Low" (L) or "High" (H). We assume that λ and θ remain constant throughout the search, so that the time to detection is exponentially distributed and the number of detections over a fixed time interval has a Poisson distribution. In an attempt to get realistic detection rates, we use representative speeds for each vehicle type. For example, for the "Low" detection rate for the search-only vehicle, we use a C-130 cruising speed of 100 knots. Assuming one person in distress in a 100 square mile search area and a sensor footprint of one square mile, we get an average detection rate of 0.00032 detections per second. Table 3 provides a test setup with the values used for each parameter.

The low value of the cueing parameter, $k = 1$, represents a "no cueing" case; i.e., either no communication or the information communicated has no

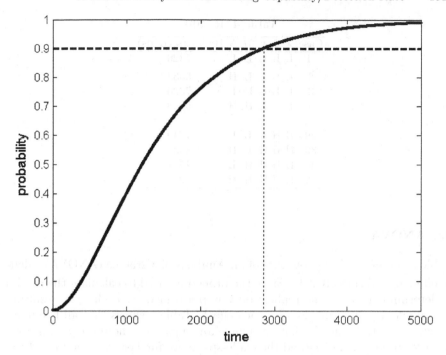

Fig. 2. Process to determine MOE.

Table 3. Test setup

Parameters	λ	θ	k	M	N
Low (L)	0.00032	0.00019	1	1	4
High (H)	0.00064	0.00038	4	2	5

value. When k is four, the cue increases the detection rate of the vehicles by a factor of four. This is equivalent to a decrease in the search area to 25% of the original area.

3 Results

Table 4 provides selected results obtained for the thirty-two cases.

The time in the result column is the earliest time at which at least two search-and-engage vehicles have found the person in distress with a 95% probability. The results range (approximately) from two hours to twenty minutes. This range seems reasonable in the context of this scenario. The next section provides an analysis of these results.

Table 4. Results

Case	λ	θ	k	M	N	Result(seconds)
1	L	L	L	L	L	7326
2	L	L	L	L	H	5638
3	L	L	L	H	L	7326
4	L	L	L	H	H	5638
29	H	H	H	L	L	1749
30	H	H	H	L	H	1438
31	H	H	H	H	L	1501
32	H	H	H	H	H	1247

3.1 ANOVA

Table 5 presents selected outputs of an Analysis of Variance (ANOVA) calculated using STATISTICA®. Since the procedure used to calculate the results is deterministic, only one replication was performed for each case, resulting in a single degree of freedom for each effect. Table 5 shows that all five main effects (λ, θ, k, M and N) have a significant impact on the response variable. In order to better understand the trade-space, we further examine four of the ten two-factor interactions in the following section.

Table 5. Analysis of variance

Effect	SS	Deg of Free	MS	F	P
λ	201612	1	201612	4032	0.010025
θ	38843298	1	38843298	776866	0.000722
k	58206655	1	58206655	1164133	0.000590
M	201613	1	201613	4032	0.010025
N	5649841	1	5649841	112997	0.001894
λ by q	30135	1	30135	603	0.025917
λ by k	201612	1	201612	4032	0.010025
θ by k	8611250	1	8611250	172225	0.001534
λ by M	325	1	325	7	0.237922
θ by M	30135	1	30135	603	0.025917
k by M	201613	1	201613	4032	0.010025
λ by N	2812	1	2812	56	0.084385
θ by N	553352	1	553352	11067	0.006051
k by N	1449253	1	1449253	28985	0.003739
M by N	2813	1	2813	56	0.084385

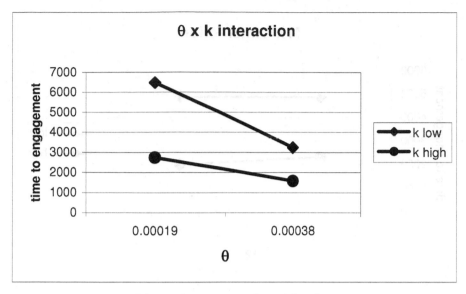

Fig. 3. Interaction effect of θ and k.

3.2 Analysis Plots

Figure 3 displays the interaction effect between the search-and-engage vehicle detection rate, θ, and the cueing factor, k. θ is plotted on the horizontal-axis and the time to engage is plotted on the vertical-axis. There are two plots, one for each level of k. Both plots have a downward slope, indicating that an increase in the search-and-engage vehicle detection rate decreases the time to engage. The k low plot (cueing not present) has a greater slope than the k high plot, showing that q has more impact on the time to engage when cueing is not present. In other words, if the vehicles are not communicating, the search capability of each search-and-engage vehicle is more critical.

We plot the interaction effect between the number of search-only vehicles, M, and cueing, k, in Figure 4. Figure 4 indicates that increasing the number of search-only vehicles does not significantly affect the time for two search-and-engage vehicles to arrive on scene, with or without cueing. There is an effect due to cueing, as shown by the gap between the two lines, but there is essentially no effect from increasing the number of search-only vehicles. This could be a result of using a relatively small search area.

Figure 5 is a plot of the interaction effect between the number of search-and-engage vehicles and cueing. There is a more pronounced downward slope in the case when cueing is not present (k low), illustrating that when there is no communication between the vehicles, an additional search-and-engage vehicle significantly decreases the time to engage.

Figure 6 is a plot of the interaction effect between the search-only vehicle detection rate, λ , and the search-and-engage vehicle detection rate, θ .

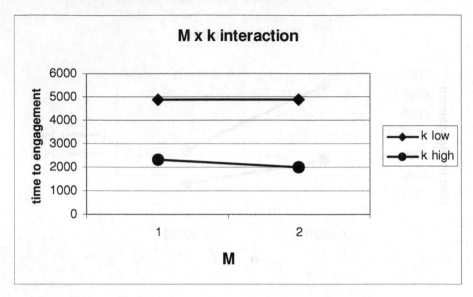

Fig. 4. Interaction effect of M and k.

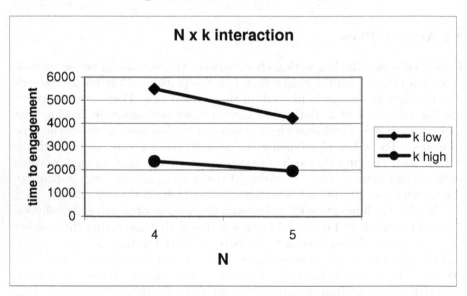

Fig. 5. Interaction effect of N and k.

The plots indicate that the search-and-engage vehicle detection rate is more significant than the search-only vehicle detection rate.

These plots are useful for analyzing the trade-space between cueing capability and detection rates in a search and rescue scenario.

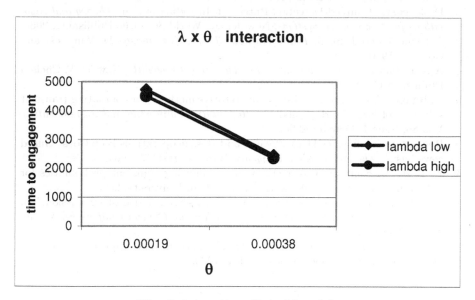

Fig. 6. Interaction effect of λ and θ.

4 Summary

This chapter analyzes the trade-space between cueing capability and detection rates in a search and rescue scenario using a Markov chain formulation. Matrix exponentiation is used to solve for the transient state probabilities. An analysis is conducted to examine the effects of cueing capability, detection rates, and numbers of vehicles on the time to engage. We conclude that the number of search-and-engage vehicles and their detection rate are the most influential parameters for this specific scenario. We also find that cueing significantly decreases the time to engage. The Markov chain model provided a useful tool for the analysis of a search and rescue scenario. We believe this approach could be used in the analysis of other cooperative search problems.

References

1. M. Baum and K. Passino, "A Search-Theoretic Approach to Cooperative Control for Uninhabited Air Vehicles," In *Proceedings of the AIAA Guidance, Navigation and Control Conference*, Monterey, California, August 2002.

2. A. Bharucha-Reid, *Elements of the Theory of Markov Processes and Their Applications*, New York: McGraw-Hill Book Company Inc, 1960.
3. J. Curtis and R. Murphey, "Simultaneous Area Search and Task Assignment for a Team of Cooperative Agents," in *Proceedings of the AIAA Guidance, Navigation and Control Conference*, Austin, Texas, August 2003.
4. D. Jeffcoat, "Coupled Detection Rates: An Introduction," in *Theory and Algorithms for Cooperative Systems,* New Jersey: World Scientific Publishing, 2004.
5. J. Kemeny and J. Snell, *Finite Markov Chains*, New Jersey: D. Van Nostrand Co., Inc, 1960.
6. A.N. Kolmogorov, *Foundations of the Theory of Probability*, New York: Chelsea Publishing Co, 1956.
7. A. Rauzy, "An experimental study on iterative methods for computing transient solutions of large Markov models," *Reliability Engineering and System Safety*, Vol. 86, Issue 1, October 2004.
8. Regulatory Intelligence Data, "Search and Rescue operations keep Coast Guard cews busy," *The America's Intelligence Wire*, August 27, 2004.
9. A. Ross, "A comparison of methods for computing transient probabilities for the Erlang loss system," Working paper, Lehigh University, 2003.
10. G. Slater, "Cooperation between UAV's in a search and destroy mission," in *Proceedings of the AIAA Guidance, Navigation and Control Conference*, Austin, Texas, August 2003.
11. L. Stone, *Theory of Optimal Search, 2nd Edition*, Military Applications Section, Operations Research Society of America, 1989.

Challenges in Building Very Large Teams

Paul Scerri[1], Yang Xu[2], Jumpol Polvichai[2], Bin Yu[1], Steven Okamoto[1], Mike Lewis[2] and Katia Sycara[1]

[1] Carnegie Mellon University
   {pscerri,byu,sokamoto,katia}@cs.cmu.edu
[2] University of Pittsburgh
   {xuy,jumpol,ml}@sis.pitt.edu

Summary. Coordination of large numbers of unmanned aerial vehicles is difficult due to the limited communication bandwidth available to maintain cohesive activity in a dynamic, often hostile and unpredictable environment. We have developed an integrated coordination algorithm based on the movement of *tokens* around a network of vehicles. Possession of a token represents exclusive access to the task or resource represented by the token or exclusive ability to propagate the information represented by the token. The movement of tokens is governed by a local decision theoretic model that determines what to do with the tokens in order to maximize expected utility. The result is effective coordination between large numbers of UAVs with very little communication. However, the overall movement of tokens can be very complex and, since it relies on heuristics, configuration parameters need to be tuned for a specific scenario or preferences. We have developed a neural network model of the relationship between configuration and environment parameters and performance, that an operator uses to rapidly configure a team or even reconfigure the team online, as the environment changes.

1 Intro

Efficient, effective automated or semi-automated coordination of large numbers of cooperative heterogeneous software agents, robots and humans has the potential to revolutionize the way complex tasks are performed in a variety of domains. From military operations, to disaster response [26, 58] to commerce [35] to space [18], automated coordination can decrease operational costs, risk and redundancy while increasing efficiency, flexibility and success rates. To achieve this promise, scalable algorithms need to be found for the key problems in coordination. Unfortunately, these problems, including deciding how to allocate tasks and resources, deciding when and what to communicate and planning, have NP-complete or worse computational complexity and thus require approximate solutions.

While automated coordination is a very active research area (e.g., [56, 61, 19]), previous work has failed to produce algorithms or implementations that meet the challenges of coordinating large numbers of heterogeneous actors in complex environments. Most algorithms developed for key problems do not scale to very large problems (e.g., optimal search techniques [32, 33]), though some scale better than others (e.g., markets [18]). The rare algorithms that do scale effectively typically either make unreasonable assumptions or are very specialized for a specific problem (e.g., emergent behavior [45]). Algorithms that have been shown to be scalable often rely on some degree of centralization [31], which may not always be desirable or feasible.

The chapter presents an integrated solution to the problem of automated coordination for large teams of unmanned aerial vehicles. The solution relies on three novel ideas. The first novel idea is to use *tokens*, encapsulate both information and control, as the basis for all coordination. For each aspect of the coordination, e.g., each task to be assigned, there is a token representing that aspect. Control information, included with the token, allows actors to locally decide what to do with the token. For example, a task token contains control information allowing an actor to decide whether to perform the task or pass it off for another actor. Movement of tokens from actor to actor, implements the coordination. The effect of encapsulating the piece of the overall coordination with its control information is to ensure that everything required for a particular decision, e.g., whether a particular actor should perform a particular task, is localized to the actor currently holding the respective token. This reduces required communication and avoids many of the problems that arise when required information is distributed. The key is to be able to find control rules and control information that can be encapsulated in the token and allow decisions about that token to be made relatively independantly of decisions about other tokens. We have developed such algorithms for a range of key coordination problems including task allocation [53], reactive plan instantiation [52], resource allocation, information sharing [63] and sensor fusion.

The efficiency of the token based coordination algorithms depends on the routing of the tokens around the network of actors. Individual actors build up local models of what types of things their neighbors in the network are most interested in and use these models to decide where to send a token (if at all) [64]. Previous work has shown that even relatively poor (i.e., often inaccurate) local routing models can dramatically improve overall performance of a particular coordination algorithm [63]. The second novel idea in this work is to exploit the homogeneity of tokens, i.e., the fact that all aspects of coordination are represented with tokens, to allow actors to exploit the movement of tokens for one coordination algorithm to improve the flow of tokens for another coordination algorithm. The inituition is that, e.g., knowing something about a task allocation should help the resource allocation which should in turn help dissemination of important information. For example, if actor A knows that actor B is performing a fire fighting task in Pittsburgh, it can infer that resources physically located in Phillidelphia will not be of interest to actor B and save the overhead of involving actor B in algorithms for allocating those resources. Our results show that exploiting synergies between coordination algorithms dramatically improves overall coordination performance.

The third novel idea in this work is to develop a general purpose *meta-level reasoning* layer, distributed across the team. The meta-reasoning is conceptually "above" the token flows and manipulates the movement of tokens in one of two ways. First, the meta-reasoning layer can extract and manipulate particular tokens when behavior is not as required. For example, the meta-reasoning layer may notice that a particular task allocation token is unable to find any actor to perform the task it represents and bring that unfilled task assignment to the attention of a human who can decide what to do. Tokens can be safely extracted because everything related to that token is encapsulated within the token. Second, the meta-reasoning layer can configure control parameters on all tokens to manipulate the token flows to maximize current performance requirements. For example, when communication bandwidth is tightly constrained the meta-reasoning can configure the tokens to move less (at a cost of quality of performance.) We use a neural network model of the relationships between control parameters and performance to allow rapid search for parameter settings to meet current performance requirements and online reconfiguration to adapt to changing requirements.

In previous work, we have shown that by leveraging a logical, static network connecting all coordinated actors, some coordination algorithms can be made to effectively scale to very large problems [49]. The approach used the network to relax some of the requirements to communicate and made it possible to apply theories of teamwork [9, 56] in large groups. The key was the *small worlds* property of the network [59] which requires that any two agents are connected by a small number of intermediate agents, despite having a relatively small number of direct neighbors. The token-based algorithms leverage this network when moving around the team

The integrated token-based approach has been implemented both in abstracted simulation environments and as a part of domain independent coordination software called Machinetta [50]. The abstract simulation environments show the effectiveness of the token based ideas across a very wide range of situations. In this simulation environment, the token-based algorithms have been shown to be extremely scalable, comfortably coordinating groups of up to 5000 actors. Machinetta is a public domain software module that has been used in several domains [50, 49, 54], including for control of unmanned aerial vehicles. Machinetta uses the concept of a *proxy* [42, 23] which gives each actor a semi-autonomous module encapsulating the coordination reasoning. The proxies work together in a distributed way to implement the coordination. Experiments with upto 200 Machinetta proxies running the token based algorithms have shown that fully distributed coordination feasible.

2 Problem Description

In this section, we describe the target application and the general coordination problem that must be addressed.

Target Application: Coordinated Wide Area Search Munitions

Our current domain of interest is coordination of large groups of Wide Area Search Munitions (WASMs). WASMs are a cross between an unmanned aerial vehicle and a standard munition. The WASM has fuel for about 30 minutes of flight, after being launched from an aircraft. The WASM cannot land, hence it will either end up hitting a target or self destructing. The sensors on the WASM are focused on the ground and include video with automatic target recognition, ladar and GPS. It is not currently envisioned that WASMs will have an ability to sense other objects in the air. WASMs will have reliable high bandwidth communication with other WASMs and with manned aircraft in the environment. These communication channels will be required to transmit data, including video streams, to human controllers, as well as for the WASM coordination.

The concept of operations for WASMs are still under development, however, a wide range of potential missions are emerging as interesting [7, 12]. A driving example for our work is for teams of WASMs to be launched from AC-130 aircraft supporting special operations forces on the ground. The AC-130 is a large, lumbering aircraft, vulnerable to attack from the ground. While it has an impressive array of sensors, those sensors are focused directly on the small area of ground where the special operations forces are operating making it vulnerable to attack. The WASMs will be launched as the AC-130s enter the battlespace. The WASMs will protect the flight path of the manned aircraft into the area of operations of the special forces, destroying ground based threats as required. Once an AC-130 enters a circling pattern around the special forces operation, the WASMs will set up a perimeter defense, destroying targets of opportunity both to protect the AC-130 and to support the soldiers on the ground. Even under ideal conditions there will be only one human operator on board each AC-130 responsible for monitoring and controlling the WASMs. Hence, high levels of autonomous operation and coordination are required of the WASMs themselves. However, because the complexity of the battlefield environment and the severe consequences of incorrect decisions, it is expected that human experience and reasoning will be extremely useful in assisting the team in effectively and safely achieving their goals.

Many other operations are possible for WASMs, if issues related to coordinating large groups can be adequately resolved. Given their relatively low cost compared to Surface-to-Air Missiles (SAMs), WASMs can be used simply as decoys, finding SAMs and drawing fire. WASMs can also be used as communication relays for forward operations, forming an adhoc network to provide robust, high bandwidth communications for ground forces in a battle zone.

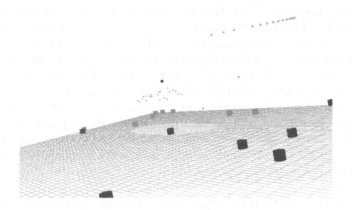

Fig. 1. A screenshot of the WASM coordination simulation environment. A large group of WASMS (small spheres) are flying in protection of a single aircraft (large sphere). Various SAM sites (cylinders) are scattered around the environment. Terrain type is indicated by the color of the ground.

Since a WASM is "expendible", it can be used for reconnasiance in dangerous areas, providing real-time video for forward operating forces. While our domain of interest is teams of WASMs, the issues that need to be addressed have close analogies in a variety of other domains. For example, coordinating resources for disaster response involves many of the same issues [26], as does intelligent manufacturing [44] and business processes.

2.1 Team Oriented Plans and Joint Activities

The problem of coordination we are dealing with here can be informally described as determining who does what at which time and with which shared resources and information. In the following, we provide a formal description of this coordination problem.

Each member of the team $a \in A$ has a copy of the *Team Oriented Plan* templates, *Templates* that describe the joint activities that need to be undertaken in particular situations [43]. These templates are defined offline by a domain expert. Each template, *template* $\in$ *Templates* has preconditions, $template_{pre}$ under which it should be instantiated into a *joint activity*, α_i. The template may also have parameters, $template_{param}$ that encode specifics of a particular instance. The same template may be instantiated multiple times when with different parameters. The joint activity should be terminated when certain postconditions, $template_{post}$ are met. These postconditions may be a function of $template_{param}$. The templates whose preconditions but not postconditions are satisfied at time t are written $JointActs(t, Templates)$.

A joint activity, α_i, breaks a complex activity down into tasks, $Tasks(\alpha_i) = \{task_i^1, \ldots, task_i^n\}$, each intended to be performed by a single actor. Constraints, $constraints(\alpha_i)$, exist between the tasks including constraints on the sequencing of tasks, the simultaneous (or not) execution of tasks and whether tasks are alternative ways of doing the same thing. The set of roles that should be executed at time t to achieve α_i, given the constraints, at time t is written $CurrTasks(\alpha_i, t) = f(Tasks(\alpha_i), Constraints(\alpha_i))$. Each team member has a capability to perform each task, which is written as $capability(a, task, t)$. This capability may change over time as, e.g., an agent moves around the environment. Notice that the actual value of assigning a particular actor to a particular task depends also on which resources and information that actor has, as described beow. A $template_i$ does not specify which actor should perform which task nor which resources will be used nor what what coordination must take place [43].

For example, a UAV team might have a template investigating a potential target, $template_{imaging}$. The template will be instantiated when there is a sensor reading indicating a vehicle at a location X, i.e., $template_{pre} = SensedVehicleX$ and parameterized with the specific location to investigate, i.e., $template_{param} = LocationX$. The template breaks the investigation task down into IR, EO and ladar imaging to be performed by three UAVs with relevant sensing capabilities.

Associates Network

The *associates network* arranges the whole team into a small worlds network defined by $N(t) = \bigcup_{a \in A} n(a)$, where $n(a)$ are the *neighbors* of agent a in the network. The minimum number of *agents* a message must pass through to get from one agent to another via the associates network is the *distance* between those agents. For example, if agents a_1 and a_3 are not neighbors but share a neighbor $distance(a_1, a_3) = 1$. We require that the network be a small worlds network, which imposes two constraints. First, $\forall a \in A, |n(a)| < K$, where K is a small integer, typically less than 10. Second, $\forall a_i, a_j \in A, distance(a_i, a_j) < D$ where D is a small integer, typically less than 10.

2.2 The Value of Information

Events and circumstances in the environment are represented discretely as beliefs, $b \in Beliefs$. Individual actors will not necessarily know all current beliefs, but typically some subset $K_a \in Beliefs$. If the environment is fully observable then $\bigcup_{a \in A} K_a = Beliefs$ otherwise $\bigcup_{a \in A} K_a \subset Beliefs$. When an actor is assigned to a task (see below) having knowledge of particular beliefs can improve how well the actor can perform the task. The value of a piece of information is dependant on the environment, the task, the actor and time and is written $value(b, a, task, time) \rightarrow \mathcal{R}$. For example, when a robot with vision based sensing is assigned to search a building for trapped civilians, knowledge of where smoke is in the building is less important than to a robot using infrared or acoustic sensors (listening for voices) than using vision. While in general the mapping between tasks, agents, information and value is very complex, in practical applications it is often straightforward to find reasonable, compact approximations.

2.3 Resources

To perform assigned tasks, actors may need *resources*, $Resources = \{r_1, \ldots, r_n\}$. In this work, resources are modeled as being freely assignable to any actor and never being exhausted, however only one actor can have access to a resource at any one time[3]. A task's need for a resource is modeled as being independant of which actor is performing the task. Often there is a set of resources that are interchangable, in so far as any one of the resources is just as effective as any of the others. Such sets are written $IR = \{r_i, \ldots, r_n\}$. Some interchangable resources, $task_{need} = \{IR_1, \ldots, IR_m\}$, are necessary for execution of the task. This means that without access to at least one resource from each $IR \in task_{need}$ no actor can execute this task. Another set of interchangable resources, $task_{useful} = \{IR_1, \ldots, IR_k\}$ are useful to the execution of the task, although the task can be executed without access to one of the interchanable resources.

Assignment of a resource, r, to actor, a, is written $assigned(r, a)$. The resources assigned to an actor are $resouces(a)$. Since a resource cannot be assigned to more than one agent, we require $\forall a, b \in A, a \neq b, resources(a) \cap resources(b) = \emptyset$.

Consider a task for a UAV to provide a video image of a potential target. To perform this task the UAV must have appropriate sensors and have access to some airspace from which it can take the an image. The airspace can be modelled as a necessary, interchangable resource, with the UAV typically having a range of options about which airspace to use to take the video.

2.4 Assignments and Optimization

This first step toward effective coordination is to determine what templates should be instantiated into joint activities, $\alpha_1, \ldots, \alpha_n$. The joint activities define the current set of tasks that need to be assigned to actors. Any templates that are not instantiated, but should be, because their preconditions are satisfied or should be terminated because their postconditions are satisfied, cost the team value. Performance of any tasks for joint activities that should have been terminated provide no value to the team. As described above, the joint activities that should be executed at time t are $JointActs(t, Templates)$.

$Tasks!(t) = \bigcup_{\alpha \in JointActs(t, Templates)} CurrTasks(\alpha, t)$ defines the set of tasks that give value to the team if assigned to capable team members at time

---

[3] If multiple actors can access the same resources simultaneously, we represent this is being multiple resources

t. The set of tasks that are assigned to an actor, a, is written $tasks(a)$. As with resources, tasks should be assigned to only one actor, hence $\forall a, b \in A, a \neq b, tasks(a) \cap tasks(b) = \emptyset$. The value an actor provides to the team is a function of the tasks assigned to it, the resources assigned to it and the information it knows, $contrib(a, tasks(a), resources(a), K_a, t) \to \mathcal{R}$. This function can be a complex function since interactions between tasks and knowledge can be very intricate, but in practice simple linear functions are often used to approximate it. In the case that the task $t \notin Tasks!(t)$ the agent team can recieve no value for the execution of the task. The overall coordination problem can be described as:

$$\int_{t=0}^{\infty} \sum_{a \in A} contrib(a, tasks(a), resources(a), K_a, t) - communicationCost \quad (1)$$

Typically, communication between actors is not free or is limited in some way (e.g., total volume). Optimization of Equation 1 should be performed taking into account these communications limitations.

3 Algorithms

To implement coordination in a large team we encapsulate anything that needs to be shared in a *token*. Specifically, tokens represent any belief that needs to be shared, any assignable task or any shared resource. Tokens cannot be copied or duplicated, but can be passed from actor to actor along links in the network connecting them. A token, Δ, contains two types of information *content* and *control*. The content component describes the belief, task or resource represented by the token. The control component captures the information that is required to allow each agent decide whether to keep or pass on the token to maximise the expected value to the team. The precise nature of the control component depends on the type of token, e.g., role or resource, and is discussed in more detail below. However, common to all is the path the token has followed through the team, denoted $\Delta.path$. In the remainder of this section, we describe how key coordination algorithms are implemented via the use of tokens. Notice, below when an actor decides to move a token to another actor it calls *Pass*. In the next section, we describe how the *Pass* function sends the token from a to the $a \in n(a)$ that is most likely to benefit from reciept of the token, e.g., most likley to be able to use the resource represented by a resource token.

Information Sharing

Members of a large team will commonly locally sense information that is useful to the execution of another agents tasks. The value of this information to a team member executing a task was formally defined in Section 2.2. However, it is not necessarily the case that the agent locally sensing the information will know which of its teammates needs information or even that a teammate needs it at all. We have developed a token-based algorithm for proactive sharing of such information that efficiently gets the information to any agent that needs it. The algorithm is described in detail in [63]. The control information for an information token is simply the number of hops through the team that the information token will be allowed to make before it is assumed that any team mate that needs the information actually has it. Algorithm 1 provides the pseudo code for local processing of an information token.

Algorithm 1: Information Token Algorithm
(1)
       if $token.TTL > 0$
(2) $token.TTL --$
(3)   PASS($token$)

Where $token.TTL$ is the number of remaining hops the token can take. Efficient values for $token.TTL$ are determined emperically and tend to be approximately $log(|A|)$.

Template Instantiation and Joint Activity Deconfliction

To instantiate a plan template, $template_i$, into a joint activity, α_i, requires that some team member know that the preconditions, $template_{pre}$, for the plan are satisfied. Since preconditions for a particular template may be sensed locally by different team members, belief sharing via information tokens is required to ensure that at least one actor knows all the preconditions. However, if multiple actors get to know the same preconditions, it may happen that the template is instantiated multiple times and the team's effort is wasted on multiple executions of the same plan. Our approach to this problem is described in detail in [29], in this chapter we just provide a brief overview. The approach to avoiding plan duplicates uses two ideas. First, an actor can choose not to instantiate a template (at least for some time), despite knowing the preconditions hold and not knowing of another instantiation. For example, in some cases an actor might wait a random amount of time to see if it hears about another instance before instantiating a plan. Second, once it does instantiate the template into a joint activity it informs each of its neighbours in the associates network. Any actor accepting a role in the joint activity must also inform its neighbors about the joint activity. It turns out that despite only a relatively small percentage of the team knowing about a particular joint activity instance, there is very high probability of at least one team member knowing about both copies of any duplicated team activity. A team member detecting a duplicate plan instantiation is obliged to inform the actors that instantiated the duplicate plans (this information is kept with the information token informing of the initiation of the joint activity) who can then initiate a straightforward deconfliction process.

Task Allocation

Once joint activities are instantiated from templates, the individual tasks that make up the joint activity must be assigned to individual actors. Our algorithm for task allocation is extensively described and evaluated in [53]. A task token is created for each task, $t \in Tasks(\alpha)$. The holder of the task token has the exclusive right to execute the task and must either do so or pass the token to a teammate. First, the actor must decide whether it is in the best interests of the team for it to perform the task represented by the token (Alg 2, line 6). A task tokens control information is the minimum capability ($capability(a, task, t)$) an actor must have in order to perform the task, $task$. This threshold is the control component of the token. The token is passed around the team until it is held by an actor with capability above threshold for the task and without other tasks that would prevent it from performing the task. Computing thresholds that maximize expected utility is a key part of this algorithm and is described in [53]. The threshold is calculated once (Alg 2, line 5), when the task arises due to team plan instantiation. A token's threshold therefore reflects the state of the world when it was created. As the world changes, actors will be able to respond by changing the threshold for newly-created tokens. This allows the team flexibility in dealing with dynamics by always seeking to maximize expected utility based on the most current information available.

Once the threshold is satisfied, the actor must check whether it can perform the task give other responsibilities (Alg. 2, line 9). If it cannot, it must choose a task(s) to reject and pass the respective tokens to a neighbor in the network (Alg. 2, lines 10 and 12). The actor keeps the tasks that maximize the use of its capabilities (performed in the MAXCAP function, Alg. 2, line 10), and so acts in a locally optimal manner. Extensions to this algorithm allow efficient handling of constraints between tasks.

Resource Allocation

Efficient teams must be able to assign resources to actors that can make best use of those resources. As described above tasks have both necessary and useful resources. Since there is no global view of which actor is doing which task, the process of allocating resources to tasks must be fully distributed. Each shareable, discrete resource is represented by an individual token. As with task tokens, control information on the token is in the form of a threshold. An actor can hold the resource token, and thus have exclusive access to the resource, if it computes its need for the token as being above the threshold.

Algorithm 2: Task Token
```
(1)      V ← ∅, PV ← ∅
(2)      while true
(3)          token ← getMsg()
(4)          if token.threshold = NULL
(5)              token.threshold ← CALCTHRESHOLD(token)
(6)          if token.threshold < Cap(token.value)
(7)              V ← V ∪ token.value

(9)              if ∑_{v∈V} Resources(v) ≥ agent.resources
(10)                 out ← V − MAXCAP(V)
(11)                 foreach v ∈ out
(12)                     PASSON(new token(v))
(13)                     V ← V − out

(15)         else
(16)             PASSON(token) /* threshold < Cap */
```

However, unlike task tokens, thresholds for resource tokens are dynamic. When an actor has a resource it slowly increases the threshold (up to some maximum value) and continues checking whether its need for the resource is above that threshold. When the token moves around the team, the threshold is slowly descreases until it is accepted by some agent. The combination of moving the threshold up and down ensures that whichever actor needs the resource most at a particular point in time gets that resource.

Algorithm 3: ProcessResourceToken
```
(1)      while true
(2)          token ← getMsg()
(3)          if token.threshold < Req(token.resource)
(4)              // Keep the token
(5)              MONITORRESOURCETOKEN(token)
(6)          else
(7)              PASS(token)
```

Algorithm 4: MonitorResourceToken
```
(1)
             haveToken ← true while haveToken
(2)              SLEEP()token.threshold ← token.threshold + inc if
                 token.threshold > Req(token.resource)
(3)              PASS(token)
(4)              haveToken ← true
```

Sensor Fusion

Individual sensors of individual actors may not be sufficient to determine the state of some part of the environment. For example, a single UAV may not have a sufficiently high fidelity sensor suite to independantly determine that an enemy tank is concealed in a forest. However, multiple sensor readings by multiple actors can result in the team having sufficiently high confidence in a determination of the state to take an action. However, in a cooperative *mobile* team an actor will not always have accurate knowledge about where other actors are and hence will not know which team mates might have readings to confirm or refute its own. We encapsulate each sensor reading in an information token and forward the token across the team. Each actor individually performs sensor fusion on the information that it has and creates a new information token with the fused belief when it is able to combine multiple low confidence readings into a single high confidence reading. The key to this algorithm is that despite each token visiting a relatively small number of team members there is high probability that some team member will get to see multiple sensor readings for the same event, if they exist. As with information tokens, the control information for sensor-reading tokens (i.e., information tokens) is the number of additional hops a token should make before assuming it cannot be fused at the current time (i.e., TTL).

4 Synergies Between Algorithms

Efficient token-based coordination depends on how well tokens are routed, i.e., how efficiently they pass from actor to actor to where the are most needed. Since routing decisions are made locally, actors must build local models of the state of the team to make appropriate routing decisions. Notice that *whether* to pass a token on is a function of the control information on the token, but *where* to route a token, if that is the decision, is a function of the local model of state. In this section, we describe an algorithm to maintain the localized decision model by utilizing previously received tokens.

We assume that there is a known relationship between tokens, called *relevance*. We define the relevance relationship between tokens Δ_i and Δ_j as $Rel(\Delta_i, \Delta_j)$. $Rel(\Delta_i, \Delta_j) > 1$ indicates that an agent interested in Δ_i will also be interested in Δ_j, while $Rel(\Delta_i, \Delta_j) < 1$ indicates that an agent interested in Δ_i is unlikely to be interested in Δ_j. If $Rel(\Delta_i, \Delta_j) = 1$ then nothing can be inferred. When an agent receives two tokens for which $Rel(\Delta_i, \Delta_j) > 1$ they are more likely to be usable in concert to obtain a reward for the team. For example, when an actor gets a task token Δ_t and resource token Δ_r representing a necessary resource for the task, $Rel(\Delta_t, \Delta_r) > 1$ and passing them to the same acquaintance is more likely to gain reward for the team than passing them to different acquaintances.

4.1 Updating Decision Model according to Previous Tokens

Each actor maintains a matrix $P[\Delta, a] \rightarrow \mathcal{R}$ that estimates that for each possible token, the probability that each of its associates would be the best to pass that token to. For example, $P[\Delta_k, a] = 0.2$ indicates that the actor estimates that the probability associate a is the best of its associates to pass token Δ_k to is 0.2. Notice that in the implementation we do not actually store the entire matrix but calculate it as needed, but in the following we assume so for clarity.

The update function of agent α's P_α based on an incoming token Δ_j, written as $\delta_P(P_\alpha[\Delta_i, b], \Delta_j)$ leverages Bayes' Rule as follows:

$$\forall b \in n(\alpha), \forall \Delta_i, d = first(n(a), \Delta_j.path)$$

$$\delta_P(P_a[\Delta_j, b], \Delta_i) = \begin{cases} P_a[\Delta_j, b] \times Rel(\Delta_i, \Delta_j) & \text{if } \Delta_i \neq \Delta_j, b = d \\ P_\alpha[\Delta_j, b] & \text{if } \Delta_i \neq \Delta_j, b \neq d \\ \varepsilon & \text{if } \Delta_i = \Delta_j, b \in \Delta_j.path \cap n(\alpha) \end{cases} \tag{2}$$

first extracts from the recorded path of the token the acquataince of the actor that earliest had the token. P is then normalized to ensure $\sum_{b \in N(\alpha)} P_\alpha[\Delta_j, b] = 1$. The first case in Eqn. 2 is the most important. The probability that d is the best agent to receive Δ_i is updated according to $Rel(\Delta_i, \Delta_j)$. The second case in the equation changes the probability of sending that token to agents other than the sender in a way that ensures the subsequent normalization has the desired effect. Finally, the third case encodes the idea that an actor should typically not pass a token back from where it came. Details about how Rel is computed to ensure appropriate behavior can be found in [64].

5 Human-in-the-Loop

The token-based process described above works effectively at controlling large teams. However, for real-world teams it is essential to have a human-in-the-loop, controlling the behavior of the team. The need for such control stems from two key reasons. First, the heuristics used to coordinate the team will not always work effectively and sometimes human "common sense" will be required to ensure appropriate behavior. Second, the human may have preferences for tradeoffs given the current situation, e.g., a willingness to trade off the quality of task allocation provided bandwidth is reduced. These two rationales for human control imply an ability for control at both a high and low level and over a wide range of aspects of behavior. Fortunately, the homogeneity of the token-based algorithms allows an effective and general control layer to be built on top of the control flows providing powerful control for a human user (or users.) The effect is to allow meta-reasoning over the token-based coordination. The specific approach we have developed has two components, one for high level control and another for more detailed control.

5.1 High-Level Control

The high level control component allows the user to tradeoff high level performance measures such as the message bandwidth versus performance task allocation versus resource allocation. To do this we need a model of the interaction between the environment algorithm configuration and performance. However, the relationship turns out to be very complex, denying straightforward means of modeling it. Moreover, non-determining leads to a relatively high standard deviation in performance. To represent the highly non-linear relationship between the environment, configuration and performance of the team, we used multilayer feed-forward neural networks, capable of representing any arbitrary function [38]. With inspirations from the idea of dynamic rearrangement [13], we use the concept, called *Dynamic Neural Networks* [39, 40], which allows all internal nodes in the network to act stochastically and independently even though all external input data remain unchanged.

We trained the multilayer feed-forward neural network using genetic algorithms because of the high standard deviation of the function being modelled. Moreover, in genetic algorithms, the unit of adaptation is not an individual agent, but a population of agents, which is excellent for dealing with very huge and noisy training data set. The fitness function was the average of square error between target output and actual output as follows:

$$\sum_{d \in D} \sum_{p \in P} (O_{p,t}^d - O_{p,a}^d)^2 / sizeof(D).$$

Where D is the set of training data ($d \in D$), P is the set of system performance measures ($p \in P$), $O_{p,t}^d$ is the target output of the p th performance parameter of the data entry d, and $O_{p,a}^d$ is the actual output of the p th performance parameter of the data entry d. The genetic algorithm training function attempts to minimize this function. The learning process converged to 20 percent error quickly and slowly converged to 15 percent error after that. Future work will look at making more accurate models.

Team Control Interface

A user interface, shown in Figure 2 was developed for working with the dynamic neural network model. There are two key interaction modes: *input-to-output* where the model shows the expected performance of a particular setup; and *output-to-input* where the model shows the optimal configuration to achieve a specified performance profile. In input-to-output mode the interface simply provide inputs to the neural network and displays the output, but the output-to-input mode is more complex.

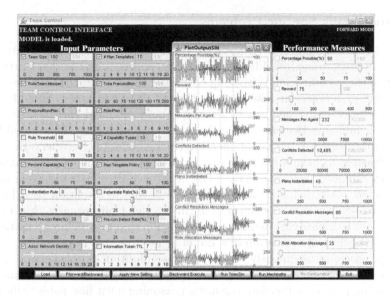

Fig. 2. The team control interface for online and offline control. Input parameters are shown on the left side, performance measures are shown on the right side. Check boxes for performance measures are used to specify the constraints for finding configurations.

Output-to-Input Mode

Using the team neural network in "reverse", the interface allows a user to change output parameters and receive a configuration that best meets some specific performance constraints both in input and output. The user specifies which performance features to constrain and what values they should have. In order to find input parameters that meet output requirements, the interface performs a search over the changeable configuration parameters to find a configuration that gives the required performance tradeoffs. Notice that this usage of the neural network allows various coordination algorithms to be traded off against each other automatically. For example, if the user requests a descrease in bandwidth usage, the neural network can determine which algorithms to limit that bandwidth usage of to have the least impact on overall performance.

The user interface can be connected to an executing team allowing the user to monitor system performance and to change configuration during execution. Special data collection tokens sample the team to determine current performance measures and the state of the environment. When the user specifies a new performance tradeoffs or the environment changes, the neural network determines the best configuration for meeting the users needs and sends information tokens to all actors to get the new configuration initiated.

5.2 Addressing Specific Problems

Because tokens completely encapsulate pieces of the overall coordination, it is feasible to examine individual tokens to determine whether that particular aspect of the coordination is working correctly. If not, or if the user has some particular preference for how that particular detaileds aspect should work, then the individual token can be extracted and modified (or its task taken over by a human.) Because of the independance of tokens, it is possible to extract any single token without effecting the behavior of the others. However, it is infeasible to have a *human* monitor all tokens and determine which are not performing to their satisfaction. Our approach is to instead have a model of expected token behavior and bring tokens to the attention of a human when the tokens behavior deviates from this model. Conceptually, this process corresponds to identifying details of coordination that may be problematic and bringing them to the attention of a human.

In practice, autonomously identifying coordination problems that might be brought to the attention of a human expert is imprecise. Rather than reliably finding poor coordination, the meta-reasoning must find *potentially* poor coordination and let the humans determine the actually poor coordination. (Elsewhere we describe the techniques that are used to ensure that this does not lead to the humans being overloaded [51].) Notice that while we allow humans to attempt to rectify problems with agent coordination, it is currently an open question whether humans can actually make better coordination decisions than the agents. For example, when a task token travels to many actor repeatedly, it may be that no actor has the required capability for the task or that the task is overloaded. A human might cancel the task or find an alternative way of acheiving the same goal.

6 Implementation

To evaluate the token-based approach we have developed both an abstracted simulator and a fully distributed implementation called Machinetta. The abstract simulator, called *TeamSim* represents tasks, information and resources as simple objects and uses simple queues for messages. It allows very rapid prototyping of algorithms and extensive testing to be performed.

Machinetta is an approach to building generic coordination software based on the concept of a *proxy* [22, 43]. Each team member is given its own proxy which encapsulates generic coordination reasoning. Plan templates are specified in XML and given to all the proxies. The proxy interacts with its team member via an abstracted interface that depends on the type of actor, e.g., for a robot it might be a simple socket while for a human it may be a sophisticated GUI. The proxies coordination together, using the token-based algorithms described above to implement the coordination. Machinetta proxies have been demonstrated to perform efficient, effective coordination with up to 500 distributed team members. They have been tested in several distinct domains and were successfully demonstrated in a U.S. Air Force flight test in October, 2005.

7 Results

In this section we present results of the individual token algorithms, the synergistic use of the algorithms and the human in the loop control. For results utilizing Machinetta refer to [50, 52, 49, 54]. Note that these results have for the most part been previously published elsewhere but are collected here to present a cohesive picture of the approach.

7.1 Task Allocation

To test the token based task allocation, we developed a simple simulator where actors are randomly given capabilities, independant of information or resources, for each of 5 types of task, with some percentage of actors being given zero capability for a type of task. For each time step that the agent has the task, the team receives ongoing reward based on the agent's capability. Message passing is simulated as perfect (lossless) communication that takes one time step. As the simulation progresses, new tasks arise spontaneously and the corresponding tokens are distributed randomly. The new tasks appear at the same rate that old tasks disappear, thus keeping the total number of tasks constant. This allows a single, fixed threshold for all tasks to be used throughout the experiment. Each data point represents the average from 20 runs.

Figure 3 shows the performance of the algorithm against two competing approaches. The first is DSA, which is shown to outperform other approximate distributed constraint optimization algorithms for problems like task assignment [32, 16]; we choose optimal parameters for DSA [65]. As a baseline we also compare against a centralized algorithm that uses a "greedy" assignment [5]. Results are shown using two different thresholds for the task tokens, T=0.0 and T=0.5. Figure 3(a) shows the relative performance of each algorithm as the number of agents is increased. The experiment used 2000 tasks over 1000 time steps. The y-axis shows the total reward, while the x-axis shows the number of agents. Not surprisingly, the centralized algorithm performs best but not dramatically better than the token based approach. The token based approach performs significantly better with a threshold of 0.5 than with no threshold. The real key to the comparison, however, is the amount of communication used, as shown in Figure 3(b). Notice that the y-axis is a logarithmic scale; thus the token based approach uses approximately four orders of magnitude fewer messages than the greedy algorithm and six orders of magnitude fewer messages than DSA. The token-based approach performs better than DSA despite using far less communication and only marginally worse than a centralized approach, despite using only a tiny fraction of the number of messages.

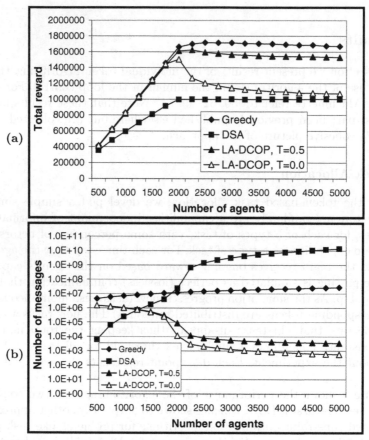

Fig. 3. (a) comparing the reward versus the number of agents. (b) the number of messages sent versus the number of agents

7.2 Information Sharing Results

To evaluate the information sharing algorithms, we arranged agents into a network and randomly picked one agent as the source of a piece of information i and another as a sink (i.e., for the sink agent $U(i)$ is very large). The sink agent sent out 30 information tokens (with $TTL = 150$) with information with a high Rel to i. Then the source agent sent out i and we measured how long it takes to get to the sink agent. In Figure 4(b) we show a frequency distribution of the time taken for a network with 8000 agents. While a big percentage of messages arrive efficiently to the sink, a small percentage get "lost" on

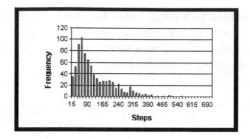

Fig. 4. Frequency distribution over the number of steps to get an information token to an unknown target location.

the network, illustrating the problem with a probabilistic approach. However, despite some messages taking a long time to arrive, they all eventually did and faster than if moved at random. We also looked in detail at exactly how many messages must be propogated around the network to make the routing efficient (Figure 5). Again using 8000 agents we varied the number of messages the sink agent would send before the source agent sent i onto the network. Notice that only a few messages are required to dramatically affect the average message delivery time.

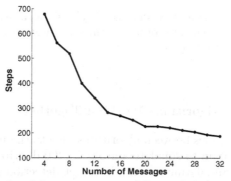

Fig. 5. The impact of training messages on delivery time for an information token in a large network.

7.3 Sensor Fusion Results

To evaluate the sensor fusion approach we use a random network of 100 nodes. Nodes are randomly chosen as the source of relevant sensor readings. Information tokens propogate the sensor readings through the network and we measure the probability of getting a successful fusion given a fixed TTL ("hops" on x-axis of graph). Figure 6 shows two cases, one where all three sensor readings must be known to a single actor for fusion to be successful (labeled 3/3) and one where three of five readings must be known to a single actor for successful fusion (labeled 3/5). Notice that a relatively small TTL is required to have high probability of successful fusion.

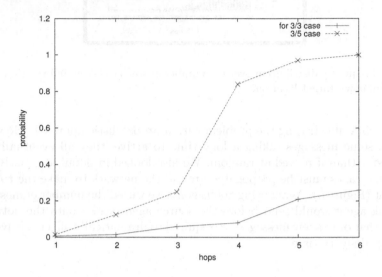

Fig. 6. The probability of a successful fusion (y-axis) by at least one actor given a specific TTL (x-axis) when three of five or three of three readings must be known by a single actor to perform fusion.

7.4 Token-Based Algorithms Working Together

To evaluate the synergies between algorithms, due to the use of the P model, we configured TeamSim to simulate a group of 400 distributed UAVs searching a hostile area. Simulating automatic target detection rates 200 pieces of information, e.g., SAM sites, were randomly "sensed" by UAVs and passed around the team. Fifty plan templates, each with four independent preconditions were used on each of 100 trials. Each plan template had four tasks to be performed. Thresholds for the tasks tokens were set such that UAVs needed to be near the target or reconnasaince site to accept the task. Shared resources were airspace that the UAVs needed to fly through to complete their tasks. One resource token was created for each "voxel" of airspace. When all four tasks in a plan were completed, the team recieved a reward of 10. A maximum reward of 500 units (10 units x 50 plans) was possible.

Five variations of the integrated algorithm were compared. The most integrated algorithm used all types of plan tokens to update P. The least integrated algorithm moved tokens randomly to associates when it was decided to move a token. Three intermediate variations of the algorithm used only one type of token, resource, role or information, tokens to update the local routing model, P. Figure 7 shows the reward received by the team on the y-axis and the time on the x-axis. The Figure shows that the team recieved more reward, faster when using the integrated algorithm. Moreover, Figure 8 shows that less messages (on the y-axis) were required to get a fixed amout of reward (on the x-axis) for the integrated approach. Both Figures show that using any type of tokens to build a routing model is better than using no routing model at all. Finally, Figure 9 shows that the algorithm scales well with increased team size. In this case, the y-axis shows the average number of messages per agent required to achieve a certain amount of reward. Notice, there is some indication that the average number of messages goes down as the team gets bigger, but more work is required to determine under what conditions this holds and what the key reasons for it are.

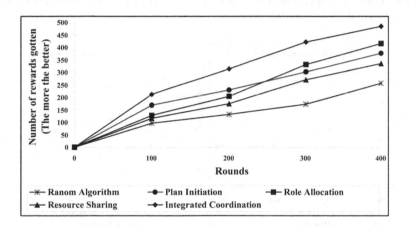

Fig. 7. The team receives considerably more reward when all previous tokens are used for local routing decisions.

7.5 Meta-Reasoning

We have evaluated both the low and high level aspects of the human-in-the-loop control of the large teams.

High Level Control

Using TeamSim we were able to verify that the user was able to reconfigure a team online and get required changes in performance tradeoffs. The interface is connected directly with TeamSim, so that users can set team configurations

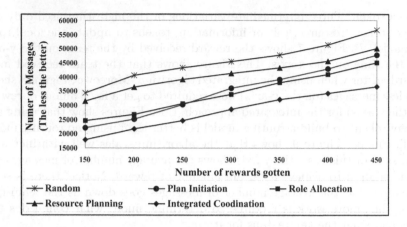

Fig. 8. The team requires far fewer messages to coordinate when all previous tokens are used to make local routing decisions.

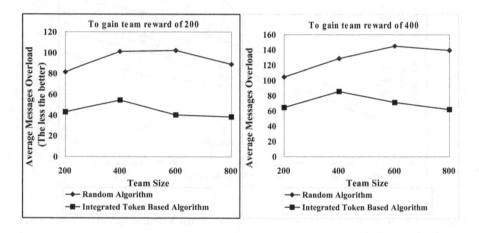

Fig. 9. The impact of team size on reward and communication with and without the use of all tokens for local routing decisions.

and monitor team performance measures online. The user configures the team at the start of the mission. When performance changes are requested the offline features of the team performance model areneural network is used to find suitable reconfigurations. The team control interface and reconfiguration assistance were evaluated over 10 scenarios. Scenarios were selected to provide situations that would require users to reconfigure their team in order to meet performance targets. For example, in a mission involving a very large team of 300 agents the user might be requested at some point in the mission to reduce the number of messages per agent or increase the number of plans instantiated. Performance measures are recorded throughout the execution. The data presented here represents 4 hours of runtime with a user in the loop. At step 1, the initial team configuration is set. At step 2, the user is asked to increase level of rewards obtained by the team disregarding other performance measures. Using the output-to-input feature of the team performance model the user finds a new coordination configuration that increases reward performance and reconfigures the team. At step 3 network communication bandwidth is reduced limiting the time-to-live for information tokens to 2 hops requiring another team reconfiguration to lessen the degradation in performance. At step 4, the user is again asked to reconfigure to improve reward performance. Results for six of the performance measures are shown in Figure 10. The bold lines show average values for the configured system while the lighter lines indicate the values predicted by the output-to-input model. The jagged lines show the moment to moment variation in the actual performance measures. Despite the high variability of team performance measures the model qualitatively predicts the effects of reconfiguration on average performance values across all six measures.

Low Level Control

To remove the need for many hours of human input, the interfaces for manipulating individual tokens were augmented with code that made decisions as if they were made by the human. These "human" decisions were made between five seconds and two minutes after control was transferred to the human. The experiments involved a team of 80 WASMs operating in a large environment. The primary task of the team was to protect a manned aircraft by finding and destroying surface-to-air missile sites spread around the environment. Half the team spread out across the environment searching for targets while the other half stayed near the manned aircraft destroying surface-to-air sites as they were found near the aircraft. Plans were simple, requiring a single WASM to hit each found target. If a target was not hit within three minutes of being found, this was considered abnormal plan execution and meta-reasoning would be invoked. Meta-reasoning was also invoked when a WASM was not allocated to hit any target for five minutes. These times are low, but reasonable since the simulation ran at approximately four times real-time. Finally, meta-reasoning was invoked when no WASM was available to hit a found target. Two human commanders were available to make meta-reasoning decisions (although, as discussed above there were not "real" human commanders).

Six different scenarios were used, each differing the number of surface-to-air missile sites. Each configuration was run ten times, thus the results below represent around 30 hours of simulation time (120 hours of real-time). As the number of missile sites increases, the team will have more to do with the same number of WASMs, thus we expected more meta-reasoning decisions.

Figure 11 shows that the total number of meta-reasoning decisions does increase with the number of targets. Over the course of a simulation, there

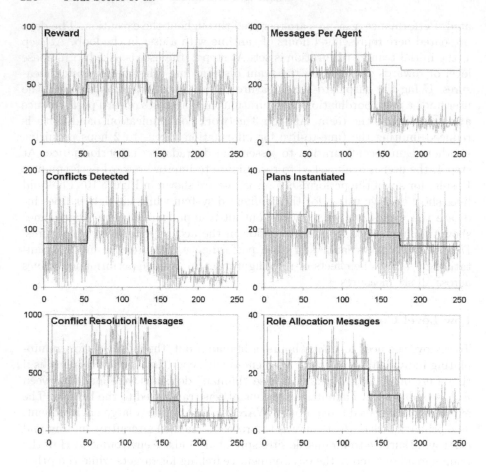

Fig. 10. Six performance measures recorded from TeamSim are ploted during the mission with 3 times of reconfiguration. Thick lines show the average values of actual performance measures of each configuration setting. Thin lines are the predicted values by the user interface.

are around 100 meta-reasoning decisions or about one per agent. However, as Figure 12 shows, only about 20% of these get transferred to a human. The large number of decisions that are made autonomously is primarily because humans are not available to make those decisions. This suggests work may need to be done to prioritize decisions for a user, to prevent high priority decisions being left to an agent, while the user is busy with low priority decisions. However, an appropriate solution is not obvious, since new decisions arrive asynchronously and it will likely not be appropriate to continually change the list of decisions the human is working on. Finally, notice in Figure 13 that a large percentage of the meta-decisions are to potentially cancel long running

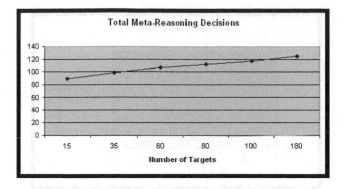

Fig. 11. The number of meta-reasoning decisions to be made as the number of targets in the environment increases.

plans. The large number of such decisions illustrates a need to carefully tune the meta-reasoning heuristics in order to avoid overloading the system with superfluous decisions. However, in this specific case, the problem of deciding whether to cancel a long running plan was the most appropriate for the human, hence the large percentage of such decisions for the human is reasonable.

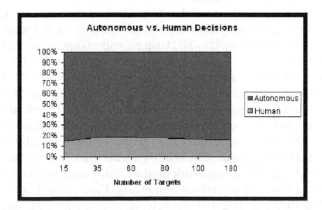

Fig. 12. The percentage of decisions transferred to humans versus the percentage made autonomously.

8 Related Work

Coordination of distributed entities is an extensively studied problem [9, 8, 24, 28, 55]. A key design decision is how the control is distributed among the group

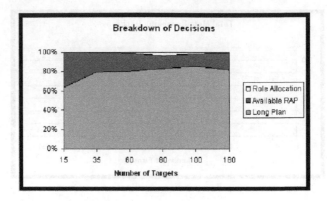

Fig. 13. Ratios of different types of meta-reasoning decisions presented to the user.

members. Solutions range from completely centralized [14], to hierarchical [11, 20] to completely decentralized [60]. While there is not yet definitive, empirical evidence of the strengths and weaknesses of each type of architecture, it is generally considered that centralized coordination can lead to behavior that is closer to optimal, but more distributed coordination is more robust to failures of communications and individual nodes [3]. Creating distributed groups of cooperative autonomous agents and robots that must cooperate in dynamic and hostile environments is a huge challenge that has attracted much attention from the research community [25, 27]. Using a wide range of ideas, researchers have had moderate success in building and understanding flexible and robust teams that can effectively act towards their joint goals [6, 10, 22, 47].

Tidhar [57] used the term "team-oriented programming" to describe a conceptual framework for specifying team behaviors based on mutual beliefs and joint plans, coupled with organizational structures. His framework also addressed the issue of team selection [57] — team selection matches the "skills" required for executing a team plan against agents that have those skills. Jennings's GRATE* [22] uses a teamwork module, implementing a model of cooperation based on the joint intentions framework. Each agent has its own *cooperation level* module that negotiates involvement in a joint task and maintains information about its own and other agents' involvement in joint goals. The Electric Elves project was the first human-agent collaboration architecture to include both proxies and humans in a complex environment [6]. COLLA-GEN [46] uses a proxy architecture for collaboration between a single agent and user. While these teams have been successful, they have consisted of at most 20 team members and will not easily scale to larger teams.

8.1 Small Worlds Networks

Research on social networks began in physics [59, 2]. Gaston [17] investigate the team formation on type of social network structures can dramatically affect team abilities to complete cooperative tasks. In particular, using a scale-free network structure for agent team will facilitate team formation by balancing between the number of skill-constrained paths available in the agent organization with the effects of potential blocking. Pujol [41] compared the merits of small world network and scale free network in the application of emergent coordination.

Task Allocation

Numerous task allocation algorithms have been proposed, although most do not consider costs (find only satisfying allocations) or scale very poorly with team size, or both. Symbolic matching techniques [57, 36] ignore costs completely which can have disastrous effects on team performance. Combinatorial auctions [21] are one approach that seek to minimize costs, but are impractical for very large teams due to the exponential number of possible bids and bottlenecks formed by centralized auctioneers. Forward-looking, decision-theoretic models [33] can exploit task decomposition to optimally allocate and reallocate tasks, but also do not scale to very large teams due to the exponential size of the search space.

Complete distributed constraint optimization algorithms [32, 31] can find optimal solutions but require impractically long running times and unacceptably high amounts of communication. Some incomplete distributed constraint optimization algorithms [65] can be scaled to large teams, but these may also suffer from a high amount of communication, and has been outperformed by our approach in previous evaluations [34].

Swarm-based approaches [48, 1, 4] provide a distributed, highly scalable way to perform task allocation and reallocation. Interestingly enough, these algorithms also rely on threshold-based computations. However, swarm algorithms rely directly on locally sensed stimuli to adjust thresholds and make decisions, while under our approach actors may use arbitrary information obtained locally or from other actors. This additional level of flexibility can be leveraged for better performance through synergistic interactions with the other algorithms presented here.

Human Control

The approach of using sensitivity analysis of multilayer neural networks to provide inverse relationship from output to input have been applied in several areas. Especially, Ming Lu et al. [30] demonstrated a simple algorithm for using a sensitivity analysis of neural networks and X. Zeng et al. [62] provide theoretical results.

Peter Eggenberger et al. [13] investigated and introduced the idea of dynamic rearrangement of biological nervous systems. Their approach allows neural networks to have an additional mechanism to dynamically change their synaptic weight modulations and neuronal states during execution. [15] presented another idea of dynamic network that dynamically modifying network structure. The algorithms start with zero or small number of hidden nodes and later the network change its structure by the number of hidden nodes to find the structure that fit well with the target system. A. Parlos et al. [37] proposed a hybrid feedforward/feedback neural network for using to identify nonlinear dynamic systems. Dynamic back propagation learning is demonstrated as the dynamic learning algorithm.

9 Conclusions and Future Work

In this chapter, we have presented a novel approach to coordination based on the concept of tokens. We have shown how such algorithms can be very effective for scalable coordination, particularly when they are combined into an integrated algorithm. The homogeneity of the token-based approach allowed us to build a general meta-reasoning layer over the top of the flows of tokens. This meta-reasoning layer gives a user powerful tools for ensuring that the team fulfills their requirements. Future work will examine how to inject fault tolerance into these algorithms and how the precise details of the associates network affect behavior.

Acknowledgements

This research has been supported by AFSOR grant F49620-01-1-0542 and AFRL/MNK grant F08630-03-1-0005.

References

1. William Agassounon and Alcherio Martinoli. Efficiency and robustness of threshold-based distributed allocation algorithms in multiagent systems. In *Proceedings of AAMAS'02*, 2002.
2. Albert-Laszla Barabasi and Eric Bonabeau. Scale free networks. *Scientific American*, pages 60–69, May 2003.
3. Johanna Bryson. Hierarchy and sequence vs. full parallelism in action selection. In *Intelligent Virtual Agents 2*, pages 113–125, 1999.
4. M. Campos, E. Bonabeau, G. Therauluz, and J.-L. Deneubourg. Dynamic scheduling and division of labor in social insects. *Adaptive Behavior*, 2001.
5. C. Castelpietra, L. Iocchi, D. Nardi, M. Piaggio, A. Scalzo, and A. Sgorbissa. Coordination among heterogenous robotic soccer players. In *Proceedings of IROS'02*, 2002.

6. Hans Chalupsky, Yolanda Gil, Craig A. Knoblock, Kristina Lerman, Jean Oh, David V. Pynadath, Thomas A. Russ, and Milind Tambe. Electric Elves: Agent technology for supporting human organizations. *AI Magazine*, 23(2):11–24, 2002.

7. Richard Clark. *Uninhabited Combat Air Vehicles: Airpower by the people, for the people but not with the people.* Air University Press, 2000.

8. D. Cockburn and N. Jennings. *Foundations of Distributed Artificial Intelligence*, chapter ARCHON: A Distributed Artificial Intelligence System For Industrial Applications, pages 319–344. Wiley, 1996.

9. Philip R. Cohen and Hector J. Levesque. Teamwork. *Nous*, 25(4):487–512, 1991.

10. K. Decker and J. Li. Coordinated hospital patient scheduling. In *Proceedings of the 1998 International Conference on Multi-Agent Systems (ICMAS'98)*, pages 104–111, Paris, July 1998.

11. Vincent Decugis and Jacques Ferber. Action selection in an autonomous agent with a hierarchical distributed reactive planning architecture. In *Proceedings of the Second International Conference on Autonomous Agents*, 1998.

12. Defense Science Board. Defense science board study on unmanned aerial vehicles and uninhabited combat aerial vehicles. Technical report, Office of the Under Secretary of Defense for Acquisition, Technology and Logistics, 2004.

13. P. Eggenberger, A. Ishiguro, S. Tokura, T. Kondo, and Y. Uchikawa. Toward seamless transfer from simulated to real worlds: A dynamically-rearranging neural network approach. In *Proceeding of 1999 the Eighth European Workshop in Learning Robot (EWLR-8)*, pages 44–60, 1999.

14. T. Estlin, T. Mann, A. Gray, G. Rapideau, R. Castano, S. Chein, and E. Mjolsness. An integrated system for multi-rover scientific exploration. In *Proceedings of AAAI'99*, 1999.

15. S. E. Fahlman and C. Lebiere. The Cascade-Correlation Learning Architecture. In Touretzky (ed.), editor, *Advances in Neural Information Processing Systems 2*. Morgan-Kaufmann.

16. Stephen Fitzpatrick and Lambert Meertens. *Stochastic Algorithms: Foundations and Applications, Proceedings SAGA 2001*, volume LNCS 2264, chapter An Experimental Assessment of a Stochastic, Anytime, Decentralized, Soft Colourer for Sparse Graphs, pages 49–64. Springer-Verlag, 2001.

17. M. Gaston and M. desJardins. The communicative multiagent team decision problem: analyzing teamwork theories and models. In *Proceedings of the 18th International Florida Artificial Intelligence Research Society Conference*, 2005.

18. Dani Goldberg, Vincent Cicirello, M Bernardine Dias, Reid Simmons, Stephen Smith, and Anthony (Tony) Stentz. Market-based multi-robot planning in a distributed layered architecture. In *Multi-Robot Systems: From Swarms to Intelligent Automata: Proceedings from the 2003 International Workshop on Multi-Robot Systems*, volume 2, pages 27–38. Kluwer Academic Publishers, 2003.

19. Barbara Grosz and Sarit Kraus. Collaborative plans for complex group actions. *Artificial Intelligence*, 86:269–358, 1996".

20. Bryan Horling, Roger Mailler, Mark Sims, and Victor Lesser. Using and maintaining organization in a large-scale distributed sensor network. In *In Proceedings of the Workshop on Autonomy, Delegation, and Control (AAMAS03)*, 2003.

21. L. Hunsberger and B. Grosz. A combinatorial auction for collaborative planning, 2000.

22. N. Jennings. The archon systems and its applications. Project Report, 1995.
23. N. Jennings, E. Mamdani, I Laresgoiti, J. Perez, and J. Corera. GRATE: A general framework for cooperative problem solving. *Intelligent Systems Engineering*, 1(2), 1992.
24. David Kinny. The distributed multi-agent reasoning system architecture and language specification. Technical report, Australian Artificial intelligence institute, Melbourne, Australia, 1993.
25. Hiraoki Kitano, Minoru Asada, Yasuo Kuniyoshi, Itsuki Noda, Eiichi Osawa, , and Hitoshi Matsubara. RoboCup: A challenge problem for AI. *AI Magazine*, 18(1):73–85, Spring 1997.
26. Hiroaki Kitano, Satoshi Tadokoro, Itsuki Noda, Hitoshi Matsubara, Tomoichi Takahashi, Atsushi Shinjoh, and Susumu Shimada. Robocup rescue: Searh and rescue in large-scale disasters as a domain for autonomous agents research. In *Proc. 1999 IEEE Intl. Conf. on Systems, Man and Cybernetics*, volume VI, pages 739–743, Tokyo, October 1999.
27. John Laird, Randolph Jones, and Paul Nielsen. Coordinated behavior of computer generated forces in TacAir-Soar. In *Proceedings of the fourth conference on computer generated forces and behavioral representation*, pages 325–332, Orlando, Florida, 1994.
28. V. Lesser, M. Atighetchi, B. Benyo, B. Horling, A. Raja, R. Vincent, T. Wagner, P. Xuan, and S. Zhang. The UMASS intelligent home project. In *Proceedings of the Third Annual Conference on Autonomous Agents*, pages 291–298, Seattle, USA, 1999.
29. E. Liao, P. Scerri, and K. Sycara. A framework for very large teams. In *AAMAS'04 Workshop on Coalitions and Teams*, 2004.
30. Ming Lu, S. M. AbouRizk, and U. H. Hermann. Sensitivity analysis of neural networks in spool fabrication productivity studies. *Journal of Computing in Civil Engineering*, 15(4):299–308, 2001.
31. Roger Mailler and Victor Lesser. A cooperative mediation-based protocol for dynamic, distributed resource allocation. 2004.
32. Pragnesh Jay Modi, Wei-Min Shen, Milind Tambe, and Makoto Yokoo. An asynchronous complete method for distributed constraint optimization. In *Proceedings of Autonomous Agents and Multi-Agent Systems*, 2003.
33. R. Nair, M. Tambe, and S. Marsella. Role allocation and reallocation in multi-agent teams: Towards a practical analysis. In *Proceedings of the second International Joint conference on agents and multiagent systems (AAMAS)*, 2003.
34. Steven Okamoto. Dcop in la: Relaxed. Master's thesis, University of Southern California, 2003.
35. Committee on Visionary Manufacturing Challenges. Visionary manufacturing challenges for 2020. National Research Council.
36. M. Paolucci, T. Kawamura, T. Payne, and K. Sycara. Semantic matching of web service capabilities. In *Proceedings of the First International Semantic Web Conference*, 2002.
37. Chong K. T. Parlos, A. G. and A. F. Atiya. Application of the Recurrent Multilayer Perceptron in Modeling Complex Process Dynamics . In *IEEE Transactions on Neural Networks*, pages 255–266, 1994.
38. Leonid I. Perlovsky. *Neural Networks and Intellect: Using Model-Based Concepts*. Oxford University Press, 2001.

39. J. Polvichai and P. Khosla. An evolutionary behavior programming system with dynamic networks for mobile robots in dynamic environments. In *Proceedings of 2002 IEEE/RSJ International Conference on Intelligent Robots and System*, volume 1, pages 978–983, 2002.
40. J. Polvichai and P. Khosla. Applying dynamic networks and staged evolution for soccer robots. In *Proceedings of 2003 IEEE/RSJ International Conference on Intelligent Robots and System*, volume 3, pages 3016–3021, 2003.
41. J. Pujol and R. Sanguesa. Emergence of coordination in scale-free networks. In *Web Intelligence and Agent Systems 131-138*, 2003.
42. David V. Pynadath and Milind Tambe. An automated teamwork infrastructure for heterogeneous software agents and humans. *Journal of Autonomous Agents and Multi-Agent Systems, Special Issue on Infrastructure and Requirements for Building Research Grade Multi-Agent Systems*, page to appear, 2002.
43. D.V. Pynadath, M. Tambe, N. Chauvat, and L. Cavedon. Toward team-oriented programming. In *Intelligent Agents VI: Agent Theories, Architectures, and Languages*, pages 233–247, 1999.
44. Paul Ranky. *An Introduction to Flexible Automation, Manufacturing and Assembly Cells and Systems in CIM (Computer Integrated Manufacturing), Methods, Tools and Case Studies*. CIMware, 1997.
45. C. Reynolds. Authoring autonomous characters. Invited Talk, Distinguished Lecture Series, Georgia Institute of Technology, Fall 1995.
46. C. Rich and C. Sidner. COLLAGEN: When agents collaborate with people. In *Proceedings of the International Conference on Autonomous Agents (Agents'97)"*, 1997.
47. P. Rybski, S. Stoeter, M. Erickson, M. Gini, D. Hougen, and N. Papanikolopoulos. A team of robotic agents for surveillance. In *Proceedings of the fourth international conference on autonomous agents*, pages 9–16, 2000.
48. Pedro Sander, Denis Peleshchuk, and Barabara Grosz. A scalable, distributed algorithm for efficient task allocation. In *Proceedings of AAMAS'02*, 2002.
49. P. Scerri, E. Liao, Yang. Xu, M. Lewis, G. Lai, and K. Sycara. *Theory and Algorithms for Cooperative Systems*, chapter Coordinating very large groups of wide area search munitions. World Scientific Publishing, 2004.
50. P. Scerri, D. V. Pynadath, L. Johnson, P. Rosenbloom, N. Schurr, M Si, and M. Tambe. A prototype infrastructure for distributed robot-agent-person teams. In *The Second International Joint Conference on Autonomous Agents and Multiagent Systems*, 2003.
51. P. Scerri, K. Sycara, and M Tambe. Adjustable autonomy in the context of coordination. In *AIAA 3rd "Unmanned Unlimited" Technical Conference, Workshop and Exhibit*, 2004. Invited Paper.
52. P. Scerri, Yang. Xu, E. Liao, J. Lai, and K. Sycara. Scaling teamwork to very large teams. In *Proceedings of AAMAS'04*, 2004.
53. Paul Scerri, Alessandro Farinelli, Steven Okamoto, and Milind Tambe. Allocating tasks in extreme teams. In *AAMAS'05*, 2005.
54. N. Schurr, J. Marecki, J.P. Lewis, M. Tambe, and P.Scerri. The DEFACTO system: Training tool for incident commanders. In *IAAI'05*, 2005.
55. Munindar Singh. Developing formal specifications to coordinate hetrogeneous agents. In *Proceedings of third international conference on multiagent systems*, pages 261–268, 1998.
56. Milind Tambe. Agent architectures for flexible, practical teamwork. *National Conference on AI (AAAI97)*, pages 22–28, 1997.

57. G. Tidhar, A.S. Rao, and E.A. Sonenberg. Guided team selection. In *Proceedings of the Second International Conference on Multi-Agent Systems*, 1996.
58. T. Wagner, J. Phelps, V. Guralnik, and Ryan VanRiper. COORDINATORS: Coordination managers for first responders. In *AAMAS'04*, 2004.
59. Duncan Watts and Steven Strogatz. Collective dynamics of small world networks. *Nature*, 393:440–442, 1998.
60. Tony White and Bernard Pagurek. Towards multi swarm problem solving in networks. In *Proceedings of the International conference on multi-agent systems*, pages 333–340, Paris, July 1998.
61. Michael Wooldridge and Nicholas Jennings. *Distributed Software agents and applications*, chapter Towards a theory of cooperative problem solving, pages 40–53. Springer-Verlag, 1994.
62. D.S.Yeung Xiaoqin Zeng. Sensitivity analysis of multilayer perceptron to input and weight perturbations. *IEEE Transactions on Neural Networks*, 12(6):1358–1366, 2001.
63. Y. Xu, M. Lewis, K. Sycara, and P. Scerri. Information sharing in very large teams. In *In AAMAS'04 Workshop on Challenges in Coordination of Large Scale MultiAgent Systems*, 2004.
64. Y. Xu, P. Scerri, B. Yu, S. Okamoto, M. Lewis, and K. Sycara. An integrated token-based algorithm for scalable coordination. In *AAMAS'05*, 2005.
65. W. Zhang and L. Wittenburg. Distributed breakout revisited. In *Proceedings of AAAI'02*, 2002.

Model Predictive Path-Space Iteration for Multi-Robot Coordination*

Omar A.A. Orqueda and Rafael Fierro

MARHES Laboratory
School of Electrical and Computer Engineering
Oklahoma State University
202 Engineering South
Stillwater OK 74078, USA
E-mail: orqueda@ieee.org, rfierro@okstate.edu

Summary. In this work, two novel optimization-based strategies for multi-robot co-ordination are presented. The proposed algorithms employ a *model predictive control* (*MPC*) version of a Newton-type approach for solving the underlying optimization problem. Both methods can generate control inputs for vehicles with nonholonomic constraints moving in a configuration space cluttered by obstacles. Obstacle- and inter-collision constraints are incorporated into the optimization problem by using interior and exterior penalty function approaches. Moreover, convergence of the algorithms is studied with and without the presence of obstacles in the environment. Simulation results verify the validity of the proposed methodology.

1 Introduction

During the last years there has been an increasing interest in controlling formations of mobile robots. The main reason of this interest is the logic expected out performance of several mobile vehicles over traditionally big structures of heavily equipped vehicles. For instance, hundred of small robots could cover better a specific terrain for land-mine removal, space exploration, surveillance, search and rescue operations than a single complex robot.

Initially, robot formations were based on the imitation of animal behavior. For example, the imitation of *flocking*, that is, agents moving together in large numbers and having a common objective. Studies on flocking mechanism show that it emerges as a combination of a desire to stay in the group and yet simultaneously keep a minimum separation distance from other members of the flock [19]. The first application of flocking behavior was on computer

---

* This work is supported in part by NSF grants #0311460 and CAREER #0348637 and by the U.S. Army Research Office under grant DAAD19-03-1-0142 (through the University of Oklahoma)

graphics [18]. Reynolds addressed the problem of simulating flocks of birds and schools of fish with a simple egocentric behavioral model. This model consisted of collision avoidance, velocity matching, and formation keeping components. He introduced three heuristic rules that led to the creation of the first computer animated flocking:

- *cohesion rule* - aim of proximity keeping to nearby flock-mates,
- *separation rule* - desire of collision avoidance,
- *alignment rule* - intention of velocity matching with neighbors.

These three rules are inherently local, and give each member the possibility of navigating using only its sensing capabilities. From a mathematical point of view, they allow to pose the flocking problem as a decentralized optimization problem. The superposition of these three rules results in all agents moving in a loose (as opposite to rigid) formation, with a common heading while avoiding collisions [22]. Also, Reynolds' model includes leader-follower strategies, in which one agent acts as a group leader and other agents follow the leader accomplishing the aforementioned rules.

Other authors have addressed the coordination problem of multiple unmanned vehicles using optimization techniques [2]. Contributions in this area include work focused on autonomous vehicles performing distributed sensing tasks [4], decentralized optimization-based control algorithms to solve a variety of multi-robot problems [9], optimal motion planning [1], and *formation reconfiguration planning* (FRP) [25].

More recently, the use of model predictive control (MPC) or receding-horizon control (RHC) is becoming popular in the multi-robot system literature [3, 10, 8, 12]. Generally, MPC algorithms rely on the optimization of a predicted model response with respect to the plant input to determine the best input changes for a given state. Either hard constraints (that cannot be violated) or soft constraints (that can be violated but with some penalty) can be incorporated into the optimization problem, giving to MPC a potential advantage over passive state feedback control laws. However, there are possible disadvantages to MPC. For instance, in its traditional use for process control, the primary disadvantage is the need of a good model of the plant, but such model is not always available. In robotics applications, the foremost disadvantage is the computational cost, negligible for slow-moving systems in the process industry, but very important in real-time applications.

Another important related area of research is *motion planning* [14]. Among different approaches on motion planning, we are particularly interested in algorithms that compute the control action to be applied to the system to reach a given goal configuration [6, 13]. One of those techniques, useful for path planning of nonholonomic systems, is *path-space iteration* (PSI). PSI methods correct the control action applied to a given system to minimize the error between an initial path and an acceptable path by using Newton-Raphson or Gradient type-algorithms [6, 7, 14, 17].

In this chapter, we present two main contributions: (1) a centralized algorithm based on an MPC/PSI to solve a nonholonomic multi-robot coordination problem; and (2) a decentralized algorithm that incorporates kinematic, formation, inter- and intra-vehicle collision avoidance constraints. Specifically, we consider a team of mobile robots (*i.e.*, agents) navigating within a dynamic, unknown environment. Moreover, any two agents that are interacting in any way (*e.g.*, sensing, communication) are referred to as *neighbors*.

The rest of the chapter is organized as follows. Section 2 presents some mathematical preliminaries and definitions that are used along the chapter. Section 3 gives the details of the centralized version of the MPC/PSI algorithm. The decentralized version of the algorithm is developed in Section 4. Conclusions and future work are given in section 5.

2 Mathematical Preliminaries

2.1 Model Predictive Control

The traditional formulation of *nonlinear model predictive control* (NMPC) consists of solving an on-line finite horizon open-loop optimization problem using the current state as the initial state for the plant. The solution to the problem gives a sequence of control inputs for the entire control horizon; however, only the first element of the optimal input sequence is applied to the plant.

Let us consider a nonlinear feedback-controlled system model

$$
\begin{aligned}
\dot{x}(t) &= f(x(t), u(t)), \\
y(t) &= h(x(t)).
\end{aligned}
\tag{1}
$$

It is assumed that the vector field $f \subseteq \mathcal{X} \times \mathcal{U} \to \mathcal{X}$ is locally Lipschitz continuous, $x(t) \in \mathcal{X} \subseteq \mathbb{R}^n$, $u(t) \in \mathcal{U} \subseteq \mathbb{R}^m$, $\forall t \geq 0$, $\mathcal{X}$ is a connected subspace that represents the state constraints, and $\mathcal{U}$ is a compact subspace that represents the input constraints.

The goal is to find a sampled-data optimal control sequence to drive the sampled system (1), *i.e.*,

$$
\begin{aligned}
x_{k+1} &= f(x_k, u_k), \\
y_k &= h(x_k),
\end{aligned}
\tag{2}
$$

to an equilibrium point such that the cost function

$$
J(x_{k+1}, \ldots, x_{k+M}, u_k, \ldots, u_{k+M-1}) = \sum_{k=0}^{M-1} \eta(x_k, u_k) + \eta(x_{k+M})
\tag{3}
$$

is minimized satisfying the state and the input constraints. $\eta(\cdot)$ is a smoothing function, $t := kh$, and h is the sampling time. Usually, to ensure stability a terminal constraint is imposed

$$x_M \in \mathcal{X}_f \subset \mathcal{X}. \tag{4}$$

The optimization problem could be re-written as

$$\mathcal{P}_{k,M} : \min_{\underline{u}_M} \left\{ J_{k,M} \left(\underline{x}_{k,M}, \underline{u}_{k,M} \right) \right\} \tag{5}$$

subject to system dynamics (2),

$$\underline{x}_{k,M} \in \mathcal{X}_{k,M}, \text{ and} \tag{6}$$

$$\underline{u}_{k,M} \in \mathcal{U}_{k,M}, \tag{7}$$

with $\underline{x}_{k,M} := \{x_{k,1}, \ldots, x_{k,M}\}$, $\underline{u}_{k,M} := \{u_{k,0}, \ldots, u_{k,M-1}\}$, $\mathcal{X}_{k,M} \subseteq \mathbb{R}^{nM}$ and $\mathcal{U}_{k,M} \subseteq \mathbb{R}^{mM}$ are the state and input constraint subspaces expanded for M components. Throughout this paper, we use the notation $\omega_{k,j} := \omega\left(k+j \,|\, k\right)$ for any function $\omega\left(\cdot\right)$ evaluated at time $k+j$ with the information available at time k.

Starting from the actual state x_k, the *MPC* algorithm:

1: solves the optimal problem $\mathcal{P}_{k,M}$ finding the sequence $\underline{u}_{k,M}$,
2: applies the first element of the input sequence $\underline{u}_{k,M}$, u_k, to the system,
3: shifts the input sequence $\underline{u}_{k,M}$, and
4: repeats for x_{k+1}.

For a more detailed discussion about *MPC* and its properties, the reader is referred to the bibliography [16].

2.2 Path-space Iteration

Path planning for nonholonomic systems is a well-know research area that has attracted the attention of many researchers since the end of the 60's. There exist several techniques to solve this problem [14, 17], such as *search-based, control-theoretic*, and *iterative learning* (IL) algorithms. Path-space iteration (PSI) methods are special types of IL algorithms. The main idea is to enhance the performance of a control system through training, that is, at each new experiment, the control law is updated on the basis of the results of the previous trial. In particular, *path-space iteration* methods iterate on the control along a *whole* trajectory until a feasible trajectory is found by minimizing a performance index in each iteration. An illustration of the method is depicted in Figure 1.

Let $\phi_M\left(x_0, \underline{u}_M\right)$ denote the end-point map of the system state that results of applying a piecewise-continuous control law $u\left(t\right)$, given by the elements of the sequence $\underline{u}_M := \{u\left(0\right), \ldots, u\left(M-1\right)\}$ for $t = 0, h, 2h,, \left(M-1\right)h$, to the system (1) evolving from an initial state x_0. Let x^d be the desired final configuration of the system at time Mh, and let the final error be defined as

$$e_M\left(x_0, \underline{u}_M\right) := \phi_M\left(x_0, \underline{u}_M\right) - x^d.$$

Assuming that there are no obstacles in the configuration space, the path-space iteration method used in this chapter can be seen as a nonlinear root

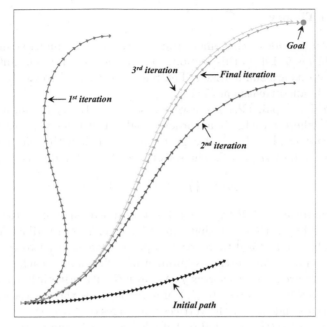

Fig. 1. Path-space iteration example. The final error is minimized in each iteration.

finding problem, where the goal is to find a control sequence $\underline{u}_M$ such that $e_M(x_0, \underline{u}_M) \to 0$. To this end we compute the derivative of the error and iterate over an iteration variable τ

$$\frac{de_M(x_0, \underline{u}_M(\tau))}{d\tau} = \nabla_{\underline{u}_M} \phi_M(x_0, \underline{u}_M(\tau)) \frac{d\underline{u}_M(\tau)}{d\tau}. \qquad (8)$$

If $\nabla_{\underline{u}_M} \phi_M(x_0, \underline{u}_M)$ is full rank, then we can choose the following update rule for $\underline{u}_M(\tau)$ minimize the final error

$$\underline{u}_M(\tau+1) = \underline{u}_M(\tau) - \alpha \left[\nabla_{\underline{u}} \phi_M(x_0, \underline{u}_M(\tau))\right]^\dagger e_M(x_0, \underline{u}_M(\tau)), \qquad (9)$$

where $\alpha > 0$ and $\left[\nabla_{\underline{u}_M} \phi_M(x_0, \underline{u}_M(\tau))\right]^\dagger$ is the Moore-Penrose pseudo-inverse of $\nabla_{\underline{u}_M} \phi_M(x_0, \underline{u}_M(\tau))$. The gradient $\nabla_{\underline{u}_M} \phi_M(x_0, \underline{u}_M(\tau))$ can be computed from the system (2) linearized about the path resulting from applying $\underline{u}_M(\tau)$

$$\delta x_{k+1} = \Phi_k \, \delta x_k + \Gamma_k \, \delta u_k, \quad \delta x(0) = 0, \qquad (10)$$

with $\Phi_k := \exp(A_k h) \in \mathbb{R}^{n \times n}$, $\Gamma_k := \int_0^h \exp(A_k \tau) \, d\tau B_k \in \mathbb{R}^{n \times m}$, $A(k) := \nabla_x f(x(t), u(t))|_{t=kh}$, and $B(k) := \nabla_u f(x(t), u(t))|_{t=kh}$. The convergence of the iterative algorithm (9) is assured if $\nabla_{\underline{u}_M} \phi_M(x_0, \underline{u}_M(\tau))$ is full rank for all τ, or equivalently, if the time-varying linearized system (10) generated by linearizing (1) about $\underline{u}_M(\tau)$ is controllable [6, 7, 15, 23].

2.3 Graph Theory

Graph theory provides a convenient framework to model multi-vehicle coordination problems[5, 19]. In this section, some basic concepts relevant to multi-robot formations are summarized. The reader is referred to the bibliography for a more detailed treatment [5, 19].

A *graph* $\mathcal{G}$ is a pair $(\mathcal{V}, \mathcal{E})$ of a vertex set $\mathcal{V} \in \{1, \ldots, n\}$ and an edge set $\mathcal{E} \in \mathcal{V} \times \mathcal{V}$, where an edge is an ordered pair of distinct vertices in $\mathcal{V}$. The *adjacency matrix* $\mathcal{A} = \{a_{ij}\}$ of a graph is a matrix with nonzero elements such that $a_{ij} \neq 0 \Leftrightarrow (i, j) \in \mathcal{E}$. The set of *neighbors* of node i is defined by

$$\mathcal{N}_i := \{j \in \mathcal{V} : a_{ij} \neq 0\}.$$

A *path* from vertex $x \in \mathcal{V}$ to vertex $y \in \mathcal{V}$ is a sequence of vertices starting with x and ending with y such that consecutive vertices are adjacent. A graph $\mathcal{G}$ is said to be *connected* if there exists a path between any two vertices of $\mathcal{G}$. An orientation in a graph is the assignment of a direction to each edge, so that edge (i, j) is an arc from i to vertex j. A graph $\mathcal{G}$ with orientation σ is denoted by $\mathcal{G}^\sigma$. The *incidence matrix* $B(\mathcal{G}^\sigma)$ of a graph $\mathcal{G}^\sigma$ is the matrix whose rows and columns are indexed by the vertices and edges of $\mathcal{G}$ respectively, such that the (i, j) entry of $B(\mathcal{G}^\sigma)$ is equal to 1 if edge (i, j) is incoming to vertex i, -1 if edge (i, j) is out coming from vertex i, and 0 otherwise.

The symmetric matrix defined as

$$L(\mathcal{G}) = B(\mathcal{G}^\sigma) B(\mathcal{G}^\sigma)^T$$

is called the *Laplacian* of $\mathcal{G}$ and is independent of the choice of orientation σ. For a connected graph, L has a single zero eigenvalue and the associated eigenvector is the n-dimensional vector of ones, $\mathbf{1}_n$.

Let $p^i = (p^i_x, p^i_y) \in \mathbb{R}^2$ denote the position of robot i, and $r^i > 0$ denote the *interaction range* between agent i and the other robots. A *spherical neighborhood* (or *shell*) of radius r^i around p^i is defined as

$$\mathcal{B}(p^i, r^i) := \{q \in \mathbb{R}^2 : \|q - p^i\| \leq r^i\}.$$

Let us define $p = \mathrm{col}(p^i) \in \mathbb{R}^{2n}$, where $n = |\mathcal{V}|$ is the number of nodes of graph $\mathcal{G}$, and $r = \mathrm{col}(r^i)$. We refer to the pair (p, r) as a *cluster* with *configuration* p and vector of radii r. A *spatial adjacency matrix* $A(p) = [a_{ij}(p)]$ induced by a cluster is defined as follows

$$a_{ij}(p) = \begin{cases} 1, & \text{if } p^j \in \mathcal{B}(p^i, r^i),\ j \neq i \\ 0, & \text{otherwise} \end{cases}$$

The spatial adjacency matrix $A(p)$ defines a *spatially induced graph* or *net* $\mathcal{G}(p)$. A node $i \in \mathcal{V}$ with a spherical neighborhood define a *neighboring graph* $\mathcal{N}_i$ as

$$\mathcal{N}^i(p) := \{j \in \mathcal{V} : a_{ij}(p) > 0\}. \tag{11}$$

We assume that $\mathcal{N}^i$ is *connected*.

An *α-lattice* [19] is a configuration p satisfying the set of constraints

$$\left\| p^j - p^i \right\| = d, \ \forall j \in \mathcal{N}^i(p).$$

A *quasi α-lattice* [19] is a configuration p satisfying the set of inequality constraints

$$-\delta \le \left\| p^j - p^i \right\| - d \le \delta, \ \forall (i,j) \in \mathcal{E}(p).$$

Throughout this chapter, mobile robots are also referred to as agents or α-agents in the sense defined in [19].

2.4 Robot Model

We are interested in coordinating a team of agents using a model predictive version of a path-space iteration algorithm. We consider a team of N_a nonholonomic robots, as the one shown in Figure 2 (left), modeled using the well-known unicycle model

$$\begin{bmatrix} \dot{p}_x^i(t) \\ \dot{p}_y^i(t) \\ \dot{\theta}^i(t) \end{bmatrix} = \begin{bmatrix} c_{\theta^i(t)} & 0 \\ s_{\theta^i(t)} & 0 \\ 0 & 1 \end{bmatrix} \begin{bmatrix} v^i(t) \\ \omega^i(t) \end{bmatrix} = F^i(t)\, u^i(t), \tag{12}$$

where $(p_x^i, p_y^i) \in \mathcal{X}^i \subseteq \mathbb{R}^2$ and $\theta^i \in [-\pi, \pi]$ denote the Cartesian position and orientation of the *i-th* vehicle, respectively, $u^i = (v^i, \omega^i) \in \mathcal{U}^i \subseteq \mathbb{R}^2$ is the velocity (linear and angular) control vector, $\mathcal{U}^i$ is a compact set of admissible inputs for robot i, $i = 1, \ldots, N_a$, and

$$F^i(t) := \begin{bmatrix} c_{\theta^i(t)} & 0 \\ s_{\theta^i(t)} & 0 \\ 0 & 1 \end{bmatrix}, \tag{13}$$

with $c_{\theta^i(t)} := \cos \theta^i(t)$ and $s_{\theta^i(t)} := \sin \theta^i(t)$.

We assume that all sets $\mathcal{U}^i$ are equal, then

$$\mathcal{U}^0 = \mathcal{U}^i := \left\{ (v^i, \omega^i) \,\middle|\, \left| v^i \right| \le v_{\max}, \left| \omega^i \right| \le \omega_{\max} \right\}.$$

We also assume that each robot is equipped with a range sensor, and has approximate information about its goal position and the number of team members.

In next sections, we describe two methodologies for multi-robot coordination that allow a team of mobile robots to reach a goal destination maintaining a desired formation and avoiding collisions.

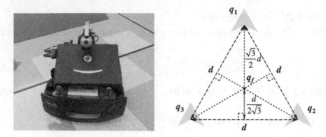

Fig. 2. ERSP Scorpion (*Evolution Robotics*) - Unicycle type robot (left). Centralized formation configuration (right).

3 Centralized MPC/Path-space Iteration

In this case, the position and orientation of the team is defined by the average of the positions and orientations of all of its members. We derive the equations for a formation of three robots (*i.e.*, $n = 9$, $m = 6$), but our approach can be easily extended to any number of robots. Moreover, it is assumed that the desired formation shape is an equilateral triangle as shown in Figure 2 (right). Thus, the group dynamics for these three robots become

$$\dot{x}(t) = F(x(t)) u(t), \tag{14}$$

$$y(t) = Cx(t), \tag{15}$$

with

$$x(t) := \left[p_x^1(t), p_y^1(t), \theta^1(t), p_x^2(t), p_y^2(t), \theta^2(t), p_x^3(t), p_y^3(t), \theta^3(t) \right]^T \in \mathbb{R}^{9 \times 1},$$

$$u(t) := \left[v^1(t), \omega^1(t), v^2(t), \omega^2(t), v^3(t), \omega^3(t) \right]^T \in \mathbb{R}^{6 \times 1},$$

$$C := \frac{1}{3} \begin{bmatrix} 1\,0\,0\,1\,0\,0\,1\,0\,0 \\ 0\,1\,0\,0\,1\,0\,0\,1\,0 \\ 0\,0\,1\,0\,0\,1\,0\,0\,1 \end{bmatrix} \in \mathbb{R}^{3 \times 9},$$

$$F(t) := \begin{bmatrix} F^1(t) & 0 & 0 \\ 0 & F^2(t) & 0 \\ 0 & 0 & F^3(t) \end{bmatrix} \in \mathbb{R}^{9 \times 6},$$

where $F(t)$ is a block-diagonal matrix, and $F^i(t)$ is given in (13), $i = 1, 2, 3$.

Let the states and the inputs computed with the information available at time k be written into the more compact *block vector form* [20]

$$\underline{x}_{k,M} := \left[x_{k,1}^T, x_{k,2}^T, \ldots, x_{k,M}^T \right]^T \in \mathbb{R}^{nM \times 1},$$

$$\underline{u}_{k,M} := \left[u_k^T, u_{k,1}^T, \ldots, u_{k,M-1}^T \right]^T \in \mathbb{R}^{mM \times 1}.$$

Let $\phi_M\left(x_k, \underline{u}_{k,M}\right)$ be the M-step-ahead-point map of the system state that results of applying the input sequence $\underline{u}_{k,M}$ with the system evolving from the initial state x_k. Then the M-step-ahead predicted formation error is defined by

$$e_{k,M}\left(x_k, \underline{u}_{k,M}\right) := C\phi_M\left(x_k, \underline{u}_{k,M}\right) - y^d, \tag{16}$$

where $C\phi_M\left(x_k, \underline{u}_{k,M}\right)$ is the position of the team in the M-step-ahead sample time, and y^d is the desired position.

The main idea of the PSI method is to iteratively refine the M-step-ahead control sequence $\underline{u}_{k,M}$ with a correcting factor $\underline{d}_{k,M}$, such that $e_{k,M} \to 0$ as $k \to \infty$. The correcting factor is computed using a Newton-type algorithm, and only the first m elements of the new control sequence $\underline{v}_{k,M}$ are used as the actual control law u_k. Then, $\underline{v}_{k,M}$ is shifted one-step ahead using the shifting matrix

$$G := \begin{bmatrix} 0_{m(M-1) \times m} & I_{m(M-1)} \\ 0_{m \times m} & 0_{m \times m(M-1)} \end{bmatrix},$$

and the process is restarted.

To obtain an expression for the correcting factor $\underline{d}_{k,M}$, we first differentiate the error vector (16) as follows

$$\frac{de_{k,M}}{d\tau} = \left[C\nabla_{\underline{u}_{k,M}}\phi_M\left(x_k, \underline{u}_{k,M}\right)\right]\frac{d\underline{u}_{k,M}}{d\tau}.$$

Then, we can use the following discrete update rule to minimize the error

$$\underline{v}_{k,M} = \underline{u}_{k,M} + \alpha_k \underline{d}_{k,M}, \tag{17}$$

with

$$\underline{d}_{k,M} = -\left[C\nabla_{\underline{u}_{k,M}}\phi_M\left(x_k, \underline{u}_{k,M}\right)\right]^\dagger e_{k,M}, \tag{18}$$

where $\nabla_{\underline{u}_{k,M}}\phi_M\left(x_k, \underline{u}_{k,M}\right) \in \mathbb{R}^{n \times mM}$ is the gradient of the predicted state $\phi_M\left(\cdot\right)$ with respect to the M-step-ahead control sequence $\underline{u}_{k,M}\left(\cdot\right)$, and $(\cdot)^\dagger$ denotes the Moore-Penrose pseudo-inverse. The gradient $\nabla_{\underline{u}_{k,M}}\phi_M\left(\cdot\right)$ can be computed from the system equations (14) linearized around a *reference trajectory* $\underline{x}_{k,M}\left(\cdot\right)$ with input sequence $\underline{u}_{k,M}\left(\cdot\right)$,

$$\delta x_{k,j+1} = \Phi_{k,j}\delta x_{k,j} + \Gamma_{k,j}\delta u_{k,j}, \quad \delta x_{k,0} = 0, \ j = 0, \ldots, M-1, \tag{19}$$

with $\Phi_{k,j} := \exp\left(A_{k,j}h\right) \in \mathbb{R}^{n \times n}$, and $\Gamma_{k,j} := \int_0^h \exp\left(A_{k,j}\tau\right)d\tau B_{k,j} \in \mathbb{R}^{n \times m}$, where $A_{k,j}$ and $B_{k,j}$ are block-diagonal matrices given by

$$A_{k,j} := [\nabla_x\left(Fu\right)]_{k,j} = \begin{bmatrix} A_{k,j}^1 & 0_{3\times3} & 0_{3\times3} \\ 0_{3\times3} & A_{k,j}^2 & 0_{3\times3} \\ 0_{3\times3} & 0_{3\times3} & A_{k,j}^3 \end{bmatrix}$$

$$B_{k,j} := [\nabla_u\left(Fu\right)]_{k,j} = F_{k,j}, \tag{20}$$

with $A_{k,j}^i := \begin{bmatrix} 0 & 0 & -v_{k,j}^i s_{\theta_{k,j}^i} \\ 0 & 0 & v_{k,j}^i c_{\theta_{k,j}^i} \\ 0 & 0 & 0 \end{bmatrix}, \ i = 1, 2, 3.$

In general, the predicted state can be expressed as

$$
\begin{bmatrix} \delta x_{k,1} \\ \delta x_{k,2} \\ \vdots \\ \delta x_{k,M} \end{bmatrix} = \begin{bmatrix} \Gamma_{k,0}, & 0, \cdots & 0 \\ \Phi_{k,1}\Gamma_{k,0}, & \Gamma_{k,1}, & \ddots & 0 \\ \vdots & \vdots & \ddots & \vdots \\ \prod_{j=1}^{M-1}\Phi_{k,j}\Gamma_{k,0}, & \prod_{j=2}^{M-1}\Phi_{k,j}\Gamma_{k,1}, & \cdots, & \Gamma_{k,M-1} \end{bmatrix} \begin{bmatrix} \delta u_{k,0} \\ \delta u_{k,1} \\ \vdots \\ \delta u_{k,M-1} \end{bmatrix}.
$$

$$(21)$$

Then, the expression for $\nabla_{\underline{u}_{k,M}}\phi_M\left(x_k,\underline{u}_{k,M}\right)$ is

$$
\nabla_{\underline{u}_{k,M}}\phi_M\left(x_k,\underline{u}_{k,M}\right) = \left[\prod_{j=1}^{M-1}\Phi_{k,j}\Gamma_{k,0}, \right.
$$

$$
\left. \prod_{j=2}^{M-1}\Phi_{k,j}\Gamma_{k,1}, \cdots, \Phi_{k,M-1}\Gamma_{k,M-2}, \Gamma_{k,M-1} \right].
$$

Proposition 1. *If the controllability matrix* $\nabla_{\underline{u}_{k,M}}\phi_M\left(x_k,\underline{u}_{k,M}\right)$ *has full rank for all* $\underline{u}_{k,M} \in \mathbb{R}^{mM}$ *and* $\underline{x}_{k,M} \in \mathbb{R}^{nM}$, *then* $e_{k,M}\left(x_k,\underline{u}_{k,M}\right) \to 0$ *as* $k \to \infty$.

Proof: Omitted here for space limitations. ∎

Algorithm 5: MPC-PSI

1: $k = 0$, $\underline{u}_{k,M}$ non singular
2: Apply u_k (first element of the sequence $\underline{u}_{k,M}$) to the system
3: Compute $e_{k.M}$
4: **while** $k < k_{\max}$ && $\|e_{k,M}\| > \epsilon$ **do**
5:     $\alpha_k = 1$
6:     $exit = false$
7:     $\underline{d}_{k,M} = -\left[C\nabla_{\underline{u}_{k,M}}\phi_M\left(x_k,\underline{u}_{k,M}\right)\right]^\dagger e_{k,M}$
8:     **while** $\alpha_k > \alpha_k^0$ && $\neg exit$ **do**
9:         $\underline{v}_{k,M} = \underline{u}_{k,M} + \alpha_k\underline{d}_{k,M}$
10:        **if** $\|C\phi_M\left(x_k,\underline{v}_{k,M}\right) - y_d\| \geq (1 - \delta\alpha_k)\|e_{k.M}\|$ **then**
11:            Choose $\sigma \in [\sigma_0,\sigma_1]$
12:            $\alpha_k = \sigma\alpha_k$
13:        **else**
14:            $exit = true$
15:        **end if**
16:    **end while**
17:    Apply v_k (first element of the sequence $\underline{v}_{k,M}$) to the system
18:    Compute $e_{k.M}$
19:    $\underline{u}_{k,M} = G\underline{v}_{k,M}$
20:    $k = k + 1$
21: **end while**

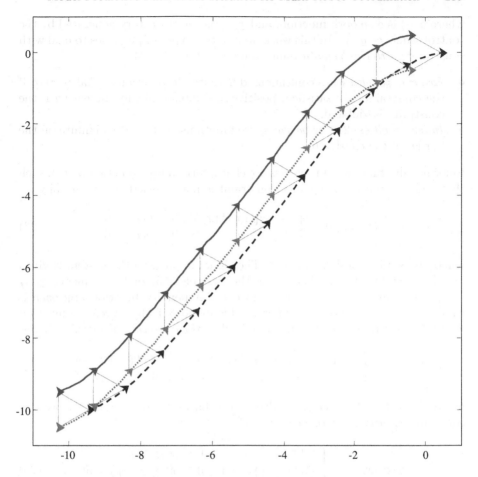

Fig. 3. Centralized multi-robot formation planning without obstacles

Algorithm 5 uses a backtracking line search over the parameter α_k to reduce the variation in the control action [2]. Figure 3 shows the convergence of the algorithm for the obstacle-free case.

3.1 Constrained MPC/PSI Problem

Formation keeping in presence of obstacles is not a trivial task. In this section, we show how to add collision avoidance and formation keeping constraints. Specifically, let us suppose that the system has to satisfy a set of p inequality constraints

$$c^i\left(x_k\right) \leq 0, \, i = 1, \ldots, p; \, \forall x_k \in \underline{x}_{k,M}, \tag{22}$$

where $c^i(\cdot)$ is a smooth function, and $\underline{x}_{k,M}$ is the trajectory generated by the control sequence $\underline{u}_{k,M}$. In this work, we use two types of penalties to deal with constraints, *exterior penalties* and *inner penalties* [15, 24]:

- *Exterior penalties* are continuous differentiable functions equal to *zero* if the constraints are satisfied, positive and monotonically increasing if the constraint is violated.
- *Inner penalties* [21], or *barriers*, are functions that tend to infinite on the border of the constraint.

Inner penalty function approaches are characterized by a function $\beta(x_k)$, such that $\beta(x_k) = 0$ if the inequality constraint is not satisfied. More formally,

$$\beta\left(c^i(x_k)\right) = \begin{cases} 1 - \exp\left[\sigma^i c^i(x_k)\right], & \text{if } c^i(x_k) \le 0 \\ 0, & \text{if } c^i(x_k) > 0 \end{cases} \tag{23}$$

where $\sigma^i > 0$ is a design constant. The signal applied to the system is given by $u'_k = \beta\left(c^i(x_k)\right) v_k$, where v_k is the first element of the sequence $\underline{v}_{k,M}$ computed using (17). In contrast, an exterior penalty function approach is characterized by a continuous differentiable monotonically increasing function when the constraint is violated, and null when the constraint is satisfied. Thus,

$$g\left(c^i(x_k)\right) = \begin{cases} \gamma^i \left(1 - \exp\left[-\kappa^i \left(c^i(x_k) + \delta^i\right)\right]\right)^2, & \text{if } c^i(x_k) > -\delta^i \\ 0, & \text{if } c^i(x_k) \le -\delta^i \end{cases}, \tag{24}$$

where $\delta^i > 0$ is the effective width of the barrier and $\kappa^i > 0$ is a positive constant. Alternatively, one can use [11]

$$g\left(c^i(x_k)\right) = \begin{cases} c^i(x_k), & \text{if } c^i(x_k) > \delta^i \\ \frac{1}{4\delta^i}\left(c^i(x_k) + \delta^i\right)^2, & \text{if } -\delta^i \le c^i(x_k) \le \delta^i \\ 0, & \text{if } c^i(x_k) < -\delta^i \end{cases} \tag{25}$$

where $\delta^i > 0$ is a design parameter. Figure 4 depicts both functions (24) and (25). Furthermore, inequality constraints in (22) are transformed into equality constraints given by

$$z_{k,M}\left(\underline{x}_{k,M}\right) = \sum_{j=1}^{M}\sum_{i=1}^{p} g\left(c^i(x_{k,j})\right).$$

Now the iterative method is applied to the composite constraint vector

$$\zeta_{k,M}\left(\underline{x}_{k,M}, \underline{u}_{k,M}\right) = \begin{bmatrix} e_{k,M}\left(x_k, \underline{u}_{k,M}\right) \\ z_{k,M}\left(\underline{x}_{k,M}, \underline{u}_{k,M}\right) \end{bmatrix},$$

to obtain a path planning solution that satisfies $\zeta_{k,M} = 0$ [6]. Then, the iteration is modified such that

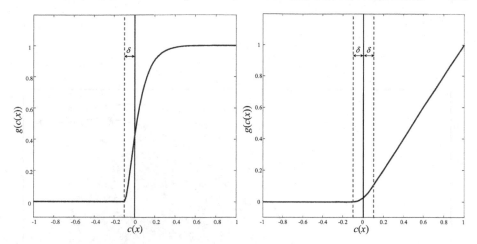

Fig. 4. Smooth exterior penalty functions.

$$\underline{d}_{k,M} = -\left[D_{k,M}\left(\underline{x}_{k,M}, \underline{u}_{k,M}\right)\right]^{\dagger} \zeta_{k,M},$$

with

$$D_{k,M}\left(\underline{x}_{k,M}, \underline{u}_{k,M}\right) := \left[\begin{array}{c} C\nabla_{\underline{u}_{k,M}}\phi_M\left(x_k, \underline{u}_{k,M}\right) \\ \nabla_{\underline{u}_{k,M}} z_{k,M}\left(\underline{x}_{k,M}, \underline{u}_{k,M}\right) \end{array}\right]. \tag{26}$$

The gradient $\nabla_{\underline{u}_{k,M}} z_{k,M}\left(\underline{x}_{k,M}, \underline{u}_{k,M}\right)$ can be computed as

$$\nabla_{\underline{u}_{k,M}} z_{k,M}\left(\underline{x}_{k,M}, \underline{u}_{k,M}\right) = \left[\sum_{i=1}^{p} \frac{dg}{dc^i}\nabla_x c^i\Bigg|_{x_{k,1}} \nabla_{\underline{u}} x_{k,1}, \cdots, \right.$$

$$\left. \sum_{i=1}^{p} \frac{dg}{dc^i}\nabla_x c^i\Bigg|_{x_{k,M}} \nabla_{\underline{u}} x_{k,M}\right], \tag{27}$$

with

$$\frac{dg\left(c^i\left(x_k\right)\right)}{dc^i} =$$

$$\begin{cases} 2\gamma^i\kappa^i \exp\left[-\kappa^i\left(c^i + \delta^i\right)\right]\left[1 - \exp\left[-\kappa^i\left(c^i + \delta^i\right)\right]\right], & \text{if } c^i\left(x_k\right) > -\delta^i, \\ 0, & \text{if } c^i\left(x_k\right) \leq -\delta^i, \end{cases}$$

or

$$\frac{dg\left(c^i\left(x_k\right)\right)}{dc^i} = \begin{cases} 1 & \text{if } c^i\left(x_k\right) > \delta^i, \\ \frac{1}{2\delta^i}\left(c^i\left(x_k\right) + \delta^i\right) & \text{if } -\delta^i \leq c^i\left(x_k\right) \leq \delta^i, \\ 0 & \text{if } c^i\left(x_k\right) < -\delta^i, \end{cases}$$

and $\nabla_x c^i\left(x_k\right)$ depends on the constraint (22). Finally, the following control law is applied to the original system

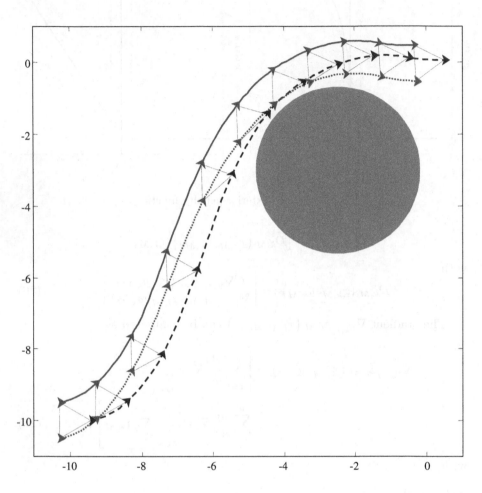

Fig. 5. Centralized algorithm with obstacle avoidance using penalty function (24). The distances between robots are between $[0.7577, 1.1229]$, the convergence is achieved in $\tau = 30$ iterations.

CASE I: Constraints due to obstacles

Let the constraint given by a circular obstacle of radius r'_o centered at $P = (x_o, y_o)$ be defined as

$$c^i (x_k) = r_o^2 - \left(p_x^i - x_o\right)^2 - \left(p_y^i - y_o\right)^2 \leq 0, \; i = 1, 2, 3,$$

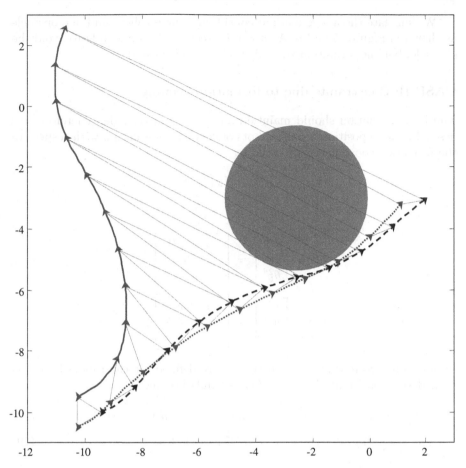

Fig. 6. Centralized algorithm with obstacle avoidance using penalty function (25). The distances between robots are between $[0.8118, 14.6445]$, the convergence is achieved in $\tau = 27$ iterations.

with $r_o = r'_o + r_{robot} + \varepsilon$, where ε is a security factor. Then, the inequality constraint (22) is transformed into the equality constraint

$$z_{k,M}\left(\underline{x}_{k,M}, \underline{u}_{k,M}\right) = \sum_{j=1}^{M} g\left(c^1\left(x_{k,j}\right)\right) + g\left(c^2\left(x_{k,j}\right)\right) + g\left(c^3\left(x_{k,j}\right)\right).$$

Then, for the given formation $c\left(x_k\right) = \left[c^1\left(x_k\right), c^2\left(x_k\right), c^3\left(x_k\right)\right]^T$, and the gradient $\nabla_x c\left(x_k\right)$ in (27) is given by

$$\nabla_x c\left(x_k\right) = -2\left[p_x^1 - p_x^0, p_y^1 - p_y^0, 0, p_x^2 - p_x^0, p_y^2 - p_y^0, 0, p_x^3 - p_x^0, p_y^3 - p_y^0, 0\right].$$

We simulate the above method considering three robots and one obstacle as shown in Figures 5 and 6. As it can be seen, the team is able to avoid the obstacle, but the formation shape is no longer maintained.

CASE II: Constraints due to formation keeping

The basic formation should maintain an equilateral triangle. Therefore, the desired relative positions of the robots of this basic formation with respect to the formation coordinates are given by

$$
q_k^{1d} = \begin{bmatrix} x_k^{1d} \\ y_k^{1d} \\ \theta_k^{1d} \end{bmatrix} = \begin{bmatrix} x_k^f + d^f \\ y_k^f \\ \theta_k^f \end{bmatrix},
$$

$$
q_k^{2d} = \begin{bmatrix} x_k^{2d} \\ y_k^{2d} \\ \theta_k^{2d} \end{bmatrix} = \begin{bmatrix} x_k^f + \frac{1}{2}d^f \\ y_k^f - \frac{\sqrt{3}}{2}d^f \\ \theta_k^f \end{bmatrix},
$$

$$
q_k^{3d} = \begin{bmatrix} x_k^{3d} \\ y_k^{3d} \\ \theta_k^{3d} \end{bmatrix} = \begin{bmatrix} x_k^f + \frac{1}{2}d^f \\ y_k^f + \frac{\sqrt{3}}{2}d^f \\ \theta_k^f \end{bmatrix}.
$$

In addition, two new sets of constraints are defined for the robots based on their desired positions. The sets of constraints become

$$
c_\rho^i \left(x_k \right) = \left(x_k^i - x^{id} \right)^2 + \left(y_k^i - y^{id} \right)^2 - \left(\rho^i \right)^2, \, i = 1, 2, 3, \tag{28}
$$

$$
c_\xi^i \left(x_k \right) = \left(\psi_k^i - \psi^{id} \right)^2 - \left(\xi^i \right)^2, \, i = 1, 2, 3, \tag{29}
$$

where ρ^i and ξ^i are design constants, $\psi_k^i := \pi + \zeta_k^i - \theta^{id}$, $\zeta_k^i = \tan^{-1}\left(\frac{y^{id} - y_k^i}{x^{id} - x_k^i} \right)$, $\psi^{1d} = 0$, $\psi^{2d} = -\frac{2\pi}{3}$, and $\psi^{3d} = \frac{2\pi}{3}$, as shown in Figure 7.

The inequality constraints in (22) are transformed into the equality constraint

$$
z_{k,M} \left(\underline{x}_{k,M} \right) = \sum_{j=1}^{M} \lambda^{M-j+1} \sum_{i=1}^{3} g \left(c_\rho^i \left(x_{k,j} \right) \right) + g \left(c_\xi^i \left(x_{k,j} \right) \right),
$$

where λ is a forgetting factor whose purpose is to allow initial errors in the formation shape. Then, the gradients $\nabla c_\rho^i \left(x_k \right)$ and $\nabla c_\xi^i \left(x_k \right)$ in (27) can be computed as follows

$$
\frac{\partial c_\rho^i \left(x_k \right)}{\partial x^j} = \begin{cases} -\frac{4}{3} \left(x^{id} - x^i \right), \, j = i \\ \frac{2}{3} \left(x^{id} - x^i \right), \quad j \neq i \end{cases}, \quad \frac{\partial c_\rho^i \left(x_k \right)}{\partial y^j} = \begin{cases} -\frac{4}{3} \left(y^{id} - y^i \right), \, j = i \\ \frac{2}{3} \left(y^{id} - y^i \right), \quad j \neq i \end{cases},
$$

$$
\frac{\partial c_\rho^i \left(x_k \right)}{\partial \theta^j} = 0, \, i, j \in \{1, 2, 3\},
$$

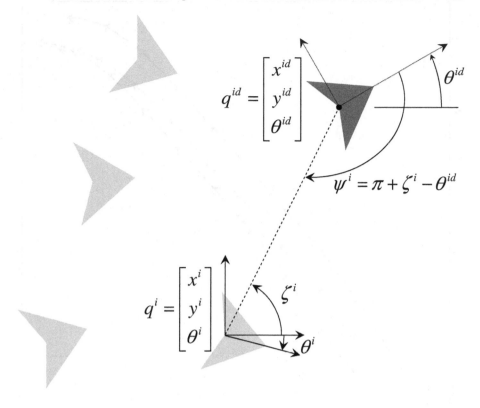

Fig. 7. Angle and distance definition between a virtual leader and a follower in a given multi-robot configuration.

$$\frac{\partial c_\xi^i(x_k)}{\partial x^i} = \frac{2\left(\gamma^i - \gamma^{id}\right)\left(y^{id} - y^i\right)}{\left(x^{id} - x^i\right)^2 + \left(y^{id} - y^i\right)^2}, \quad \frac{\partial c_\xi^i(x_k)}{\partial y^i} = -\frac{2\left(\gamma^i - \gamma^{id}\right)\left(x^{id} - x^i\right)}{\left(x^{id} - x^i\right)^2 + \left(y^{id} - y^i\right)^2},$$

$$\frac{\partial c_\xi^i(x_k)}{\partial x^j} = \frac{\partial c_\xi^i(x_k)}{\partial y^j} = 0, \; i \neq j, \; \frac{\partial c_\xi^i(x_k)}{\partial \theta^j} = 0, \; i, j \in \{1, 2, 3\}.$$

Figures 8-10 depict the result of adding formation keeping constraints to the path-space iteration problem. Note that the desired formation is achieved after a few iterations as shown in Figure 8. Furthermore, Figures 9 and 10 illustrate the case of obstacle avoidance. As it can be seen, the formation shape is successfully maintained during obstacle avoidance maneuvers.

4 Decentralized MPC/Path-space Iteration

A centralized computation of the control law for each robot in a formation is developed in Section 3. In this section, on the other hand, a decentralized

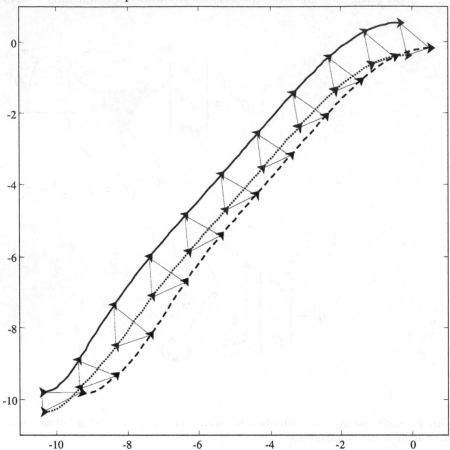

Fig. 8. Centralized multi-robot formation planning without obstacles and arbitrary initial conditions.

MPC/PSI is presented. Each agent (called a *follower* here) is assumed to have a limited sensing and communication range. In other words, it is only capable of avoiding obstacles and following (*virtual* or *real*) leaders within a neighborhood defined by its sensing range. A leader agent sends its position and the M-step ahead control law to its followers. The followers use this information to compute their desired positions with respect to the formation.

The control law to achieve a relative desired position is

$$\underline{e}_{k,M} = \underline{\phi}_{k,M}\left(x_k, \underline{u}_{k,M}\right) - \underline{y}_k^d,$$

with $\underline{e}_{k,M}$, $\underline{\phi}_{k,M}$, and $\underline{y}_k^d \in \mathbb{R}^{nM}$. Differentiating this equation, we obtain

$$\frac{d\underline{e}_{k,M}}{dt} = \nabla_{\underline{u}_{k,M}} \underline{\phi}_{k,M}\left(x_k, \underline{u}_{k,M}\right)\frac{d\underline{u}_{k,M}}{dt},$$

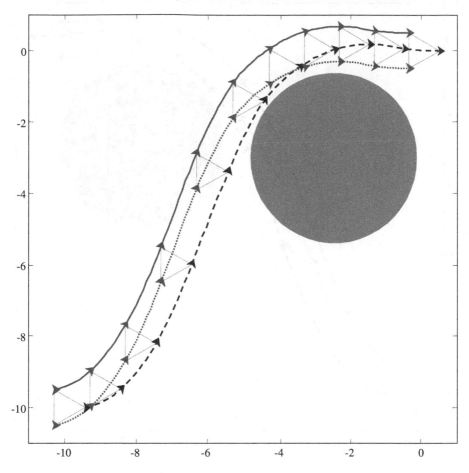

Fig. 9. Centralized algorithm with obstacle avoidance and formation constraints using penalty function (24). The distances between robots are between $[0.9817, 1.0048]$, the convergence is achieved in $\tau = 42$ iterations.

with $\nabla_{\underline{u}_{k,M}} \underline{\phi}_{k,M} \in \mathbb{R}^{nM \times mM}$ and $\underline{u}_{k,M} \in \mathbb{R}^{mM}$. Then, the control law becomes

$$\underline{v}_{k,M} = \underline{u}_{k,M} - \alpha_k \left[\nabla_{\underline{u}_{k,M}} \underline{\phi}_{k,M} \left(x_k, \underline{u}_{k,M}\right) \right]^\dagger \underline{e}_{k,M}, \tag{30}$$

where the gradients are computed in analogous fashion to the centralized case given in Section 3. Thus,

$$\nabla_{\underline{u}_{k,M}} \underline{\phi}_{k,M} \left(x_k, \underline{u}_{k,M}\right) = \begin{bmatrix} \Gamma_{k,0} & 0 & \cdots & 0 \\ \Phi_{k,1} \Gamma_{k,0} & \Gamma_k^1 & \ddots & \vdots \\ \vdots & \vdots & \ddots & 0 \\ \prod_{j=1}^{M-1} \Phi_{k,j} \Gamma_{k,0} & \prod_{j=2}^{M-1} \Phi_{k,j} \Gamma_{k,1} & \cdots & \Gamma_{k,M-1} \end{bmatrix}.$$

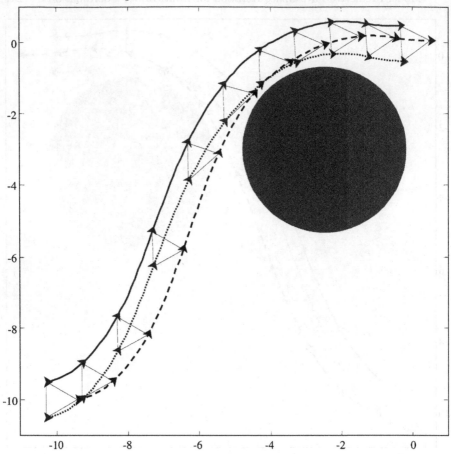

Fig. 10. Centralized algorithm with obstacle avoidance and formation constraints using penalty function (25). The distances between robots are between [0.7269, 1.0698], the convergence is achieved in $\tau = 48$ iterations

Now, a path planning algorithm is used to determine the path of the leader of the formation. The other agents are controlled by (30). In order to use this controller, a virtual leader or group of leaders has to be defined for a particular robot. In this work, the leader for each member of the formation is chosen from $\mathcal{N}_i(q)$ in equation (11). Then, formation keeping is assured by equations (29), where ψ^{id}, $i = 1, 2, 3$, depends on the choice of the leader.

Figure 11 shows the result of simulating a formation of 7 robots in an obstacle-free environment. Figure 12 presents the same formation but with the presence of an obstacle. Finally, Figure 13 depicts the results of obstacle avoidance using formation keeping constraints. Note that the controller (30) with constraints (29) allow the agents to avoid inter-collisions by adequately defining the parameters ψ^{id}, ρ^i, and ξ^i.

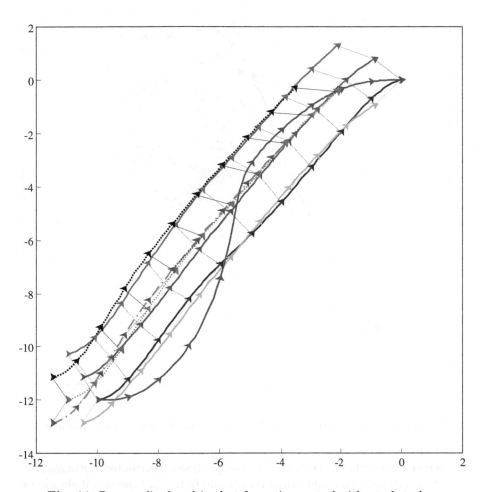

Fig. 11. Decentralized multi-robot formation control without obstacles.

5 Conclusions

In this work, we develop a centralized and a decentralized optimization-based strategies for multi-robot coordination. Both strategies use a Newton-type approach for solving a model predictive control optimization problem, avoiding the use of any optimization solver package.

The centralized strategy requires that a team leader performs all the computations and broadcasts all the resultant control sequences to its followers.

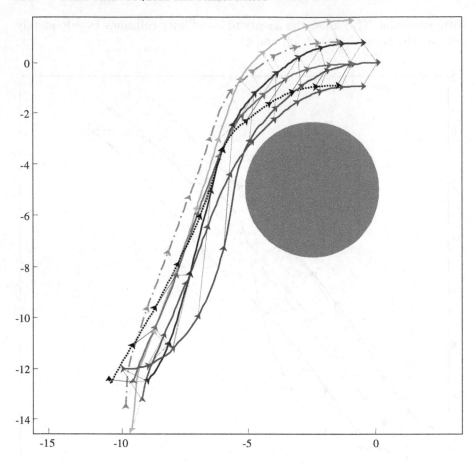

Fig. 12. Decentralized multi-robot formation control with obstacle avoidance.

On the contrary, the decentralized strategy allows distributed optimization, reducing computational and communicational burdens. However, it should be noted that the robot or robots acting as leaders must, at least, broadcast the information about their positions and projected movements to the their followers with the decentralized strategy.

The convergence of both algorithms is very fast, as shown in the simulation experiments. The main drawback of both methodologies is the need of a *feasible* control sequence in the initial setup. This problem could be solved by using a holonomic motion planner to find such a sequence. Currently, we are implemented the algorithms presented herein on the MARHES experimental testbed[2].

---

[2] http://marhes.okstate.edu

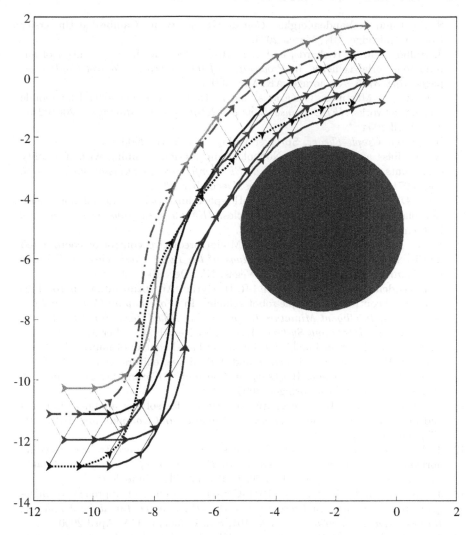

Fig. 13. Decentralized multi-robot formation control with obstacle avoidance and formation constraints.

6 Acknowledgments

The authors thank Professor Teo Kok Lay for his suggestion on the formulation of the constrained optimization problem.

References

1. C. Belta and V. Kumar. Optimal motion generation for groups of robots: A geometric approach. *ASME Journal of Mechanical Design*, 126:63–70, 2004.

2. S. Boyd and L. Vandenberghe. *Convex Optimization*. Cambridge University Press, Cambridge, UK, March 2004.
3. J. Bullingham, A. Richards, and J. P. How. Receding horizon control of autonomous aerial vehicles. In *Proceedings of the American Control Conference*, pages 3741–3746, Anchorage, AK, May 2002.
4. J. Cortés, S. Martínez, T. Karatas, and F. Bullo. Coverage control for mobile sensing networks. *IEEE Transactions on Robotics and Automation*, 20(2):243–255, April 2004.
5. R. Diestel. *Graph Theory*. Springer-Verlag, New York, 2000.
6. A. Divelbiss and J.T. Wen. Nonholonomic motion planning with inequality constraints. In *Proceedings of the IEEE Conference on Decision and Control*, pages 2712–2717, San Antonio, Texas, December 1993.
7. A.W. Divelbiss and J.T. Wen. A path space approach to nonholonomic motion planning in the presence of obstacles. *IEEE Transactions on Robotics and Automation*, 13(3):443–451, June 1997.
8. W. B. Dunbar and R. M. Murray. Model predictive control of coordinated multi-vehicle formations. In *Proceedings of the IEEE Conference on Decision and Control*, pages 4631–4636, Las Vegas, NV, Dec. 10-13 2002.
9. J. T. Feddema, R. D. Robinett, and R. H. Byrne. An optimization approach to distributed controls of multiple robot vehicles. In *Workshop on Control and Cooperation of Intelligent Miniature Robots, IEEE/RSJ International Conference on Intelligent Robots and Systems*, Las Vegas, Nevada, October 31 2003.
10. R. Fierro and K. Wesselowski. Optimization-based control of multi-vehicle systems. In V. Kumar, N.E. Leonard, and A.S. Morse, editors, *A Post-Workshop Volume 2003 Block Island Workshop on Cooperative Control Series*, volume 309 of *LNCIS*, pages 63–78. Springer, 2005.
11. C.J. Goh and K.L. Teo. Alternative algorithms for solving nonlinear function and functional inequalities. *Applied Mathematics and Computation*, 41(2):159–177, January 1991.
12. T. Keviczky, F. Borrelli, and G.J. Balas. A study on decentralized receding horizon control for decoupled systems. In *Proceedings of the American Control Conference*, volume 6, pages 4921–4926, Boston, MA, June 2004.
13. J.J. Kuffner and S.M. LaValle. RRT-Connect: An efficient approach to single-query path planning. In *Proceedings of the IEEE International Conference on Robotics and Automation*, pages 95–101, San Francisco, CA, April 2000.
14. J.C. Latombe. *Robot Motion Planning*. Kluwer Academic Publications, Boston, MA, 1991.
15. F.C. Lizarralde. *Estabilização de sistemas de controle não lineares afins por um método do tipo Newton (in Portuguese)*. PhD thesis, Universidade Federal do Rio de Janeiro, COPPE, Río de Janeiro, RJ-BRASIL, September 1998.
16. D.Q. Mayne, J.B. Rawings, C.V. Rao, and P.O.M. Scokaert. Constrained model predictive control: Stability and optimality. *Automatica*, 36(6):789–814, June 2000.
17. D-O. Popa. *Path-Planning and Feedback Stabilization of Nonholonomic Control Systems*. PhD thesis, Rensselaer Polytechnic Institute, Troy, NY 12180, April 1998.
18. C.W. Reynolds. Flocks, herds, and schools: A distributed behavioral model. *Computer Graphics*, 21(4):25–34, July 1987.

19. R. Olfati Saber. Flocking for multi-agent dynamic systems: Algorithms and theory. Technical Report CIT-CDS 2004-005, Control and Dynamical Systems, Pasadena, Cal, June 2004.
20. N. Sadegh. Trajectory learning and output feedback control of nonlinear discrete time systems. In *Proceedings of the 40th IEEE Conference on Decision and Control*, pages 4032–4037, Orlando, Florida USA, December 2001, December 2001.
21. E.D. Sontag. Control of systems without drift via generic loops. *IEEE Transactions on Automatic Control*, 40:413–440, 1995.
22. H.G. Tanner, A. Jabdabaie, and G.J. Pappas. Stable flocking of mobile agents, part I: Fixed topology. In *Proceedings of the IEEE Conference on Decision and Control*, pages 2010–2015, Maui, Hawaii, USA, December 2003.
23. J.T. Wen and S. Jung. Nonlinear model predictive control based on predicted state error convergence. In *Proceedings of the American Control Conference*, pages 2227–2232, Boston, MA, June 2004.
24. J.T. Wen and F. Lizarralde. Nonlinear model predictive control based on the best-step Newton algorithm. In *Proceedings of the 2004 IEEE Conference on Control Applications*, pages 823–829, Taipei, Taiwan, August 2004.
25. S. Zelinski, T.J. Koo, and S. Sastry. Optimization-based formation reconfiguration planning for autonomous vehicles. In *Proceedings of the IEEE International Conference on Robotics and Automation*, number 3, pages 3758–3763, September 2003.

Path Planning for a Collection of Vehicles With Yaw Rate Constraints

Sivakumar Rathinam[1], Raja Sengupta[1] and Swaroop Darbha[2]

[1] CEE Systems,
    University of California,
    Berkeley, CA 94702, USA
    E-mail: rsiva@berkeley.edu
[2] Mechanical Engineering,
    Texas A & M University,
    College Station, TX 77843-3123

Summary. Multi-vehicle systems are naturally encountered in civil and military applications. Cooperation amongst individual "miniaturized" vehicles allows for flexibility to accomplish missions that a single large vehicle may not readily be able to accomplish. While accomplishing a mission, motion planning algorithms are required to efficiently utilize a common resource (such as the total fuel in the collection of vehicles) or to minimize a collective cost function (such as the maximum time taken by the vehicles to reach their intended destination). The objective of this chapter is to present a constant factor approximation algorithm for planning the path of each vehicle in a collection of vehicles, where the motion of each vehicle must satisfy yaw rate constraints.

1 Introduction

The motivation for this work stems from the need to develop combined motion planning and resource allocation algorithms for multi-vehicle systems such as those envisioned in [1]. Reference [1] describes a collection of miniaturized, self-propelled vehicles that are capable of searching a particular area, identifying targets, and mapping areas of interest. In such a collection, it is imperative that fuel be efficiently used to accomplish missions. The vehicles considered in this work have the following constraints:

1. Each vehicle has a limited fuel capacity; this implies that the distance that can be traveled by a vehicle is limited.
2. The yaw rate of every vehicle in the collection is bounded. At a constant speed, this constraint is equivalent to a constraint on the minimum turning radius of the vehicle.

It is assumed that vehicles in the collection have the ability to communicate with each other. It is also assumed that there is no loss and/or corruption of information when vehicles communicate with each other. Given a set of vehicles, targets with yaw rate constraints on the vehicles and the approach angles at the targets, the problem **P** addressed in this chapter is

- to optimally assign each vehicle, a sequence of targets to visit, and
- to find the optimal paths of the vehicles to their respective targets that satisfy yaw rate constraints, so that a collective cost function is minimized. The cost function considered is the total distance traveled by all the vehicles to traverse their assigned targets.

Once the optimal distance path between any pair of targets is determined using the well-known result of L.E. Dubins [2] in the motion planning literature, problem **P** can be posed as a Multi Vehicle-Asymmetric Traveling Salesman Problem with the costs satisfying triangle inequality[3]. The yaw rate constraints present in the problem **P** make the costs asymmetric. For example, the distance traveled by any vehicle to travel from a target A with a heading ψ_A to another target B with a different heading ψ_B is, in general, different from the distance traveled by the vehicle from target B with a heading ψ_B to target A with a heading ψ_A.

Asymmetric Traveling Salesman Problem (ATSP) is a well known combinatorial optimization that is known to be NP-hard. Currently, there are no algorithms with a constant approximation factor available for solving ATSP problems even when the costs satisfy triangle inequality. Approximation factor $\beta(P, A)$ of using an algorithm A to solve the problem P (objective is minimize some cost function) is defined as

$$\beta(P, A) = \sup_S \left(\frac{C(S, A)}{C_o(S)} \right), \tag{1}$$

where S is a problem instance, $C(S, A)$ is the cost of the solution by applying algorithm A to the instance S and $C_o(S)$ is the cost of the optimal solution of S. The algorithm by Markus Blaser given in [22] for a single vehicle ATSP problem (visiting n targets) has an approximation factor $0.999 \log n$. Hence, the bound $\rightarrow \infty$ as $n \rightarrow \infty$. There are also other kind of algorithms where the bound $\rightarrow \infty$ due to the data but are independent of n. For example, the algorithm by Kumar and Li given in [21] has an approximation ratio which is a increasing function of $\frac{d_{max}}{d_{min}}$. Here, $d_{max} = \max_{i,j} d(i, j)$ and $d_{min} = \min_{i,j} d(i, j)$, where $d(i, j)$ denotes the costs between two targets i and j. In this chapter, we present an algorithm with an approximation factor to solve problem **P** by making certain *assumptions* about the positions of the targets. This will be discussed later in the next subsection.

---

[3] If i, j, k denote the targets to be visited and $d(i, j)$, the distance to travel from the ith to the jth target, then satisfying triangle inequality means that $d(i, j) \leq d(i, k) + d(k, j)$

1.1 Related Work

A series of papers by Chandler et al. in [3],[4],[5],[6] and [9] discusses the complexity and other related issues that arise in cooperative control of UAVs. Task allocation and multi-assignment problems are solved using network flow and auction algorithms in [7],[8] and [9]. Theju et al. in [10] present methods for solving the multi-vehicle, target assignment problem in the presence of threats with the goal of minimizing the maximum path length. A hierarchical decomposition approach involving both target assignment and feasible trajectory generation is addressed in [11]. Mixed Integer Linear Programming is used to solve task assignment problems with timing constraints in [12],[13]. Cooperative path planning for teams of vehicles with different types of timing constraints are addressed in [15]. Dynamic programming is used to solve a cooperative search problem with several UAVs in [16].

Relevant to this work is the paper by Yang et al. [17] where they consider path planning for a UAV with kinematic constraints given fixed initial and final positions in the presence of obstacles. The UAV in their work is required to tour a target and then reach the final position. This is related to the single vehicle problem addressed in this chapter. A bound on the tour distance for a single vehicle was derived in [22]. A more general version of problem **P** was also formulated in [18].

In this chapter, we present algorithms with an approximation factor of $\frac{3}{2}(1 + 1.33\pi)$ for the single vehicle and $2(1 + 1.33\pi)$ for the multiple vehicle case. The assumption is that the Euclidean distances between any two targets and the Euclidean distance between the initial position of each vehicle and a target is greater than twice the minimum turning radius of the vehicles. This is reasonable assumption in the context of unmanned aerial vehicles which carry sensors that have footprints that are greater than $2r$.

2 Problem Setting

As a first approximation, it is reasonable to ignore the dynamics of a vehicle for purposes of resource allocation. Even with this simplifying assumption, the motion planning is a non-trivial problem. Since the primary focus is on resource allocation that accounts for motion planning constraints, we are not particularly concerned that the resulting motion plans may not be flyable trajectories; however, we do assume that there are flyable trajectories close to the motion plans. We treat each vehicle as a Dubins car that travels at a constant speed and has a bound on its yaw rate. Basically, the motion of each vehicle in the collection is governed by the following equations:

$$\dot{x}(t) = v_o \cos \theta(t) \tag{2}$$

$$\dot{y}(t) = v_o \sin \theta(t) \tag{3}$$

$$\dot{\theta}(t) = \Omega \ \ where \ \ \Omega \ \epsilon \ [-\omega, +\omega] \tag{4}$$

where v_o denote the velocity of the vehicle and ω represents the bound on the yaw rate of the vehicle. The maximum yaw rate is related to the minimum turning radius, r, for the vehicle through the following relation: $r = \frac{v_o}{\omega}$.

The assumption regarding motion planning is that the fuel spent by the vehicle in moving a given distance is directly proportional to the distance covered. This assumption implies that minimal fuel path is the same as the minimal distance path. Due to the constraints on the fuel capacity, the aim here is to minimize the fuel consumed by the vehicles to arrive at their respective targets. Hence the problem in motion planning is to find out minimal distance paths that a vehicle should follow from its present position to a desired future location, e.g., a target, subject to path constraints. L.E. Dubins [2] gives the optimal path the vehicle must travel for this problem subject to the path constraints. This result transforms the above motion planning problem to an algebraic problem as follows:

The curve joining the two points (x_1, y_1, θ_1) and (x_2, y_2, θ_2) that has minimal length subject to a limit on the yaw rate, consists of at most three pieces, each of which is either a straight line or an arc of of a circle of radius r. This curve must necessarily be

1. An arc of an circle of radius r, followed by a line segment, followed by an arc of circle of radius r.

2. A sequence of three arcs of circles of radius r.

3. A sub path of a path of type 1 or 2.

Henceforth, the optimal path between any two targets will be referred to as a Dubins path in this chapter. Examples of such paths are shown in figures 1 and 2.

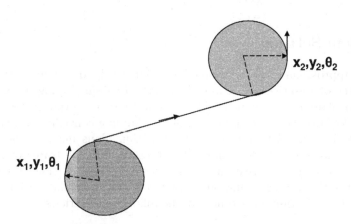

Fig. 1. Shortest path - $\{cw, s, ccw\}$.

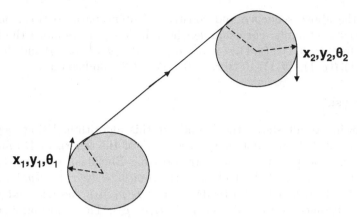

Fig. 2. Shortest path - $\{cw, s, cw\}$.

3 Resource Allocation Problem

Let $(t(x, j), t(y, j))$ denote the position coordinates of the (target) t_j. Similarly, let $(v(x, i), v(y, i))$ indicate the initial position of the vehicle v_i. Let each vehicle start at a initial heading α_i. Given a set of n vehicles, m targets and the angles of approach at each target $t(\theta, j)$, the problem is to

- Assign a sequence P_i of targets to each vehicle such that $\{t_1, t_2...t_m\} = \bigcup_i P_i$, where $P_i \cap P_j = \Phi$ if $i \neq j$. P_i is a sequence of targets $\{t_{i1}, ...t_{ik}\}$ to be visited by the ith vehicle. For example, the ith vehicle moves from $(v(x, i), v(y, i), v(\alpha, i))$ to $(t(x, i1), t(y, i1), t(\theta, i1))$ and then from $(t(x, i1), t(y, i1), t(\theta, i1))$ to $(t(x, i2), t(y, i2), t(\theta, i2))$ and so on. After reaching t_{ik}, it comes back to its initial position. The paths of each vehicle when it travels between the targets should satisfy the maximum yaw rate constraints.

The objective is to minimize $\sum_{i=1}^{n} Cost(P_i)$, where $Cost(P_i)$ is the distance traveled by the ith vehicle.

4 Single Vehicle Path Planning

In this section, we present an algorithm for the single vehicle path planning problem (i.e., when $n = 1$). The $Algorithm(SVTP)$ to solve the single vehicle problem is as follows:

1. Ignoring the yaw rate constraints on the vehicle, use Christofides algorithm [23] to find an approximate tour for the Euclidean TSP problem. The output is a sequence of vertices $\{t_1, t_2, ...t_n\}$ for the vehicle to visit.

2. Use the above sequence and construct Dubins paths between any two consecutive targets. For example, the vehicle v_1 moves along the Dubins path from $(v(x,1), v(y,1), v(\alpha,1))$ to $(t(x,1), t(y,1), t(\theta,1))$ and then from $(t(x,1), t(y,1), t(\theta,1))$ to $(t(x,2), t(y,2), t(\theta,2))$ and so on.

4.1 Analysis

The following result shows the bound for this algorithm. Using the result of Savla et al. [22], we first bound the ratio of the length of Dubins path between any two targets to the corresponding Euclidean distance between them. Let the length of the Dubins path from $p_1 = (v(x,1), v(y,1), v(\alpha,1))$ to $p_2 = (t(x,1), t(y,1), t(\theta,1))$ be denoted as $D(p_1, p_2)$. Assume that the the the Euclidean distance between p_1 and p_2 is $E(p_1, p_2)$. The following result is a simple consequence of the result given in [22].

Lemma 1. *If* $E(p_1, p_2) \geq 2r$, $\frac{D(p_1,p_2)}{E(p_1,p_2)} \leq 1 + 1.33\pi$.

Proof: $D(p_1, p_2)$ is upper bounded by $E(p_1, p_2) + 2.66\pi r$ as given in [22], where r is the minimum turning radius. Since by assumption $E(p_1, p_2) \geq 2r$, $\frac{D(p_1,p_2)}{E(p_1,p_2)} \leq 1 + 1.33\pi$.

Once the distances between the individual points are bounded, it can be combined with the Christofides' result to get an approximation for the single vehicle problem.

Theorem 1. *Algorithm(SVTP) solves the single vehicle problem with an approximation factor* $\frac{3}{2}(1 + 1.33\pi)$.

Proof: The Christofides' algorithm has an approximation factor of $\frac{3}{2}$ [23]. This is assuming all the points are on a Euclidian plane. Lemma 1 basically implies that the maximum ratio of the distance of the path constructed using the Dubins path to the Euclidian distance between any two points is $1 + 1.33\pi$. Combining these two results, the Algorithm(SVTP) has a bound $\frac{3}{2}(1 + 1.33\pi)$.

5 Multiple Vehicle Path Planning

In this section, we present an algorithm for the multiple vehicle ATSP problem $(n > 1)$. The *Algorithm(MVTP)* for the multi vehicle path planning problem is as follows:

1. Construct a complete graph with vertices corresponding to all the vehicles and targets. Assign zero cost to an edge that joins any two vehicles. Assign the Euclidean distance as the cost to each edge between any other pair of vertices.
2. Find a minimum spanning tree of the constructed graph using Prim's algorithm [23]. The minimum spanning tree will contain exactly $n - 1$ zero cost edges, where n is the number of vehicles (figure 3).

3. Remove the zero cost edges to get a tree for each vehicle.
4. For each tree corresponding to a vehicle, double its edges to construct a Eulerian graph (figure 4). Then construct a tour for each vehicle based on the Eulerian graph. A tour for each vehicle is a sequence of vertices (targets) for it to visit (figure 5). (This step is similar to Tarjan's algorithm for a single vehicle Euclidean TSP [23]).
5. Use the sequence derived from the previous step for the each vehicle and construct Dubins paths between any two consecutive targets as in the single vehicle case (figure 6).

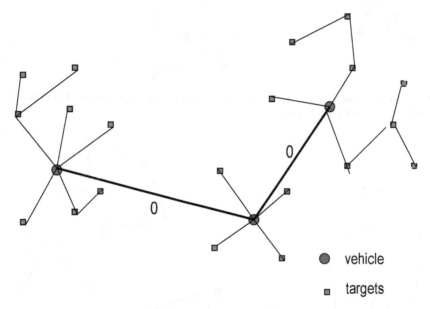

Fig. 3. Calculate the minimum spanning tree (MST). In this example, there are 3 vehicles, hence MST will have 2 zero cost edges.

5.1 Analysis

First we show that multi-vehicle problem without any yaw rate constraints has an approximation factor of 2. Then as in the single vehicle case, each edge is replaced with a path that satisfies the yaw rate constraints and a bound similar to the single vehicle problem can be obtained.

Let $G(V, E)$ be a graph with vertices $V = \{v_1, v_2...v_n, t_1, t_2...t_m\}$. The graph is complete, that is, there is a edge e_{ab} between any pair of vertices a and b. Each edge is assigned a cost $C : E \longrightarrow R^+$ such that $C(e_{ab})$ is the Euclidean distance between vertices a and b if either a or(inclusive or) b $\notin \{v_1, v_2...v_n\}$ and $C(e_{ab}) = 0$ if both a and $b \in \{v_1, v_2...v_n\}$

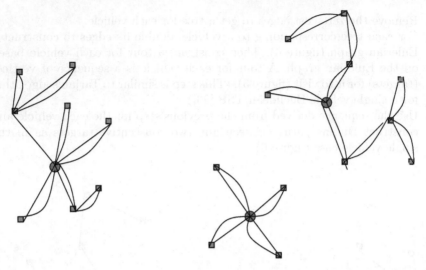

Fig. 4. After removing the zero cost edges, double the edges of the MST to get a Eulerian graph for each vehicle

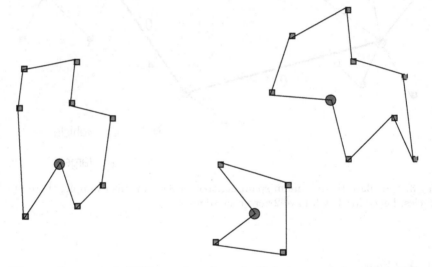

Fig. 5. Compute the TSP tour based on the Eulerian graph for each vehicle

Lemma 2. *The minimum spanning tree MST of the graph G computed using the Prim's algorithm has n − 1 zero cost edges.*

Proof: Start the Prim's algorithm at a vertex representing a vehicle. Since, the Prim's algorithm is greedy, it will add n − 1 zero costs edges before adding a target vertex. The algorithm cannot add more than n − 1 zero cost edges, because it would form a cycle otherwise. Hence there will be exactly n − 1 zero cost edges.

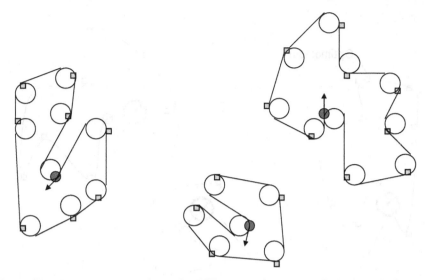

Fig. 6. Use the sequence got from the TSP tour and construct Dubins paths between the corresponding targets.

Now the following theorem gives a bound for visiting all the targets based on the *Euclidean distances*.

Lemma 3. *Step 4 of Algorithm(MVTP) produces a sequence of targets for each of the vehicle to visit and has a factor of approximation of 2.*

Proof: Consider the optimal tours for all the vehicles for the graph $G(V, E)$ based on the cost function C. From each tour, remove one of the two edges that connect to the vehicle vertex to yield a tree for each vehicle as shown in the figure 7. Now, add an appropriate set of $n-1$ zero cost edges to join all the trees connected to the vehicles to make a tree (that connects all the vertices), say T' (figure 8). Clearly the cost of T' must be greater than the cost of the minimum spanning tree. Hence the cost of the tour constructed using Step 4 of the *Algorithm(MVTP)* must be lower bounded by $Cost(MST)$.

Let MST_i represent the tree for the ith vehicle after removing the zero cost edges from the minimum spanning tree. Each tree MST_i must be a minimum spanning tree for the subset of targets connected to the corresponding vehicle. Hence doubling the edges results in a Eulerian graph for the corresponding subset of vertices with a cost $\leq 2MST_i$ (This can be done because triangle inequality is satisfied). As given in [23], a TSP tour can be constructed for each vehicle with a total cost upper bounded by $2\sum_i MST_i$ or $\leq 2MST$. Hence the tour constructed has a cost which is less than two times the cost of the optimal tour.

Now, the following is the result for the approximation factor of the *Algorithm(MVTP)*.

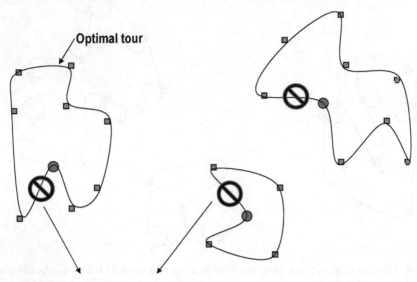

Fig. 7. Removes edges to form a set of trees with each tree containing exactly one vehicle vertex.

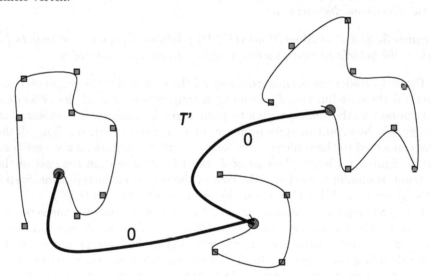

Fig. 8. Adding the zero cost edges to form a tree.

Theorem 2. *Algorithm(MVTP) solves the multiple vehicle problem with an approximation factor* $2(1 + 1.33\pi)$

Proof: Follows from lemma 1 and lemma 3.

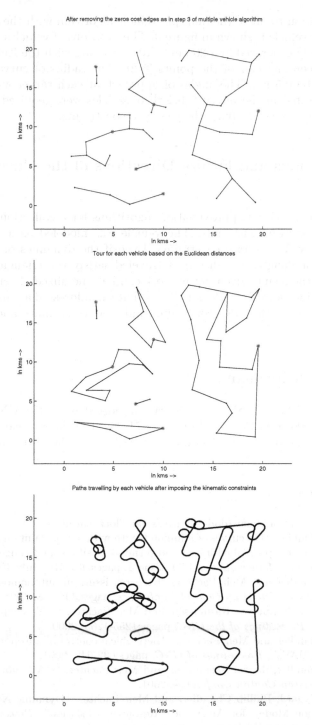

Fig. 9. A simulation result showing the application of the multiple vehicle algorithm to 6 vehicles visiting 40 targets. In the figures '*' indicates a vehicle and '.' indicates a target.

A simulation result of the multiple vehicle algorithm with the paths traveled by each vehicle is shown in figure 9. The positions of vehicles and targets were randomly generated in a 20 km × 20 km square with the Euclidean distances between any pair of the points $\geq 2r$. The radius of curvature r was assumed to be 0.5 km. The angles of approach at each target was also uniformly generated in the interval $[0, 2\pi]$. 6 vehicles were required to visit 40 targets exactly once satisfying the yaw rate constraints.

6 Conclusions and Future Directions of the Current Work

This chapter presented approximation algorithms for a collection of vehicles satisfying yaw rate constraints. The vehicle was modeled as a simple unicycle model with yaw rate constraints. Even if the dynamics of the vehicle is included, as long as the distances traveled satisfy the triangle inequality constraints, the results given in this work can be generalized. There are many future directions for this work. The issues that can addressed are multiple task targets; missions with timing constraints and order constraints; and stochastic uncertainty.

7 Acknowledgements

Rathinam and Sengupta's research was supported by ONR AINS Program - grant # N00014-03-C-0187,SPO #016671-004. Darbha's research was supported by a summer faculty fellowship at AFRL, Wright-Patterson AFB.

References

1. http://www.fas.org/man/dod-101/sys/smart/locaas.htm .
2. Lester E. Dubins, "On curves of minimal length with a constraint on average curvature, and with prescribed initial and terminal positions and tangents", *American Journal of Mathematics*, vol 79, Issue 3, pages:487-516, July 1957.
3. Phillip Chandler and Meir Pachter, "Research issues in autonomous control of tactical UAVs", *American Control Conference*, pages:394-398, 1998.
4. Phillip Chandler, Steven Rasmussen, and Meir Pachter, "UAV cooperative path planning", *Proceedings of the GNC*, pages:1255-1265, 2000.
5. Phillip Chandler and Meir Pachter, "Hierarchical control of autonomous control of tactical UAVs", *Proceedings of GNC*, pages:632-642, 2001.
6. Phillip Chandler, Steven Rasmussen, and Meir Pachter, "UAV cooperative control", *American Control Conference*, 2001.
7. Kendall Nygard, Phillip Chandler, and Meir Pachter, "Dynamic Network Flow Optimization Models for Air Vehicle Resource Allocation", *Proceedings of the American Control Conference*, Arlington, 2001.

8. Corey Schumacher, Phillip R. Chandler and Steven R. Rasmussen, "Task allocation for wide area search munitions via network flow optimization", *AIAA Guidance, Navigation, and Control Conference and Exhibit*, Montreal, Canada, August 6-9, 2001.

9. Phillip Chandler, Meir Pachter, Darba Swaroop, Jeffrey M. Fowler, Jason K. Howlett, Steven Rasmussen, Corey Schumacher and Kendall Nygard, "Complexity in UAV Cooperative Control", *Proceedings of the American Control Conference*, Anchorage, Alaska, May 8-10, 2002.

10. Theju Maddula, Ali A. Minai and Marios M. Polycarpou, "Multi-Target assignment and path planning for groups of UAVs", *S. Butenko, R. Murphey, and P. Pardalos (Eds.), Kluwer Academic Publishers*, December 4-6, 2002.

11. Randall W. Beard, Timothy W. Mclain, Michael A. Goodrich and Erik P. Anderson, "Coordinated target assignment and intercept for unmanned air vehicles", *IEEE Transactions on Robotics and Automation*, 18(6),pages:911 -922, December 2002.

12. John Bellingham, Michael Tillerson, Arthur Richards, Jonathan P. How, "Multi-Task Allocation and Trajectory Design for Cooperating UAVs", *Cooperative Control: Models, Applications and Algorithms at the Conference on Coordination, Control and Optimization*, November 2001.

13. Arthur Richards, John Bellingham, Michael Tillerson, and Jonathan P. How, "Co-ordination and Control of Multiple UAVs", *AIAA Guidance, Navigation, and Control Conference*, August 2002.

14. Mehdi Alighanbari, Yoshiaki Kuwata, and Jonathan P. How, "Coordination and Control of Multiple UAVs with Timing Constraints and Loitering", *Proceeding of the IEEE American Control Conference*, June 2003.

15. Timothy Mclain and Randal Beard, "Cooperative path planning for timing critical missions", *Proceedings of the American Control Conference*, Denver, Colorado, June 4-6, 2003.

16. Matthew Flint, Marios Polycarpou and Emmanuel Fernandez-Gaucherand, "Cooperative Control for Multiple Autonmous UAVs Searching for Targets", *41st IEEE Conference on Decision and Control*, Las Vegas, Nevada USA, December 2002.

17. Guang Yang and Vikram Kapila, "Optimal path planning for unmanned air vehicles with kinematic and tactical constraints", , *Proceedings of the 41st IEEE Conference Decision and Control*, Volume 2, pages:1301 - 1306, 10-13 December 2002.

18. Swaroop Darbha, "Teaming Strategies for a resource allocation and coordination problem in the cooperative control of UAVs", AFRL Summer Faculty Report, Dayton, Ohio, 2001.

19. Zhijun Tang and Umit Ozguner, Motion planning for multi-target surveillance with mobile sensor agents", to appear in IEEE Transactions on Robotics, 2005.

20. Markus Blaser, "A new approximation algorithm for the asymmetric TSP with triangle inequality", *Proceedings of the fourteenth annual ACM-SIAM symposium on Discrete algorithms*, pages:638 - 645, 2003.

21. Ratnesh Kumar Haomin Li, "On Asymmetric TSP: Transformation to Symmetric TSP and Performance Symmetric TSP and Performance Bound", Submitted to Operations Research.

22. Ketan Savla, Emilio Frazzoli, and Francesco Bullo, On the point-to-point and traveling salesperson problems for Dubins' vehicle, *American Control Conference*, Portland, Oregon, June 2005.

23. Christos H. Papadimitriou and Ken Steiglitz, Combinatorial optimization: algorithms and complexity, Prentice-Hall 1982, Dover publications 1998.

Estimating the Probability Distributions of Alloy Impact Toughness: a Constrained Quantile Regression Approach

Alexandr Golodnikov[1], Yevgeny Macheret[2], A. Alexandre Trindade[3], Stan Uryasev[4] and Grigoriy Zrazhevsky[4]

[1] Department of Industrial and Systems Engineering
  University of Florida, Gainesville, FL 32611, USA
[2] Institute for Defense Analysis
  Alexandria, VA 22311, USA
[3] Department of Statistics
  University of Florida, Gainesville, FL 32611, USA
[4] Department of Industrial and Systems Engineering
  University of Florida, Gainesville, FL 32611, USA

Summary. We extend our earlier work, Golodnikov *et al* [3] and Golodnikov *et al* [4], by estimating the entire probability distributions for the impact toughness characteristic of steels, as measured by Charpy V–Notch (CVN) at $-84°$C. Quantile regression, constrained to produce monotone quantile function and unimodal density function estimates, is used to construct the empirical quantiles as a function of various alloy chemical composition and processing variables. The estimated quantiles are used to produce an estimate of the underlying probability density function, rendered in the form of a histogram. The resulting CVN distributions are much more informative for alloy design than singular test data. Using the distributions to make decisions for selecting better alloys should lead to a more effective and comprehensive approach than the one based on the minimum value from a multiple of the three test, as is commonly practiced in the industry.

1 Introduction

In recent work, Golodnikov *et al* [3] developed statistical models to predict the tensile yield strength and toughness behavior of high strength low alloy (HSLA-100) steel. The yield strength was shown to be well approximated by a linear regression model. The alloy toughness (as evaluated by a Charpy V-notch, CVN, at $-84°$C test), was modeled by fitting separate quantile regressions to the 20th, 50th, and 80th percentiles of its probability distributions. The toughness model was shown to be reasonably accurate. Ranking of the alloys and selection of the best composition and processing parameters based on the strength and toughness regression models, produced similar results to the experimental alloy development program.

Models with the capability to estimate the effect of processing parameters and chemical composition on toughness, are particularly important for alloy design. While the tensile strength can be modeled with reasonable accuracy by, for example, Neural Networks (Metzbower and Czyryca [10]), the prediction of CVN values remains a difficult problem. One of the reasons for this is that experimental CVN data often exhibit substantial scatter. The Charpy test does not provide a measure of an invariant material property, and CVN values depend on many parameters, including specimen geometry, stress distribution around the notch, and microstructural inhomogeneities around the notch tip. More on the CVN test, including the reasons behind the scatter and statistical aspects of this type of data analysis, can be found in McClintock and Argon [9], Corowin and Houghland [2], Lucon et al [8], and Todinov [13].

Developing alloys with minimum allowable CVN values, therefore, results in multiple specimens for each experimental condition, leading to complex and expensive experimental programs. In addition, it is possible that optimum combinations of processing parameters and alloy compositions will be missed due to the practical limitations on the number of experimental alloys and processing conditions. This issue was addressed by Golodnikov et al [4], in a follow-up paper to Golodnikov et al [3]. The statistical models developed therein, could be used to simulate a multitude of experimental conditions with the objective of identifying better alloys on the strength vs. toughness diagram, and determining the chemical composition and processing parameters of the optimal alloys. The optimization was formulated as a linear programming problem with constraints. The solution (the efficient frontier) plotted on the strength-toughness diagram, could be used as an aid in successively refining the experimental program, directing the characteristics of the resulting alloys ever closer to the efficient frontier.

The objective of this paper is to build on our previous two papers, by estimating the entire distribution function of CVN at $-84°C$ values for all specimens of steel analyzed in Golodnikov et al [3], and for selected specimens on the efficient frontiers considered by Golodnikov et al [4]. As a tool we use quantile regression, simultaneously fitting a model to several percentiles, while constraining the solution in order to obtain sensible estimates for the underlying distributions. To this end, Section 2 outlines the details and reasoning behind the methodology to be used. The methodology is applied to the steel dataset in Section 3.

2 Estimating the Quantile Function With Constrained Quantile Regression

For a random variable Y with distribution function $F_Y(y) = P(Y \leq y)$ and $0 \leq \theta \leq 1$, the θth quantile function of Y, $Q_Y(\theta)$, is defined to be

$$Q_Y(\theta) = F_Y^{-1}(y) = \inf\{y \mid F_Y(y) \geq \theta\}.$$

For a random sample $Y_1, \ldots, Y_n$ with empirical distribution function $\hat{F}_Y(y)$, we define the θth empirical quantile function as

$$\hat{Q}_Y(\theta) = \hat{F}_Y^{-1}(y) = \inf\{y \mid \hat{F}_Y(y) \geq \theta\},$$

which can be determined by solving the minimization problem

$$\hat{Q}_Y(\theta) = \arg\min_y \left\{ \sum_{i|Y_i \geq y} \theta|Y_i - y| + \sum_{i|Y_i \leq y} (1-\theta)|Y_i - y| \right\}.$$

Introduced by Koenker and Bassett [5], the θth **quantile regression** function is a generalization of the θth quantile function to the case when Y is a linear function of a vector of $k+1$ explanatory variables $\boldsymbol{x}' = [1, x_1, \ldots, x_k]$ plus random error, $Y = \boldsymbol{x}'\boldsymbol{\beta} + \varepsilon$. Here, $\boldsymbol{\beta}' = [\beta_0, \beta_1, \ldots, \beta_k]$ are the regression coefficients, and ε is a random variable that accounts for the surplus variability or scatter in Y that cannot be explained by $\boldsymbol{x}$. The θth quantile function of Y can therefore be written as

$$Q_Y(\theta|\boldsymbol{x}) = \inf\{y \mid F_Y(y|\boldsymbol{x}) \geq \theta\} \equiv \boldsymbol{x}'\boldsymbol{\beta}(\theta), \tag{1}$$

where $\boldsymbol{\beta}(\theta)' = [\beta_0(\theta), \beta_1(\theta), \ldots, \beta_k(\theta)]$. The relationship between the ordinary regression and the quantile regression coefficients, $\boldsymbol{\beta}$ and $\boldsymbol{\beta}(\theta)$, is in general not a straightforward one. Given a sample of observations $y_1, \ldots, y_n$ from Y, with corresponding observed values $\boldsymbol{x}_1, \ldots, \boldsymbol{x}_n$ for the explanatory variables, estimates of the quantile regression coefficients can be obtained as follows:

$$\hat{\boldsymbol{\beta}}(\theta) = \arg\min_{\boldsymbol{\beta}(\theta) \in \mathcal{R}^{k+1}} \left\{ \sum_{i|y_i \geq \boldsymbol{x}_i'\boldsymbol{\beta}(\theta)} \theta|y_i - \boldsymbol{x}_i'\boldsymbol{\beta}(\theta)| + \right.$$
$$\left. + \sum_{i|y_i \leq \boldsymbol{x}_i'\boldsymbol{\beta}(\theta)} (1-\theta)|y_i - \boldsymbol{x}_i'\boldsymbol{\beta}(\theta)| \right\}. \tag{2}$$

This minimization can be reduced to a linear programming problem and solved via standard optimization methods. A detailed discussion of the underlying optimization theory and methods is provided by Portnoy and Koenker [11]. The value $\hat{Q}_Y(\theta|\boldsymbol{x}) = \boldsymbol{x}'\hat{\boldsymbol{\beta}}(\theta)$ is then the estimated θth quantile (or 100θth percentile) of the response variable Y at $\boldsymbol{x}$ (instead of the estimated mean value of Y at $\boldsymbol{x}$ as would be the case in ordinary – least squares – regression).

Instead of restricting attention to a single quantile, θ, one can in fact solve (2) for all $\theta \in [0, 1]$, and thus recover the entire conditional quantile function (equivalently, the conditional distribution function) of Y. Efficient algorithms for accomplishing this have been proposed by Koenker and d'Orey [7], who show that this results in H_n distinct quantile regression hyperplanes, with $\mathbf{E}(H_n) = O(n \log n)$. Although Bassett and Koenker [6] show that the estimated conditional quantile function at the mean value of $\boldsymbol{x}$ is a monotone jump function on the interval $[0, 1]$, for a general design point the quantile hyperplanes are not guaranteed to be parallel, and thus $\hat{Q}_Y(\theta|\boldsymbol{x})$ may not be a proper quantile function. This also usually results in multimodal estimated probability density functions, which may not be desirable in certain applications.

In order to ensure proper and unimodal estimated distributions are obtained from the quantile regression optimization problem, we introduce additional constraints in (2), and simultaneously estimate $\boldsymbol{\beta}(\theta_j)$, $j = 1, \ldots, m$, over a grid of probabilities, $0 < \theta_1 < \cdots < \theta_m < 1$. That is, with $Q(\theta_j|\boldsymbol{x}_i) = \boldsymbol{x}_i'\boldsymbol{\beta}(\theta_j)$ denoting the θ_jth quantile function of Y at $\boldsymbol{x}_i$, we solve the following expression for the $((k+1) \times m)$ matrix whose jth column is $\boldsymbol{\beta}(\theta_j)$:

$$\arg\min_{\boldsymbol{\beta}(\theta_j) \in \mathcal{R}^{k+1},\ j=1,\ldots,m} \sum_{i=1}^{n} \sum_{j=1}^{m} \left\{ (1 - \theta_j) \left[Q(\theta_j|\boldsymbol{x}_i) - y_i \right]^+ + \right.$$
$$\left. + \theta_j [y_i - Q(\theta_j|\boldsymbol{x}_i)]^+ \right\}, \qquad (3)$$

where for a real number z, $(z)^+$ denotes the positive part of z (equal to z itself if it is positive, zero otherwise). The following additional constraints are imposed on (3).

- **Monotonicity.** This is obtained by requiring that quantile hyperplanes corresponding to larger probabilities be larger than those corresponding to smaller ones, i.e.

$$Q(\theta_{j+1}|\boldsymbol{x}_i) \geqslant Q(\theta_j|\boldsymbol{x}_i), \quad j = 1,\ldots,m-1, \quad i = 1,\ldots,n. \qquad (4)$$

- **Unimodality.** Suppose the conditional probability density of Y, $f_Y(\cdot|\boldsymbol{x})$, is unimodal with mode occurring at the quantile θ^*. Let $\theta_{m_1} < \theta^* < \theta_{m_2}$, for some indices $m_1 < m_2$ in the chosen grid. Then, the probability density is monotonically increasing over quantiles $\{\theta_1,\ldots,\theta_{m_1}\}$, and monotonically decreasing over quantiles $\{\theta_{m_2},\ldots,\theta_m\}$. Over $\{\theta_{m_1},\ldots,\theta_{m_2}\}$, the probability density first increases monotonically up to θ^*, then decreases monotonically. The following expressions, to be satisfied for all $i = 1,\ldots,n$, formalize these (approximately) sufficient conditions for unimodality:

$$Q(\theta_{j+2}|\boldsymbol{x}_i) - 2Q(\theta_{j+1}|\boldsymbol{x}_i) + Q(\theta_j|\boldsymbol{x}_i) < 0, \quad j = 1,\ldots,m_1 - 2, \qquad (5)$$

$$Q(\theta_{j+3}|\boldsymbol{x}_i) - 3Q(\theta_{j+2}|\boldsymbol{x}_i) + 3Q(\theta_{j+1}|\boldsymbol{x}_i) - Q(\theta_j|\boldsymbol{x}_i) > 0,$$
$$j = m_1,\ldots,m_2 - 3, \qquad (6)$$

$$Q(\theta_{j+2}|\boldsymbol{x}_i) - 2Q(\theta_{j+1}|\boldsymbol{x}_i) + Q(\theta_j|\boldsymbol{x}_i) > 0, \quad j = m_2,\ldots,m - 2. \qquad (7)$$

Requiring monotonicity leads immediately to (4). Substantiation of (5)-(7) as sufficient conditions for unimodality (in the limit as $m \to \infty$) is not so straightforward, and is deferred to A. We summarize these requirements formally as an estimation problem.

Estimation Problem (Unimodal Conditional Probability Density). *Conditional quantile estimates of Y based on in-sample data $y_1,\ldots,y_n$ and*

$x_1, \ldots, x_n$, can be obtained by solving (3) over the grid $0 < \theta_1 < \cdots < \theta_m < 1$, subject to constraints (4) and (5)-(7). The resulting θth quantile estimate at any given in-sample design point x_i, $\hat{Q}(\theta|x_i) = x_i'\hat{\beta}(\theta)$, is by construction monotone in θ for $\theta \in \{\theta_1, \ldots, \theta_m\}$. Constraints (5)-(7) are approximate sufficient conditions for unimodality of the associated density.

Estimation Problem 1 can be solved in two steps. First omit the unimodality constraints by solving (3) subject only to (4). Analyzing the sample of m empirical quantile estimates thus obtained, determine indices m_1 and m_2 that define a probable quantile interval, $\theta_{m_1} < \hat{\theta}^* < \theta_{m_2}$, around the mode of the empirical density function, $\hat{\theta}^*$. In the second step the full problem (3)-(7) is solved.

Although the literature on nonparametric unimodal density estimation is vast (see for example Cheng et al [1]), we are not aware of any work that approaches the problem from the quantile regression perspective, as we have proposed it. In fact, we are not proposing a method for density estimation per se, rather quantile construction with a view toward imparting desirable properties on the associated density. As far as we are able to ascertain, Taylor and Bunn [12] is in fact the only paper to date dealing with constrained quantile regression, albeit in the context of combining forecast quantiles for data observed over time. Focusing on the effects of imposing constraints on the quantile regression problem, they conclude that this leads in general to a loss in the unbiasedness property for the resulting estimates.

3 Case Study: Estimating the Impact Toughness Distribution of Steel Alloys

In Golodnikov et al [3], we developed statistical regression models to predict tensile yield strength (Yield), and fracture toughness (as measured by Charpy V-Notch, CVN, at $-84°C$) of High Strength Low Alloy (HSLA-100) steel. These predictions are based on a particular steel's chemical composition, namely C (x_1), Mn (x_2), Si (x_3), Cr (x_4), Ni (x_5), Mo (x_6), Cu (x_7), Cb (x_8), Al (x_9), N (x_{10}), P (x_{11}), S (x_{12}), V (x_{13}), measured in weight percent, and the three alloy processing parameters: plate thickness in mm (Thick, x_{16}), solution treating (Aust, x_{14}), and aging temperature (Aging, x_{15}). As described in the analysis of Golodnikov et al [3], the yield strength, chemical composition, and temperature data, have been normalized by their average values.

Finding that the CVN data had too much scatter to be usefully modeled via ordinary regression models, Golodnikov et al [3] instead fitted separate quantile regression models, each targeting a specific percentile of the CVN distribution. In particular, the 20%th percentile of the distribution function of CVN is of interest in this analysis, since it plausibly models the smallest of the three values of CVN associated with each specimen (which is used as the minimum acceptability threshold for a specimen's CVN value). Letting $\hat{Q}(0.2)$ denote the estimated 20th percentile of the conditional distribution function of log CVN, the following model was obtained:

$$\hat{Q}(0.2) = 0.000 - 0.1x_2 + 0.04x_4 - 0.419x_6 + 0.608x_7 - 0.144x_{10}$$
$$- 0.035x_{13} - 0.693x_{14} + 1.692x_{15} - 0.004x_{16}. \quad (8)$$

(CVN was modeled on the logarithmic scale in order to guarantee that model-predicted values would always be positive.) Although not quite as successful as the model for Yield, the $R^1(0.2) = 52\%$ value and further goodness of fit analyses showed this model to be sufficiently useful for its intended purpose, the selection and ranking of good candidate steels (Golodnikov et al [3]).

Using the method described in Section 2, we now seek to extend (8) by estimating the CVN conditional quantiles functions for several quantiles simultaneously, and for all specimens of steels described in Golodnikov et al [3] and some points on the efficient frontier determined by Golodnikov et al [4]. This dataset had an overall sample size of $n = 234$. We used a grid of $m = 99$ quantiles, $\{\theta_j = j/100\}$, $j = 1, \ldots, 99$. The resulting histograms estimate the conditional densities, and are presented in B. Each histogram is based on the 99 values, $\{\hat{Q}(0.01), \ldots, \hat{Q}(0.99)\}$.

We discuss three of these histograms, that illustrate distinct types of behavior of interest in metallurgy. The three dots appearing in each histogram identify the location of the observed values of CVN at $-84°C$[5]. The most decisive cases are those where the entire probability density falls either above or below the minimum acceptability threshold of 2.568 for CVN (0.943 on the log scale). The histogram on Figure 2 corresponding to steel #16, is an example of the former, while the first histogram of Figure 5 (steel #28 with Thick= 51, Aust= 1.05 and Aging= 0.84) is an example of the latter case. The first histogram of Figure 1 (steel #1) illustrates the intermediate case, less desirable from a metallurgy perspective, since a positive probability for CVN values to straddle the minimum acceptability threshold leads to greater uncertainty in the screening of acceptable specimens.

4 Summary

We extended our earlier work, Golodnikov et al [3] and Golodnikov et al [4], by estimating the entire probability distributions for the impact toughness characteristic of steels, as measured by Charpy V-Notch at $-84°C$. Quantile regression, constrained to produce monotone quantile function and unimodal density function estimates, was used to construct empirical quantiles which were subsequently rendered in the form of a histogram. The resulting CVN distributions are much more informative for alloy design than singular test

---

[5] If less than three dots are displayed, then two or more CVN values coincide.

data. Using the distributions to make decisions for selecting better alloys should lead to a more effective and comprehensive approach than the one based on the minimum value from a multiple of the three test, as is commonly practiced in the industry. These distributions may be also used as a basis for subsequent reliability and risk modeling.

References

1. M. Cheng, T. Gasser and P. Hall (1999). "Nonparametric density estimation under unimodality and monotonicity constraints", *Journal of Computational and Graphical Statistics*, 8, 1-21.
2. W.R. Corowin and A.M. Houghland, (1986). "Effect of specimen size and material condition on the Charpy impact properties of 9Cr-1Mo-V-Nb steel", in: *The Use of Small-Scale Specimens for Testing Irradiated Material*, ASTM STP 888, (Philadelphia, PA,) 325-338.
3. A. Golodnikov, Y. Macheret, A. Trindade, S. Uryasev and G. Zrazhevsky, (2005). "Modeling Composition and Processing Parameters for the Development of Steel Alloys: A Statistical Approach", *Research Report # 2005-1*, Department of Industrial and Systems Engineering, University of Florida.
4. A. Golodnikov, Y. Macheret, A. Trindade, S. Uryasev and G. Zrazhevsky, (2005). "Optimization of Composition and Processing Parameters for the Development of Steel Alloys: A Statistical Approach", *Research Report # 2005-2*, Department of Industrial and Systems Engineering, University of Florida.
5. R. Koenker and G. Bassett (1978). "Regression Quantiles", *Econometrica*, 46, 33-50.
6. G. Bassett and R. Koenker (1982). "An empirical quantile function for linear models with iid errors", *Journal of the American Statistical Association*, 77, 407-415.
7. R. Koenker, and V. d'Orey (1987; 1994). "Computing regression quantiles", *Applied Statistics*, 36, 383-393; and 43, 410-414.
8. E. Lucon *et al*(1999). "Characterizing Material Properties by the Use of Full-Size and Sub-Size Charpy Tests", in: *Pendulum Impact Testing: A Century of Progress*, ASTM STP 1380, T. Siewert and M.P. Manahan, Sr. eds., American Society for Testing and Materials, (West Conshohocken, PA) 146-163.
9. F.A. McClintock and A.S. Argon (1966). *Mechanical Behavior of Materials*, (Reading, Massachusetts), Addison-Wesley Publishing Company, Inc.
10. E.A. Metzbower and E.J. Czyryca (2002). "Neural Network Analysis of HSLA Steels", in: T.S. Srivatsan, D.R. Lesuer, and E.M. Taleff eds., *Modeling the Performance of Engineering Structural Materials*, TMS.
11. S. Portnoy and R. Koenker (1997). "The Gaussian hare and the Laplacian tortoise: Computability of squared-error versus absolute-error estimators", *Statistical Science*, 12, 279-300.
12. J.W. Taylor and D.W. Bunn (1998). "Combining forecast quantiles using quantile regression: Investigating the derived weights, estimator bias and imposing constraints", *Journal of Applied Statistics*, 25, 193-206.
13. M.T. Todinov (2004). "Uncertainty and risk associated with the Charpy impact energy of multi-run welds", *Nuclear Engineering and Design*, 231, 27-38.

A Substantiation of the Unimodality Constraints

Let $F(x) = P(X \le x)$ and $f(x) = F'(x)$ be respectively, the cumulative distribution function (cdf), and probability density function (pdf), for random variable X, and assume the cdf to be continuously differentiable on $\mathbf{R}$. Suppose $f(x) > 0$ over the open interval (x_l, x_r), whose endpoints are quantiles corresponding to probabilities $\theta_l = F(x_l)$ and $\theta_r = F(x_r)$. Then for $\theta \in (\theta_l, \theta_r)$, the quantile function of X is uniquely defined as the inverse of the cdf, $Q(\theta) \equiv F^{-1}(\theta) = x$. Now, differentiating both sides of the identity $F(Q(\theta)) = \theta$, we obtain for $x \in (x_l, x_r)$,

$$f(x) = \frac{1}{dQ/d\theta}\bigg|_{\theta=F(x)}. \tag{9}$$

This implies that since $f(x) > 0$ on (x_l, x_r), $dQ/d\theta > 0$, and therefore $Q(\theta)$ is monotonically increasing on (θ_l, θ_r). Differentiating (9) once more gives

$$f'(x) = -\frac{d^2Q/d\theta^2}{(dQ/d\theta)^2}\bigg|_{\theta=F(x)}. \tag{10}$$

With the above in mind, we can now state our main result.

Proposition 1 *If $Q(\theta)$ is three times differentiable on (θ_l, θ_r) with $d^3Q/d\theta^3 > 0$, then $f(x)$ is unimodal (has at most one extremum) on $[x_l, x_r]$.*

Proof: Since $d^3Q/d\theta^3 > 0$, it follows that $d^2Q/d\theta^2$ is continuous and monotonically increasing on (θ_l, θ_r). Therefore, $d^2Q/d\theta^2|_{\theta_l} < d^2Q/d\theta^2|_{\theta_r}$, and there are 3 cases to consider:

(i) $d^2Q/d\theta^2|_{\theta_l} < d^2Q/d\theta^2|_{\theta_r} < 0$. Recalling that $dQ/d\theta > 0$ for all $\theta \in (\theta_l, \theta_r)$, it follows from (10) that $f'(x) > 0$. Therefore $f(x)$ is monotonically increasing and has no extrema on (x_l, x_r). Consequently, the maximum value of $f(x)$ on $[x_l, x_r]$ is attained at $x = x_r$.

(ii) There exists a $\theta^* \in (\theta_l, \theta_r)$ such that $d^2Q/d\theta^2 < 0$ on (θ_l, θ^*), and $d^2Q/d\theta^2 > 0$ on (θ^*, θ_r). Then $f'(x) > 0$ on $[x_l, x^*)$, and $f'(x) < 0$ on $(x^*, x_r]$. Therefore, $f(x)$ has exactly one extremum on $[x_l, x_r]$, a maximum occurring at $x^* = Q(\theta^*)$.

(iii) $0 < d^2Q/d\theta^2|_{\theta_l} < d^2Q/d\theta^2|_{\theta_r}$. As in case (i), but the numerator of (10) is now positive, which implies $f'(x) < 0$. Therefore $f(x)$ is monotonically decreasing and has no extrema on (x_l, x_r). Consequently, the maximum value of $f(x)$ on $[x_l, x_r]$ is attained at $x = x_l$. $\square$

The three unimodality constraints (5)-(7), are the discretized empirical formulation of the three cases in this proof. To see this, let $\delta = 1/(m+1)$ be the inter-quantile grid point distance, and $Q(\theta)$ denote the conditional quantile function of Y, where for notational simplicity we suppress the dependence of $Q(\cdot)$ on the x_i. We then have that for large m,

$$\frac{dQ(\theta)}{d\theta}\bigg|_{\theta_j} \approx \frac{Q(\theta_{j+1}) - Q(\theta_j)}{\delta}$$

$$\frac{d^2Q(\theta)}{d\theta^2}\bigg|_{\theta_j} \approx \frac{Q'(\theta_{j+1}) - Q'(\theta_j)}{\delta} \approx \frac{Q(\theta_{j+2}) - 2Q(\theta_{j+1}) + Q(\theta_j)}{\delta^2} \quad (11)$$

$$\frac{d^3Q(\theta)}{d\theta^3}\bigg|_{\theta_j} \approx \frac{Q''(\theta_{j+1}) - Q''(\theta_j)}{\delta} \approx$$

$$\approx \frac{Q(\theta_{j+3}) - 3Q(\theta_{j+2}) + 3Q(\theta_{j+1}) - Q(\theta_j)}{\delta^3}. \quad (12)$$

Now, recall for example that the pdf of Y is required to be monotonically increasing over quantiles $\{\theta_1, \ldots, \theta_{m_1}\}$. Since this corresponds to case (i) in the Proposition, from (11) a sufficient condition (in the limit as $m \to \infty$) is that $Q(\theta_{j+2}) - 2Q(\theta_{j+1}) + Q(\theta_j) < 0$, for all $j = 1, \ldots, m_1$. Similarly, the pdf of Y is required to be monotonically decreasing over $\{\theta_{m_2}, \ldots, \theta_m\}$ which corresponds to case (iii), and can be obtained by having $Q(\theta_{j+2}) - 2Q(\theta_{j+1}) + Q(\theta_j) > 0$, for all $j = m_2, \ldots, m$. Finally, over $\{\theta_{m_1}, \ldots, \theta_{m_2}\}$ the pdf of Y has at most one mode. From the statement of the Proposition, this is ensured if $Q(\theta_j)$ has a positive 3rd derivative, which from (12) corresponds to $Q(\theta_{j+3}) - 3Q(\theta_{j+2}) + 3Q(\theta_{j+1}) - Q(\theta_j) > 0$, for all $j = m_1, \ldots, m_2$.

B Histograms for Selected Distributions of log CVN at −84°C

Fig. 1. Estimated distribution of log CVN at −84°C for Steels #1-12. The location of each of the three observed CVN values is indicated with a dot.

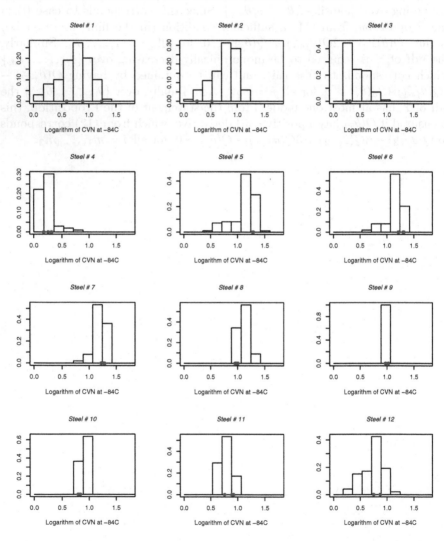

Fig. 2. Estimated distribution of log CVN at −84°C for Steels #13-18. The location of each of the three observed CVN values is indicated with a dot.

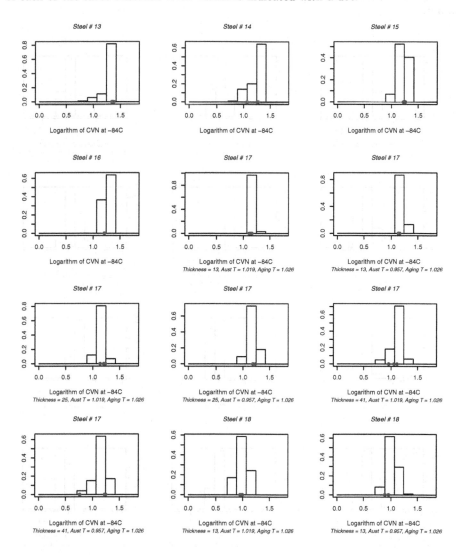

Fig. 3. Estimated distribution of log CVN at −84°C for Steels #18 (continued) and #24. The location of each of the three observed CVN values is indicated with a dot.

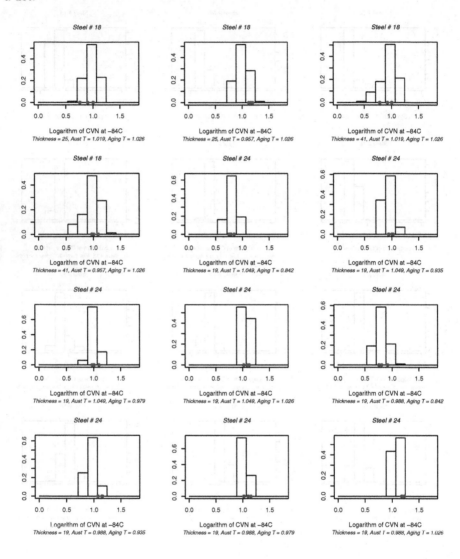

Fig. 4. Estimated distribution of log CVN at −84°C for Steels #19-20. The location of each of the three observed CVN values is indicated with a dot.

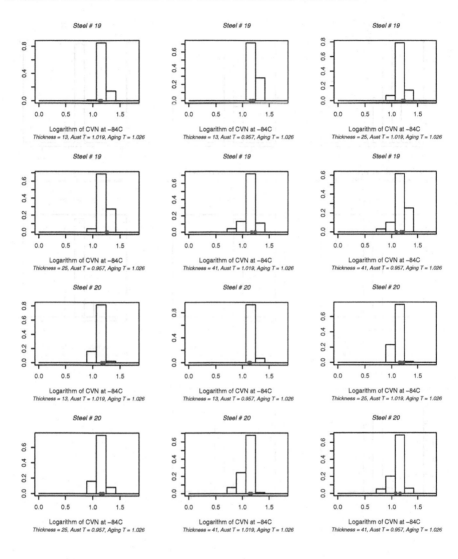

Fig. 5. Estimated distribution of log CVN at −84°C for Steel #28. The location of each of the three observed CVN values is indicated with a dot.

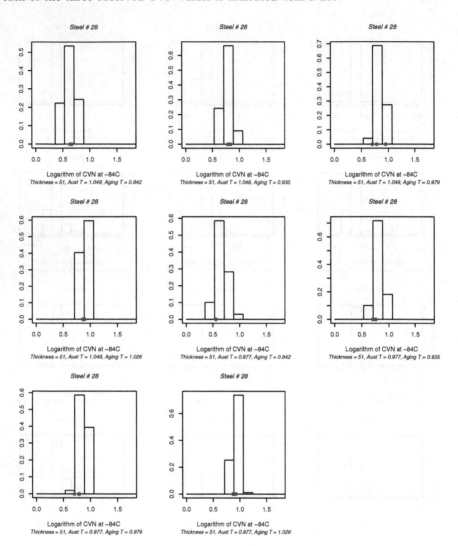

Fig. 6. Estimated distribution of log CVN at $-84°C$ for 6 points on the efficient frontier.

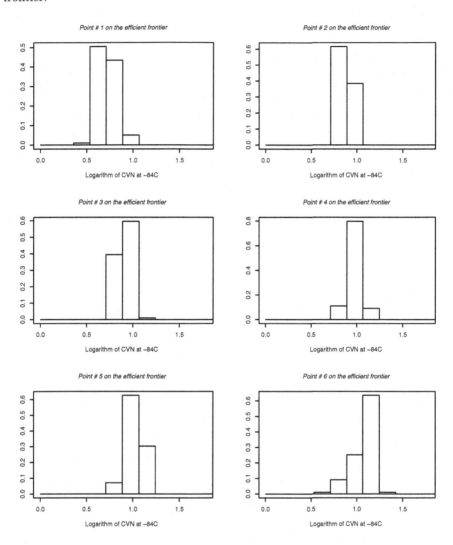

A One-Pass Heuristic for Cooperative Communication in Mobile Ad Hoc Networks

Clayton W. Commander[1,2], Carlos A.S. Oliveira[3], Panos M. Pardalos[2], and Mauricio G.C. Resende[4]

[1] Air Force Research Laboratory, Munitions Directorate, Eglin AFB, FL USA
   `clayton.commander@eglin.af.mil`
[2] Department of Industrial and Systems Engineering, University of Florida,
   Gainesville, FL USA `pardalos@ufl.edu`
[3] School of Industrial Engineering and Management Oklahoma State University,
   Stillwater, OK USA `coliv@okstate.edu`
[4] Internet and Network Systems Research Center AT&T Labs Research, Florham
   Park, NJ USA `mgcr@research.att.com`

Summary. Ad hoc networks have been used in the last few years to provide communications means among agents that need to accomplish common goals. Due to the importance of communication for the success of such missions, we study the problem of maximizing communication among a set of agents. As a practical tool to solve such problems, we introduce a one-pass randomized algorithm that maximizes the total communication, as measured by the proposed objective function. Agents in this problem are routed along the edges of a graph, connecting their individual starting nodes to their respective destination nodes. This problem, known as the *Cooperative Communication Problem in Mobile Ad Hoc Networks*, is known to be NP-hard. We present a new heuristic and motivate the need for more advanced methods for the solution of this problem. In particular, we describe 1) a construction algorithm and 2) a local improvement method for maximizing communication. Computational results for the proposed approach are provided, showing that instances of realistic size can be efficiently solved by the algorithm.

1 Introduction

Advances in wireless communication and networking have lead to the development of new network organizations based on autonomous systems. Among the most important example of such networks systems are mobile ad hoc networks (MANETs). MANETs are composed of a set of loosely coupled mobile agents which communicate using a wireless medium via a shared radio channel. Agents in the network act as both clients and as servers and use various multi-hop protocols to route messages to other users in the system[17, 18]. Unlike traditional cellular systems, ad hoc networks have no fixed topology.

Moreover, in a MANET the topology changes each time an agent changes its location. Thus, the communication between the agents depends on their physical location and their particular radio devices.

Interest in MANETs has surged in the recent years, due to their numerous civilian and military applications. MANETs can be successfully implemented in situations where communication is necessary, but no fixed telephony system exists. Real applications abound, especially when considering adversarial environments, such as the coordination of unmanned aerial vehicles (UAVs) and combat search and rescue groups. Other examples include the coordination of agents in a hostile environment, sensing, and monitoring. More generally, the study of protocols and algorithms for MANETs is of high importance for the successful deployment of sensor networks – which are themselves composed of a large number of autonomous processors that can coordinate to achieve some higher level task such as sensing and monitoring.

The lack of a central authority in MANETs leads to several problems in the areas of routing and quality assurance[6]. Many of these problems can be viewed as combinatorial in nature, since they involve finding sets of discrete objects satisfying some definite property, such as for example connectedness or minimum cost. Among the challenging problems encountered in MANETs, we can cite routing as one of the most difficult to solve, because of the temporary nature of communication links in such a system. In fact, as nodes move around, they dynamically define topologies for the entire network. In such an environment, it is difficult to determine if two nodes are connected, since any of the intermediate nodes may leave the network at any time.

This scenario makes clear the importance of close coordination among groups of nodes if a definite goal needs to be attained. If at all possible, a plan must be devised such that communication among nodes is maintained for as long as possible. With this objective in mind, we study in this chapter algorithms that would allow a set of agents to accomplish a mission, with definite starting and ending positions, but at the same time maximizing the communication time among the agents involved.

This chapter is organized as follows. We first present a graph formulation and discuss the computational complexity of the problem in Section 2. Next, in Section 3 we discuss some of the previous work on areas related to cooperative communication in wireless systems. Potential solution techniques are presented in the following sections. In Section 4, a simple construction algorithm for the maximum communication problem is proposed. In Section 5, a hill-climbing method for improving an initial solution for the problem is presented. Then, in Section 6 a one-pass local search heuristic is presented, combining the ideas discussed on the preceding sections. The numerical results from computational experiments are analyzed on Section 7. Finally, concluding remarks and future research ideas are presented in Section 8.

2 Problem Statement

In MANETs, bandwidth and communication time are usually severely constrained resources. To allow for successful interaction among nodes, we propose the solution of the *cooperative communication problem on ad hoc networks* (CCPM). In this problem, the objective is to determine a set of routes to be followed by mobile agents who must cooperate to accomplish some preassigned tasks. The function we try to maximize represents the total communication time for all agents along the computed trajectories. As described in this section, the *Cooperative Communication Problem* in MANETs can be modeled as a combinatorial optimization problem on graphs.

Consider a graph $G = (V, E)$, where $V = \{v_1, v_2, \ldots, v_n\}$ represents the set of candidate positions for the wireless agents. Suppose that a node in G is connected only to those nodes that can be reached in one unit of time. Let U represent the set of agents, $S = \{s_1, s_2, \ldots, s_{|U|}\} \subseteq V$ represent the set of initial positions, and $D = \{d_1, d_2, \ldots, d_{|U|}\} \subseteq V$ the set of destination nodes. Let $N(v) \subseteq 2^V$, for $v \in V$, represent the set of neighbors of node v in G. Given a time horizon T, the objective of the problem is to determine a set of routes for the agents in U, such that each agent $u_i \in U$ starts at a source node s_i and finishes at the destination node $d_i \in D$ after at most T units of time.

For each agent $u \in U$, the function $p_t : U \to V$ returns the position of the agent at time $t \in \{1, 2, \ldots, T\}$, where T is the time limit by which the agents must reach their destinations. Then at each time instant t, an agent $u \in U$ can either remain in its current location, i.e., $p_{t-1}(u)$, or move to a node in $N(p_{t-1}(u))$.

We can represent a route for an agent $u \in U$ as a path $\mathcal{P} = \{v_1, v_2, \ldots, v_k\}$ in G where $v_1 = s_u$, $v_k = d_u$, and, for $i \in \{2, \ldots, k\}$, $v_i \in N(v_{i-1}) \cup \{v_i\}$. Finally, if $\{\mathcal{P}_i\}_{i=1}^{|U|}$ is the set of trajectories for the agents, we are given a corresponding vector $\mathcal{L}$ such that $\mathcal{L}_i$ is a threshold on the size of path $\mathcal{P}_i$. This value is typically determined by fuel or battery life constraints on the wireless agents.

We assume that the agents have omnidirectional antennas and that two agents in the network are connected if the distance between them is less than some radius r. More specifically, let $\delta : V \times V \to \mathbb{R}$ represent the Euclidean distance between a pair of nodes in the graph. Then, we can define a function $c : V \times V \to \{0, 1\}$ such that

$$c(p_t(u_i), p_t(u_j)) = \begin{cases} 1, & \text{if } \delta(p_t(u_i), p_t(u_j)) \leq r \\ 0, & \text{otherwise.} \end{cases} \tag{1}$$

With this, we can define the CCPM as the following optimization problem:

$$\max \sum_{t=1}^{T} \sum_{u,v \in U} c(p_t(u), p_t(v)) \tag{2}$$

$$\text{s.t.} \sum_{j=2}^{n_i} \delta(v_{j-1}, v_j) \leq \mathcal{L}_i \quad \forall \, \mathcal{P}_i = \{v_1, v_2, \dots, v_{n_i}\}, \tag{3}$$

$$p_1(u) = s_u \quad \forall \, u \in U, \tag{4}$$

$$p_T(u) = d_u \quad \forall \, u \in U, \tag{5}$$

where constraint (3) ensures that the length of each path $\mathcal{P}_i$ is less than or equal to its maximum allowed length $\mathcal{L}_i$.

Oliveira and Pardalos[17] have shown that CCPM is $\mathcal{NP}$-hard via a reduction to 3SAT[1]. Furthermore, to find an optimal solution at each time $t \in \{1, 2, \dots, T\}$ remains $\mathcal{NP}$-hard. This can be shown by a reduction to MAXIMUM CLIQUE[1], another well-known $\mathcal{NP}$-hard problem. The computational complexity of the problem does not allow for real-world instances to be solved exactly. This motivates the need for efficient heuristics to solve real-world instances within reasonable computing times. In the upcoming sections, we describe one such algorithm and present the framework for a different, non deterministic heuristic.

3 Previous Work

Communication is an important measure of collaboration between entities involved in a mission. It allows different agents to perform the set of tasks that have been planned, and at the same time to implement changes in the case that an unexpected event occurs. Moreover, high communication levels are necessary in order to perform complicated tasks, where several agents must be coordinated. We describe in this section the main concepts found in the literature related to optimizing communication time in ad hoc network systems.

One of the main difficulties concerning the maintenance of communication is an ad hoc network is determining the location of agents at a given moment in time. Several methods have been proposed for improving localization in this situation. Moore et al.[16], for example, presented a linear time algorithm for determining the location of nodes in an ad hoc network in the presence of noise. Other algorithms for the same problem have been suggested by Capkun et al.[7], Doherty et al.[13], and Priyantha et al.[19].

While such algorithms can be useful in determining the correct location of nodes, they are only able to provide information about current positions, and are not meant to optimize locations for a specific objective. Packet routing, on the other hand, has been previously studied with the goal of optimizing some common parameters, such as latency, cost of the resulting route, and energy consumed. For example, Butenko et al.[6] proposed a new algorithm for computing a backbone for wireless networks with minimum size, based on a number of related algorithms for this problem[5, 8, 15].

Another problem involving the minimization of an objective function over all feasible positions of agents in an ad hoc network is the so-called *location error minimization* problem. In the location error minimization problem, given a set of measurements of node positions (taken from different sources), the goal is to determine a set of locations for wireless nodes such that errors in the given measurements are minimized. This problem has been formulated and solved using mathematical programming techniques, by the use of a relaxation for the general problem into a semi-definite programming model[20, 3, 4].

In this chapter we consider a different optimization problem (CCPM) over the relative position of nodes in a wireless network. The CCPM has the objective of maximizing the *total communication time* of a given set of wireless agents. The problem has been proposed by considering military situations where a set of agents needs to accomplish a mission. It has been proved[17] that the problem is NP-hard, which makes clear the necessity of fast algorithms for the practical solutions of instances with large size. We present a method for constructing and optimizing solutions for the CCPM in the next sections.

4 Construction Heuristic

In this section, we propose a construction heuristic to create high quality solutions for the CCPM. Our objective is to provide a fast way of constructing a set of paths, connecting wireless agents from their initial positions S to the destinations D such that the resulting routes are feasible for the problem. The union of such sequences of nodes will uniquely determine the cost of the solution, which is calculated using equation (2). The algorithm also tries to create solutions that have as large a value as possible for the objective function.

The pseudo-code for the construction heuristic is showed in Figure 1. The algorithm starts initializing the cost of the solution to zero. The incumbent solution, represented by the variable **solution**, is initialized with the empty set.

The next step consists of finding shortest paths connecting each source $s_i \in S$ to a destination $d_i \in D$. Standard minimum cost flow algorithms can be used to calculate these shortest paths. For example, the Floyd-Warshall algorithm[14, 22] can be used to compute the shortest path between all pairs of nodes in a graph. The Dijkstra algorithm[12] can also be used to perform this step of the algorithm (with the only difference that, being a single-source shortest path algorithm, it must be run for $|U|$ iterations, one for each of the $|U|$ source-destination pairs).

In the loop from lines 4 to 10, the algorithm performs the assignment of new paths to the solution, using the shortest path algorithm described above.

First, a source-destination path s_i-d_i is selected, and based on this a shortest path $\mathcal{P}_i$ corresponding to this pair is generated. Notice that, if the length (number of edges) of the shortest path $\mathcal{P}_i$ is more than T there is not feasible solution for the problem, since the destinations cannot be reached at the end of the requested time horizon. The algorithm checks for this condition on line 6.

If all source-destination pairs are found to be feasible, then a solution is generated by the union of all $\mathcal{P}_i$. Notice that once agent i reaches node d_i it can simply loiter at d_i during all remaining time (until instant T), as shown in line 7. The sequence of nodes found as a result of this process is then added to the solution in line 8 of the algorithm, and the optimum objective value is updated (line 9). Finally, a complete solution is returned on line 11, along with the value of that solution.

```
procedure ShortestPath(solution)
1      c ← 0;
2      solution ← ∅;
3      Compute all shortest paths for each (s_i, d_i) pair;
4      for i = 1 to |U| do→
5          P_i ← SP(s_i, d_i);
6          if length of P_i > T then return ∅;
7          let agent i stay in d_i until time T is reached;
8          solution ← solution ∪ P_i;
9          c ← c + number of new connections generated by P_i;
10     rof
11     return (c, solution);
end ShortestPath
```

Fig. 1. Pseudocode for the Shortest Path Constructor.

Theorem 1. *The construction algorithm presented above finds a feasible solution for the* CCPM *in* $\mathcal{O}(|V|^3)$ *time.*

Proof: A feasible solution for this problem is given by a sequence of positions starting at s_i and ending at d_i, for each agent $u_i \in U$. Clearly, the union of the shortest paths provide the required connection between each source-destination pair, according to the remarks in the preceding paragraph; therefore the solution is feasible. Suppose that, in line 3, we use the Floyd-Warshall algorithm for all-pairs shortest path[14, 22]. This algorithm runs in $\mathcal{O}(|V|^3)$ time. Then, at each step of the **for** loop we need only to refer to the solution calculated by the Floyd-Warshall algorithm and add it to the variable **solution**. This can be done in time $\mathcal{O}(|V|)$, and therefore the **for** loop will run in at most $\mathcal{O}(|V||U|)$ time. Thus, the step with highest time complexity is the one appearing in line 3, which implies that the total complexity of the algorithm is $\mathcal{O}(|V|^3)$. ∎

5 Local Search Heuristic

A construction algorithm is a good starting point in the process of solving a combinatorial optimization problem. However, due to the $\mathcal{NP}$-hard nature of the CCPM, such an algorithm provides no guarantee that a good solution will be found. In fact, it is possible that for some instances the solution found by the construction heuristic is far from the optimum, and not even a local optimal solution.

To guarantee that the solution found is at least locally optimal, we propose a local search algorithm for the CCPM. A local search algorithm receives as input a feasible solution and, given a neighborhood structure for the problem, returns a solution that is guaranteed to be optimal with respect to that neighborhood.

For the CCPM, the neighborhood structure is defined as follows. Given an instance Π of the CCPM, let $\mathcal{S}$ be the set of feasible solutions for that instance. Then, if $s \in \mathcal{S}$ is feasible for Π, the neighborhood $\mathcal{N}(s)$ of s is the set of all solutions $s' \in \mathcal{S}$ that differ from s in exactly one route. Obviously, considering this neighborhood, there are $|U|$ positions where a new path could be inserted; moreover, the number of feasible paths between any source-destination pair is exponential.

Thus, in our algorithm, instead of exhaustively searching the entire neighborhood for each point, we probe only $|U|$ neighbors at each iteration (one for each source-destination pair). Also, because of the exponential size of the neighborhood, we limit the maximum number of iterations performed to a constant MaxIter.

We use randomization to select a new route, given a source destination pair. This is done in our proposed implementation using a modified version of the depth-first-search algorithm[11]. A *randomized depth-first-search* is identical to a depth-first-search algorithm, but at each step the node selected to explore is uniformly chosen among the available children of the current node. Using the randomized depth-first-search we are able to find a route that may improve the solution, while avoiding being trapped at a local optimum after only a few iterations.

A description of the local search procedure in form of pseudo-code is given in Figure 2. The algorithm used can be described as follows. Initially, the algorithm receives as input the basic feasible solution generated on phase 1 (the construction phase). A neighborhood for this solution is then defined to be the set of feasible solutions that differ from the current solution by one route, as previously described.

Given the basic feasible solution obtained from the construction subroutine, the neighborhood is explored in the following manner. For each agent $u_i \in U$, we reroute the agent on an alternate feasible path from s_i to d_i (lines 3 to 13). Recall that a path $\mathcal{P}_i$ is feasible if the total length of this path is less than $\mathcal{L}_i$ and the agent reaches its target node by time T. This alternate path is created on line 5 using a modified depth-first-search algorithm[2]. The mod-

```
procedure HillClimb(solution)
1        Compute cost c of solution;
2        while solution not locally optimal and iter < MaxIter do→
3            for all agent pairs (sᵢ, dᵢ) do→
4                Remove 𝒫ᵢ from solution;
5                Find alternate feasible path 𝒫ᵢ′
                     using the randomized DFS algorithm;
6                compute cost c' of new soluton
7                if new solution is feasible and c' > c then
8                    c ← c';
9                        iter ← 0;
11               else
12                   Restore path 𝒫ᵢ;
10               fi
11           rof
12           iter ← iter + 1;
13       elihw
end HillClimb
```

Fig. 2. Pseudocode for the Hill Climbing intensification procedure.

ification to the DFS is a randomization which selects the child node uniformly during each iteration. This procedure is efficient in that it can be implemented in polynomial time, as shown bellow.

Theorem 2. *The time complexity of the algorithm above is* $\mathcal{O}(kTu^2m)$, *where* $k = \texttt{MaxIter}$, T *is the time horizon,* $u = |U|$ *and* $m = |E|$.

Proof: Notice the the most time consuming step of the algorithm is the construction of a new path (line 5). However, using a randomized depth-first-search procedure this can be done in $\mathcal{O}(m)$ time[2]. Each iteration of the while loop (lines 2 to 13) will perform local improvements in the solution using the re-routing procedure to improve the objective function. An upper bound on the best solution for an instance of this problem is $Tu(u-1)/2$ (the time horizon multiplied by maximum number of connections). Each improvement can require at most MaxIter iterations to be achieved. Therefore, in the worst case this heuristic will end after $\mathcal{O}(kTu^2m)$ time.   ∎

6 Combining Algorithms into a One-Pass Heuristic

The two algorithms described in Sections 4 and 5 can be combined into a single one-pass heuristic for the CCPM. The pseudo-code for the complete algorithm used can be seen in Figure 3. The new algorithm now behaves as a single-start, diversification and intensification heuristic for the CCPM.

The total time complexity of this heuristic can be determined from Theorems 1 and 2. Taking the maximum of the two time complexities determined previously, we have a total time of $O(\max\{n^3, kTu^2m\})$, where T is the time horizon, $u = |U|$, $n = |V|$, $m = |E|$, and $k = $ MaxIter is the maximum number of iterations allowed on the local search phase.

---

procedure OnePass(Instance)
1   **Input:** Instance of the CCPM, with n nodes, m edges, and a set U of agents;
2   solution $\leftarrow$ ConstructionHeuristic(Instance);
3   solution $\leftarrow$ HillClimbHeuristic(solution);
4   **return solution** ;
end OnePass

Fig. 3. Pseudocode for the One-Pass Heuristic

7 Computational Results

The algorithm proposed above was tested to verify the quality of the solutions produced, as well as the efficiency of the resulting method. The test instances employed in the experiments were composed of 60 random unit graphs, distributed into groups of 20, each group having graphs with 50, 75, and 100 nodes. The communication radius of the wireless agents was allowed to vary from 20 to 50 units. This has provided us with a greater base for comparison, resulting in random graphs and wireless units that more closely resemble real-world instances.

The graphs used in the experiment were created with the algorithm proposed by Butenko et. al.[10, 9] in the context of the BROADCAST SCHEDULING problem. The routines were coded in FORTRAN. Random numbers were generated using Schrage's algorithm[21]. In all experiments, the random number generator was started with the seed value 270001.

Results obtained in our preliminary experiments are reported in Table 1. In this table, the results of the one-pass algorithm (OnePass column) are compared to a simple routing scheme where only the construction phase is explored (the SP Soln column). The solutions shown in the table represent the average of the objective function values from the 5 instances in each class.

The numerical results provided in the table demonstrate the effectiveness of the proposed heuristic when the improvement phase is added to the procedure. The proposed heuristic increased the objective value of the shortest path solutions by an average of 38%. One reason for this is the fact that, when agents are routed solely according to a shortest path, they are not taking advantage of the remaining time they are allotted (i.e., the time horizon T) and the values from the distance limit given by $\mathcal{L}$.

Our heuristic, on the other hand, allows wireless agents to take full advantage of these bounds. The algorithm can do this by adjusting the paths

Instance	Nodes	Radius	Agents	OnePass	SP Soln	Agents	OnePass	SP Soln	Agents	OnePass	SP Soln
1	50	20	10	63.6	52.4	15	152	120.8	25	414.66	353.6
2	50	30	10	83.8	58.4	15	182.2	124.4	25	516.2	415.6
3	50	40	10	95.4	67.4	15	228.6	171.8	25	695	474.8
4	50	50	10	115.4	64.4	15	275.8	167.4	25	797.4	485.4
5	75	20	10	76.8	59	20	270.2	228.6	30	575.2	464
6	75	30	10	85.8	56	20	299.6	241.2	30	725.4	554
7	75	40	10	96.4	64.4	20	386	261	30	862.6	595.4
8	75	50	10	125	67.8	20	403.2	246.8	30	1082.4	670.8
9	100	20	15	113.6	100.4	25	333.4	269.4	50	1523.2	1258.8
10	100	30	15	166.2	124.4	25	511.2	365	50	1901.4	1515.8
11	100	40	15	203.4	141	25	600.6	389.8	50	2539.2	1749.4
12	100	50	15	255.8	151.8	25	756.8	479.6	50	3271.2	2050.6

Table 1. Comparative results between shortest path solutions and heuristic solutions.

to include those nodes within the range of other agents. In addition, at any given time an agent is allowed to loiter in its current position, possibly waiting for other agents to come into its range. This cannot occur in the phase 1 algorithm because, according to the shortest path routing protocol, loitering is forbidden.

We notice that our method provides solutions that are better than the shortest path protocol. The time spent on the algorithm has always been less than a few seconds, therefore the computational time is small enough for the problem sizes explored in our experiments. We believe, however, that the quality of the solutions and computational time can be further improved using a better implementation, and more sophisticated data structures to handle the information stored during the algorithm.

8 Conclusions and Future Research

In this chapter we presented a heuristic approach to solve the *cooperative communication problem on ad hoc networks*. This problem, known to be NP-hard, is of importance in the planning of operations involving high levels of collaboration among team members. The proposed algorithm creates a high quality solution for the problem using two phases: 1) a construction heuristic, which uses shortest path algorithms to create a feasible solution, and 2) a local search algorithm, which improves the solution previously found in order to guarantee local optimality.

This chapter reflects the current stage of our research in this problem. We plan to extend the algorithmic methods presented in this chapter using more efficient optimization strategies. We will use the two phase algorithm described as the starting point for a greedy randomized procedure (GRASP metaheuristic). This will allow us to escape local minima inherent to the approach used in this chapter, and find solutions closer to the desired global optimum.

Acknowledgments

The first and third authors have been partially supported by NSF and U.S. Air Force grants. The second author has been partially supported by a grant from the Measurement & Control Engineering Center/NSF.

References

1. M.R. Garey, D.S. Johnson. *Computers and Intractability: A Guide to the Theory of NP-Completeness.* W.H. Freeman and Company, 1979.
2. R.K. Ahuja, T.L. Magnanti, J.B. Orlin. *Network Flows: Theory, Algorithms, and Applications.* Prentice-Hall, 1993.
3. P. Biswas and Y Ye. A distributed method for solving semidefinite programs arising from ad hoc wireless sensor network localization. Technical report, Dept of Management Science and Engineering, Stanford University, 2003.
4. P. Biswas and Y. Ye. Semidefinite programming for ad hoc wireless sensor network localization. In *Proceedings of the third international symposium on Information processing in sensor networks*, pages 46–54. ACM Press, 2004.
5. S. Butenko, X. Cheng, D.-Z. Du, and P. M. Pardalos. On the construction of virtual backbone for ad hoc wireless network. In S. Butenko, R. Murphey, and P. M. Pardalos, editors, *Cooperative Control: Models, Applications and Algorithms*, pages 43–54. Kluwer Academic Publishers, 2002.
6. S.I. Butenko, X. Cheng, C.A.S. Oliveira, and P.M. Pardalos. A new algorithm for connected dominating sets in ad hoc networks. In S. Butenko, R. Murphey, and P. Pardalos, editors, *Recent Developments in Cooperative Control and Optimization*, pages 61–73. Kluwer Academic Publishers, 2003.
7. S. Capkun, M. Hamdi, and J. Hubaux. Gps-free positioning in mobile ad-hoc networks. In *HICSS '01: Proceedings of the 34th Annual Hawaii International Conference on System Sciences (HICSS-34)-Volume 9*, page 9008, Washington, DC, USA, 2001. IEEE Computer Society.
8. X. Cheng, X. Huang, D. Li, W. Wu, and D.Z. Du. A polynomial-time approximation scheme for the minimum-connected dominating set in ad hoc wireless networks. *Networks*, 42(4):202–208, 2003.
9. C.W. Commander, S.I. Butenko, and P.M. Pardalos. On the performance of heuristics for broadcast scheduling. In D. Grundel, R. Murphey, and P. Pardalos, editors, *Theory and Algorithms for Cooperative Systems*, pages 63–80. World Scientific, 2004.
10. C.W. Commander, S.I. Butenko, P.M. Pardalos, and C.A.S. Oliveira. Reactive grasp with path relinking for the broadcast scheduling problem. In *Proceedings of the 40th Annual International Telemetry Conference*, pages 792–800, 2004.

11. T. H. Cormen, C. E. Leiserson, and R. L. Rivest. *Introduction to Algorithms*. MIT Press, Cambridge, MA, 1992.
12. E. W. Dijkstra. A note on two problems in connexion with graphs. *Numer. Math.*, 1:269–271, 1959.
13. L. Doherty, K. S. J. Pister, and Ghaoui L. E. Convex position estimation in wireless sensor networks. In *Proc. IEEE INFOCOM*, Anchorage, AK, 2001.
14. R.W. Floyd. Algorithm 97 (shortest path). *Communications of the ACM*, 5(6):345, 1962.
15. M.V. Marathe, H. Breu, H.B. Hunt III, S.S. Ravi, and D.J. Rosenkrantz. Simple heuristics for unit disk graphs. *Networks*, 25:59–68, 1995.
16. David Moore, John Leonard, Daniela Rus, and Seth Teller. Robust distributed network localization with noisy range measurements. In *SenSys '04: Proceedings of the 2nd international conference on Embedded networked sensor systems*, pages 50–61, New York, NY, USA, 2004. ACM Press.
17. C.A.S. Oliveira and P.M. Pardalos. An optimization approach for cooperative communication in ad hoc networks. Technical report, School of Industrial Engineering and Management, Oklahoma State University, 2005.
18. C.A.S. Oliveira, P.M. Pardalos, and M.G.C. Resende. Optimization problems in multicast tree construction. In *Handbook of Optimization in Telecommunications*, pages 701–733. Springer, New York, 2006.
19. N.B. Priyantha, H. Balakrishnan, E.D. Demaine, and S. Teller. Mobile-assisted localization in wireless sensor networks. In *INFOCOM 2005. 24th Annual Joint Conference of the IEEE Computer and Communications Societies*, volume 1, pages 172–183, 2005.
20. C. Savarese, J. Rabay, and K. Langendoen. Robust positioning algorithms for distributed ad-hoc wireless sensor networks. In *USENIX Technical Annual Conference*, 2002.
21. L. Schrage. A more portable FORTRAN random number generator. *ACM Transactions on Mathematical Software*, 5:132–138, 1979.
22. S. Warshall. A theorem on boolean matrices. *Journal of the ACM*, 9(1):11–12, 1962.

Mathematical Modeling and Optimization of Superconducting Sensors with Magnetic Levitation *

Vitaliy A. Yatsenko[1] and Panos M. Pardalos[2]

[1] Space Research Institute of NASU and NSAU
   Kiev, Ukraine
   E-mail: vitaliy_yatsenko@yahoo.com
[2] Center for Applied Optimization,
   Department of Industrial and Systems Engineering
   University of Florida
   Gainesville, FL 32601, USA
   E-mail: pardalos@ufl.edu

Summary. Nonlinear properties of a magnetic levitation system and an algorithm of a probe stability are studied. The phenomenon, in which a macroscopic superconducting ring chaotically and magnetically levitates, is considered. A nonlinear control scheme of a dynamic type is proposed for the control of a magnetic levitation system. The proposed controller guarantees the asymptotic regulation of the system states to their desired values. We found that if a non-linear feedback is used then the probe chaotically moves near an equilibrium state. An optimization approach for selection of optimum parameters is discussed.

1 Introduction

The suspension of objects with no visible means of support is a fascinating phenomenon [1],[2], [14]. To deprive objects of the effects of gravity is a dream common to generations of thinkers from Benjamin Franklin to Robert Goddard, and even to mystics of the East. This modern fascination with superconducting levitation stems from four singular technical and scientific achievements:(i) the creation of superconducting gravity meters; (ii) the creation of high-speed vehicles to carry people at 500 km/hr; (iii) the creation of digitally controlled magnetic levitation turbo molecular pump and (iv) the discovery of new superconducting materials.

The modern development of super high-speed transport systems, known as maglev, started in the late 1960s as a natural consequence of the development

---

* This work is partially supported by Airforce, CRDF and STCU grants

of low-temperature superconducting wire, the transistor and chip-based electronic control technology [6],[8]. Maglev provides high-speed running, safety, reliability, low environmental impact and minimum maintenance. In the 1980s, maglev matured to the point, when Japanese and German technologists were ready to market these new high-speed levitated machines.

At the same time, Paul C.W. Chu of the University of Houston and co-workers in 1987 discovered a new, higher-temperature superconductivity (HTS) in the non-inter-metallic compounds (nyttrium-barium-copper oxide). Those premature promises of superconducting materials have been tempered by the practical difficulties of development. First, bulk YBCO (yttrium barium copper oxide) was found to have a low current density, and early samples were found to be too brittle to fabricate into useful wire [7],[8]. Scientists are interested in YBCO because when it is cooled below around 90 Kelvin, which can be accomplished with liquid nitrogen, it becomes a superconductor. The two most important properties of YBCO are that it has no electrical resistance and that it expels a magnetic field.

However, from the very beginning, the hallmark of these new superconductors was their ability to levitate small magnets. This property, captured on the covers of both scientific and popular magazines, inspired a group of engineers and applied scientists to envision a new set of levitation applications based on superconducting magnetic bearings.

In the past few years, the original technical obstacles of YBCO have gradually been overcome, and new superconducting materials such as bismuth-strontium-calcium-copper oxide (BSCCO) have been discovered. Higher current densities for practical applications have been achieved, and longer wire lengths have been produced with good superconducting properties. At this juncture of superconducting technology, we can now envisage, in the coming decade, the levitation of large machine components as well as the enhancement of existing maglev transportation systems with new high-temperature superconducting magnets.

A levitation phenomenon is created by opposing magnetic fluxes. Commonly it refers to levitated high-speed trains equipped with superconducting magnets, proposed by James R. Powell and Gordon T. Danby of Brookhaven National Laboratory in the late 1960s. Pursued since 1970 by the Japan Railway Technical Research Institute, which is presently building a second maglev test track 40-km long. In the 1980s demonstration maglevs were built in Germany. We can imagine the relative velocity of 100–200 m/sec between moving bodies with no contact, no wear, no need for fluid or gas intervention, and no need for active controls.

The superconductivity phenomenon was a significant step to improve suspensions. Most, but not all, conductors of electrical current, when cooled sufficiently in the direction of absolute zero, become superconductors. The superconducting state itself is one in which there is zero electrical resistance and perfect diamagnetism. Free suspension of a probe of a superconducting gravimeter is realized by the Braunbeck-Meisner phenomenon. Here we concentrate on a new high sensitive cryogenic-optical sensor and a method of estimation of the gravitational perturbation acting on the levitated probe [13], [14].

In this chapter we describe basic properties of a magnetic levitation, theoretical background, and control algorithms of a probe stability. Bilinear control schemes of the static and dynamic types are proposed for the control of a magnetic levitation system. The proposed controllers guarantee the asymptotic regulation of the system states to their desired values. We also describe a simple superconducting gravity meter, its mathematical model, and design of nonlinear controllers that stabilize it at an equilibrium state. Furthermore, an accurate mathematical model of asymptotically stable estimation of a weak noisy signal using the stochastic measurement model is proposed.

2 Stability and Levitation

Levitation can be achieved using electric or magnetic forces or by using air pressure, though some purists would argue whether flying or hovering is levitation. However, the analogy of magnetic levitation with the suspension of aircraft provides insight into the essential requirements for levitation; that is, *lift alone is not levitation*. The success of the Wright machine in 1903 was based, in part, on the invention of a mechanism on the wings to achieve *stable* "levitated" flight. The same can be said about magnetic bearing design – namely, that an understanding of the nature of mechanical stability is crucial for the creation of a successful levitation device.

Simple notions of stability often use the paradigm of the ball in a potential well or on the top of a potential hill. This idea uses the concept of potential energy, which states that physical systems are stable when they are at their lowest energy level.

The minimum potential energy definition of stability is good to begin with, but is not enough in order to understand magnetic levitation. Not only one must consider the stability of the center of mass of the body, but it is also necessary to achieve the stability of the orientation or an angular position of the body. If the levitated body is deformable, the stability of the deformed shape may also be important.

The second difficulty with the analogy with particles in gravitational potential wells is that we have to define what we mean by the magnetic or electric potential energy [6]. This is straightforward if the sources of the levitating magnetic or electric forces are fixed. But when magnetization or electric currents are induced due to changes in the position or orientation of our levitated body, then the static concept of stability using potential energy can involve pitfalls that can yield the wrong conclusion regarding the stability of the system.

To be rigorous really in magneto-mechanics, one must discuss stability in the context of dynamics. For example, in some systems one can have static instability but dynamic stability. This is especially true in the case of time-varying electric or magnetic fields as in the case of actively controlled magnetic bearings. However, it is also important when the forces (mechanical or magnetic) depend on generalized velocities.

In general, the use of concepts of dynamic stability in the presence of modeling error due to uncertainties, rooted in modern nonlinear dynamics, must be employed in order to obtain a robust position control of a magnetic levitation system. This theory not only requires the knowledge of how magnetic forces and torques change with the position and orientation (i.e., magnetic stiffness), but also the knowledge of how these forces change with both linear and angular velocities.

Earnshaw's Theorem. It is said that a collection of point charges cannot be maintained in an equilibrium configuration solely by the electrostatic interaction of the charges. Early in the nineteenth century (1839) a British minister and natural philosopher, Samuel Earnshaw (1805–1888), examined this question and stated a fundamental proposition known as Earnshaw's theorem. The essence of this theorem is that a group of particles governed by inverse square law forces cannot be in stable equilibrium. The theorem naturally applies to charged particles and magnetic poles and dipoles. A modern statement of this theorem can be found in Jeans [[3], [5]: *"A charged particle in the field of a fixed set of charges cannot rest in stable equilibrium."* This theorem can be extended to a set of magnets and fixed circuits with constant current sources. To the chagrin of many a would-be inventor, and contrary to the judgment of many patent officers or lawyers, the theorem rules out many clever magnetic levitation schemes. This is especially the case of levitation with a set of permanent magnets as any reader can verify. Equilibrium is possible, but stability is not.

Later we will address the question of how and why one can achieve stable levitation of a superconducting ring using an active feedback. However, here we will try to motivate why superconducting systems appear to violate or escape the consequences of Earnshaw's theorem. One of the first to show how diamagnetic or superconducting materials could support stable levitation was Braunbeck (Braunbeck, 1939).

Earnshaw's theorem is based on the mathematics of inverse square force laws. Particles which experience such forces must obey a partial differential equation known as Laplace's equation. The solutions of this equation do not admit local minima or maxima, but only saddle-type equilibria. However, there are circumstances under which electric and magnetic systems can avoid the consequences of Earnshaw's theorem:

- time-varying fields (e.g., eddy currents, alternating gradient);
- active feedback;
- diamagnetic systems;
- ferrofluids;
- superconductors.

The theorem is easily proved if the electric and magnetic sources are fixed in space and time, and one seeks to establish the stability of a single free-moving magnet or charged particle. However, in the presence of polarizable, magnetizable, or superconducting materials, the motion of the test body will induce changes in the electric and magnetic sources in the nearby bodies. In general magnetic flux attractors such as ferromagnetic materials still obey Earnshaw's theorem, whereas for flux repellers such as diamagnetic or Type I superconductors, stability can sometimes be obtained. Superconductors, however, have several modes of stable levitation:

- Type I or Meissner repulsive levitation based on complete flux exclusion;
- Type II repulsive levitation based on both partial flux exclusion and flux pinning;
- Type III suspension levitation based on flux pinning forces;
- Type IV suspension levitation based on magnetic potential well.

In the case of Meissner repulsive levitation superconducting currents in the bowl-shaped object move in response to changes in the levitated magnet. The concave shape is required to achieve an energy potential well.

In the case of Type II levitation, both repulsive and suspension (or attractive) stable levitation forces are possible without shaping the superconductor. Magnetic flux exclusion produces equivalent magnetic pressures which result in repulsive levitation whereas flux attraction creates magnetic tensions (similar to ferromagnetic materials) which can support suspension levitation. Flux penetration into superconductors is different from ferromagnetic materials, however.

In Type III superconductors, vortex-like supercurrent structures in the material create paths for the flux lines. When the external sources of these flux lines move, however, these supercurrent vortices resist motion or are pinned in the superconducting material. This so-called *flux-pinning* is believed to be the source of stable levitation in these materials (Brandt, 1989, 1990).

Type IV suspension levitation described in details in the next section.

Finally, from a fundamental point of view, it is not completely understood why supercurrent-based magnetic forces can produce stable attractive levitation while spin-based magnetic forces in ferromagnetic materials produce unstable attractive or suspension levitation. Given the restricted assumptions upon which Earnshaw's theorem is based, the possibility that some new magnetic material will be discovered, which supports stable levitation, cannot be entirely ruled out.

3 Dynamics of Magnetically Levitated Systems

This section considers mathematical models of a sensor based on the principle of magnetic levitation. The sensor consists of two superconducting current rings and a levitated probe placed between them (see subsection 5.1).

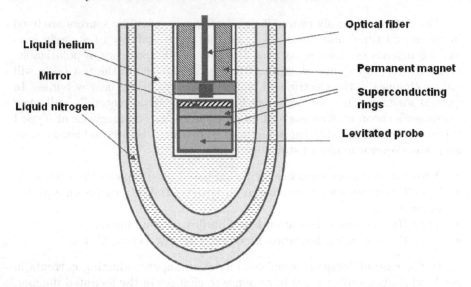

Fig. 1. Construction of the sensitive element.

The stability is provided by a set of superconducting short-circuited loops placed around the floating ring. A novel method using short-circuited superconducting loops as stabilizers has been proposed. We showed that for a given magnetic configuration there exists a minimum current in the levitated ring below which the system is unstable.

The newly developed superconducting gravimeter (Fig. 1) represents a spring type device [13], [14]. An analogue of the mechanical spring of our device accomplishes the magnetic returning force acting on a superconducting probe in a non-uniform magnetic field of superconducting rings or a permanent magnet (in another variant). Due to the high stability of superconducting currents of rings a highly stable non-dissipative spring is created.

As shown by [10], a set of variables uniquely defining an energy state can be determined for any electromechanical system possessing power function and storing energy in the form of magnetic field energy. In this case such variables will be mechanical displacements (mechanical degrees of freedom) q_j, $j = 1, 2, \ldots, l$ (l is the number of degrees of freedom), as well as total magnetic fluxes Ψ_m and currents I_m, $m = 1, 2, \ldots, n$ (n is the number of superconducting rings).

There are inner couplings between magnetic variables

$$\Psi_m = \sum_{i=1}^{n} L_{im} I_i, \quad m = 1, 2, \ldots, n, \tag{1}$$

where L_{im} are mutual inductances, L_{ii} are internal conductances. In the case of superconducting current rings, magnetic-flux linkages Ψ_m retain constant value independently of variations of a ring position. This circumstance allows us to consider the relations (1) as a system of n equations for currents I_i, $i = 1, 2, \ldots, n$, where Ψ_m ($m = 1, 2, \ldots, n$) are constants.

We will assume that the determinant Δ of (1) is not equal to zero. Then it can be solved for currents

$$I_m = \frac{\Delta_m}{\Delta}, \quad m = 1, 2, \ldots, n, \tag{2}$$

where Δ_m is the determinant of the current I_m.

If we place the solution of (2) into the following formula for the energy of magnetic field of the current loop system

$$W = \frac{1}{2} \sum_{i,m=1}^{n} L_{im} I_i I_m, \tag{3}$$

then the energy will be expressed in terms of magnetic-flux linkages Ψ_i, Ψ_m and the inductance L_{im}

$$W = \frac{1}{2} \Delta^{-2} \sum_{i,m=1}^{n} L_{im} \Delta_i(\Psi_i, \Psi_m, L_{im}) \Delta_m(\Psi_i, \Psi_m, L_{im}), \tag{4}$$

$$\Delta = \Delta(\Psi_i, \Psi_m, L_{im}), \quad i, m = 1, 2, \ldots, n. \tag{5}$$

It follows from (4) that energy W depends only on mechanical coordinates q_j, which are incorporated in mutual inductances L_{im} ($i \neq m$). Because of this, in "pure mechanical" terms it is either power function or potential energy. The formula for magnetic force (White, 1959) prompts which precisely mechanical function will be the energy of magnetic field. This formula appears as

$$\Delta = \Delta(\Psi_i, \Psi_m, L_{im}), \quad , i, m = 1, 2, \ldots, n,$$

i.e. the magnetic force is a partial derivative with respect to magnetic energy expressed in terms of magnetic-flux linkages and coordinates, taken with opposite sign.

But this is precisely the definition of force as a function of potential energy of any power field. Therefore, energy of a magnetic field in form (4), where $\Psi_i, \Psi_m = \text{const}$, is the potential energy of magnetic interaction of n ideal currents, i.e.

$$W = U_m = U_m(L_{im}(q_j)), \quad i, m = 1, 2, \ldots, n. \tag{6}$$

If the system is located in an external power field, e.g. in gravitational one, gravitational energy U_G should be added to magnetic potential energy. Then the total energy of the system is

$$U = U_m + U_G. \tag{7}$$

For the system of circuits inner linkages between magnetic variables take the form

$$\Psi_1 = LI_1 + L_{12}I_2 + L_{13}I_3 + L_{14}I_4,$$
$$\Psi_2 = L_{12}I_1 + LI_2 + L_{23}I_3 + L_{24}I_4,$$
$$\Psi_3 = L_{13}I_1 + L_{23}I_2 + LI_3 + L_{24}I_4,$$
$$\Psi_4 = L_{14}I_1 + L_{24}I_2 + L_{34}I_3 + LI_4, \tag{8}$$

and energy of magnetic field (3) with respect to (8) can be written as

$$W = U_m = \frac{1}{2}(LI_1^2 + L_{12}I_1I_2 + L_{13}I_1I_3 + L_{14}I_1I_4 + L_{12}I_1I_2 +$$
$$+ LI_2^2 + L_{23}I_2I_3 + L_{24}I_2I_4 + L_{13}I_1I_3 + L_{23}I_2I_3L_{14}I_1I_4 +$$
$$+ LI_3^2 + L_{34}I_3I_4 + L_{24}I_2I_4 + L_{34}I_3I_4 + LI_4^2) =$$
$$= \frac{1}{2}(\Psi_1I_1 + \Psi_2I_2 + \Psi_2I_3 + \Psi_1I_4). \tag{9}$$

All the coils of the sensor are modeled by thin short-circuited ring-shaped loops of similar radius, therefore internal inductances of the loops are $L_1 = L_{22} = L_{33} = L_{44} = L$.

By solving the system of equations (8), we find expressions for currents and substitute them into (9) thus defining dependence of magnetic potential energy on mechanical coordinates:

$$U_m = \Psi_1^2(2L)^{-1}\{2(1 - y_{14})(1 - y_{23}^2) - (y_{13} - y_{24})^2 - (y_{13} - y_{34})^2 +$$
$$+ 2y_{23}(y_{12} - y_{24})(y_{13} - y_{34}) +$$
$$+ 2p[(y_{12}y_{34} - y_{13}y_{24})(y_{12} - y_{13} - y_{24} + y_{34}) -$$
$$- (1 - y_{14})(1 - y_{23})(y_{12} + y_{13} + y_{24} + y_{34})] + p^2[2(1 - y_{14}^2)(1 - y_{23}) -$$

$$- (y_{12} - y_{13})^2 - (y_{24} - y_{34})^2 + 2y_{14}(y_{12} - y_{13})(y_{24} - y_{34})]\} \times$$
$$\times [(1 - y_{14}^2)(1 - y_{23}^2) - y_{12}^2 - y_{13}^2 - y_{24}^2 - y_{34}^2 +$$
$$+ (y_{12}y_{34} - y_{13}y_{24})^2 + 2y_{14}(y_{12}y_{24} - y_{13}y_{34}) +$$
$$+ 2y_{23}(y_{12}y_{13} - y_{24}y_{34}) - 2y_{14}y_{23}(y_{12}y_{34} - y_{13}y_{24})]^{-1} =$$
$$= \Psi_1^2(2L)^{-1}(M + Np + Qp^2)D^{-1}, \tag{10}$$

where $y_{im} = L_{im}L^{-1}$; $p = \Psi_2\Psi_1^{-1}$, and relative mutual inductances $y_{im} = y_{im}(q_1, \ldots, q_0)$ are functions of coordinates.

In order to define the explicit relation $y_{im}(q)$ we will introduce the inertial coordinate system $O\xi\eta\zeta$, whose $O\eta$ axis coincides with the axis of stationary loops 1, 4 of the sensor; i_1, i_2, i_3 are basis vectors of the system $O\xi\eta\zeta$. We place in the center of mass of the sensor the origin of the coordinate system associated with it, with basis vectors i_{11}, i_{12}, i_{13} and with its O_1, ζ_1 axis coinciding with the axis of the loops 2, 3. We will describe position of the center of mass of the probe by cylindrical coordinates ρ_2, α, ζ and orientation of trihedron $O_1\xi_1\eta_1\zeta_1$ with respect to system $O\xi\eta\zeta$ will be described by Euler angles (v is a nutation angle, ψ is a precession angle, φ is a proper rotation angle).

As it is seen from formula (10), the potential energy depends on all six mutual inductances y_{im}, but y_{14}, $y_{23} = \text{const}$. Therefore, only four inductances y_{12}, y_{13}, y_{24}, y_{34} are to be determined. All of them are calculated in a similar way.

Let us define the following notation: $\mathbf{R}_i$ ($i = 1, 2$) are radius-vectors of centers of rings mass in the system $O\xi\eta\zeta$; dl_1, dl_2 are elements of arcs of rings 1, 2; $\mathbf{R}_{12}$ is a radius-vector connecting the center of the i-th ring with the respective element dl_i; $\mathbf{e}$ is a radius-vector of the center of the ring 2 in the system $O_1\xi_1\eta_1\zeta_1$. Then mutual inductance can be calculated by Neumann formula

$$y_{12} = \frac{L_{12}}{L} = \frac{1}{20\pi} \oint \oint \frac{dl_1 dl_2}{|\mathbf{R}_{12}|}, \tag{11}$$

$$\mathbf{R}_{12} = \mathbf{R}_2 + \mathbf{e} + \mathbf{a}_2 - \mathbf{a}_1 - \mathbf{R}_1, \tag{12}$$

where

$$\mathbf{R}_i = \rho_i \cos\alpha i_1 + \rho_i \sin\alpha i_2 + \zeta_i i_3,$$
$$\mathbf{a}_i = a(\cos\lambda_i i_{11} + \sin\lambda_i i_{12}), \quad i = 1, 2,$$
$$\mathbf{e} = e i_{13}, \quad dl_i = a(\sin\lambda_i i_{i1} - \cos\lambda_i i_{i2})d\lambda_i.$$

Since the ring 1 is fixed and coordinate system $O\xi\eta\zeta$ is selected such that its axis $O\eta$ coincides with the axis of ring 1, all coordinates describing position of the ring 1 have the following values:

$$\rho'\rho_i, \quad \alpha_1 = \frac{\pi}{2}, \quad \zeta_1 = 0, \quad \vartheta_1 = \frac{\pi}{2}, \quad \Psi_1 = 0, \quad \phi_1 = 0. \tag{13}$$

Then for fixed ring 1

$$\mathbf{R}_1 = \rho_1 i_2,$$
$$\mathbf{a}_1 = a(\cos\lambda_1 i_1 + \sin\lambda_1 i_3),$$
$$dl_1 = a(\sin\lambda_1 i_1 - \cos\lambda_1 i_3)d\lambda_1,$$

and for the sensor:

$$\mathbf{R}_2 = \rho_2 \cos \alpha i_1 + \rho_2 \sin \alpha i_2 + \zeta i_3,$$

$$\mathbf{a}_z = \{[\cos(\lambda_2 + \lambda) \cos \Psi - \sin(\lambda_2 + \phi) \sin \Psi -$$
$$- \sin(\lambda_2 + \phi) \sin \Psi \cos \vartheta]i_2 + \sin(\lambda_2 + \phi) \sin \vartheta i_3\},$$

$$\mathbf{e} = e \sin \phi \sin \vartheta i_1 - e \cos \phi \sin \vartheta i_2 + e \cos \vartheta i_3,$$

$$dl_2 = a\{[\sin(\lambda_2 + \phi) \cos \Psi + \cos(\lambda_2 + \phi) \sin \Psi \cos \theta]i_1 -$$
$$- [\sin(\lambda_2 + \phi) \sin \Psi - \cos(\lambda_2 + \phi) \sin \Psi \cos \vartheta]i_2 -$$
$$- \cos(\lambda_2 + \phi) \sin \vartheta i_3\}.$$

By performing elementary transformations, we obtain

$$y_{12} = \frac{1}{40\pi} \int_0^{2\pi} d\lambda_1 \int_0^{2\pi} [\sin x_4 \cos \lambda_1 \cos(\lambda_2 + x_6) +$$
$$+ \sin \lambda_1 \sin(\lambda_2 + x_6) + \cos x_4 \sin x_5 \, sin\lambda_1 \cos(\lambda_2 + x_6)] \times$$
$$\times \left\{ \frac{1}{2} + e^2 + \rho_1^2 + x_1^2 + x_3^2 + \frac{1}{2}[\cos x_4 \sin x_5 - \right.$$
$$- \cos \lambda_1 \sin(\lambda_2 + \varphi_6) - \cos x_5 \cos \lambda_1 \cos(\lambda_2 + x_6)] -$$
$$- 2\rho_1 x_1 \sin x_2 + x_1[\cos(x_2 - x_5) \cos(\lambda_2 + x_6) +$$
$$+ \sin(x_2 - x_5) \cos x_4 \sin(\lambda_2 + x_6) - \cos x_2 \cos \lambda_1] -$$
$$- \rho_1[\sin x_5 \cos(\lambda_2 + x_6) + \cos x_4 \cos x_5 \sin(\lambda_2 + x_6)] -$$
$$- e(\sin x_4 \sin x_5 \cos \lambda_1 + \cos x_4 \sin \lambda_1) -$$
$$- 2e x_1 \sin(x_2 - x_5) \sin x_4 + 2e\rho_1 \sin x_4 \cos x_5 -$$
$$\left. - x_3[\sin \lambda_1 - \sin x_4(\lambda_2 + x_6)] + 2e x_3 \cos x_4 \right\}^{-1/2} d\lambda_2,$$

where dimensionless variables are introduced

$$x_1 = \frac{\rho_2}{2a}; \quad x_2 = \alpha; \quad x_3 = \frac{\zeta}{2a}; \quad x_1 = v; \quad x_5 = \psi; \quad x_6 = \varphi. \tag{14}$$

Thus, the collection of formulas (10) and (11) determines dependence of magnetic potential energy on coordinates of the sensor, and total potential energy

$$U = U_m - mg\rho \tag{15}$$

provided that direction of a gravitational force coincides with the direction of $O\eta$ axis.

The integrals in the formula (11) are not taken in general form. Only in the case, where axes of fixed loops coincide, integral relationship can be reduced to linear combinations of complete elliptic integrals. In our case, where magnetic forces are large as compared with perturbing ones, the potential energy can be expanded into the following power series:

$$U = U_0 + \sum_{j=1}^{6} \left(\frac{PU}{Pq} \right)\bigg|_0 (q_j - q_{j0}) + \tag{16}$$

$$+ \frac{1}{2} \sum_{j,n=1}^{6} \frac{\partial^2 U}{\partial q_j \partial q_n}\bigg|_0 (q_j - q_{j0})(q_n - q_{n0}),$$

where derivatives are calculated at the point q_{j0}:

$$x_{10} = x_{10}; \quad x_{20} = \frac{\pi}{2}; \quad x_{30} = 0; \quad x_{40} = \frac{\pi}{2}; \quad x_{50} = x_{60} = 0. \tag{17}$$

After simple manipulations the final expression for potential energy can be rewritten as

$$U = \sum_{\substack{i,m=1 \\ i \neq m}} \left(\frac{\partial U_m}{\partial y_{im}} \frac{\partial y_{im}}{\partial x_1}\bigg|_0 - mg \right) (x_2 - x_{10}) +$$

$$+ \frac{1}{2} \sum_{j,n=1}^{5} \sum_{\substack{r,s=1 \\ r \neq s}}^{4} \sum_{\substack{i,m=1 \\ i \neq m}}^{4} \left(\frac{\partial^2 U_m}{\partial y_{im} \partial y_{rs}} \frac{\partial y_{im}}{\partial x_j} \frac{\partial y_{rs}}{\partial x_n} + \frac{\partial U_m}{\partial y_{im}} \frac{\partial^2 y_{im}}{\partial x_j \partial x_n} \right)\bigg|_0 \times$$

$$\times (x_j - x_{j0})(x_n - x_{n0}), \tag{18}$$

where

$$\frac{\partial U_m}{\partial y_{im}} = \frac{\Psi_1}{2L}[(M_{im} + N_{im}p + Q_{im}p^2)D - (M + Np + Qp^2)D_{im}]D^{-2};$$

$$\frac{\partial U_m^2}{\partial y_{im} \partial y_{rs}} = \frac{\Psi_1}{2L}[(M_{im,rs}D^2 - M_{im}DD_{rs} - M_{rs}DD_{im} - MDD_{im,rs} +$$

$$+ 2MD_{im}D_{rs}) + p(N_{im}DD_{rs} - N_{rs}DD_{im}NDD_{im,rs} + 2ND_{im}D_{rs}) +$$

$$+ p^2(Q_{im,rs}D^2 - Q_{im}DD_{rs} - Q_{rs}DD_{im} - QDD_{im,rs} + 2QD_{im}D_{rs})]D^{-3}$$

(expressions for M, N, Q, D are clear from formula (10), symbol M_{im} denotes derivative of M with respect to y_{im});

$$M_{12} = -M_{24} = [y_{12} - y_{24} - y_{23}(y_{13} - y_{34})];$$
$$M_{13} = -M_{34} = [y_{13} - y_{34} - y_{23}(y_{12} - y_{24})];$$
$$N_{12} = 2[y_{34}(y_{12} - y_{13} - y_{24} + y_{34}) + (y_{12}y_{34} - y_{13}y_{24}) -$$
$$- (1 - y_{14})(1 - y_{23})];$$
$$N_{13} = -2[y_{24}(y_{12} - y_{13} - y_{24} + y_{34}) + (y_{12}y_{34} - y_{13}y_{24} +$$
$$+ (1 - y_{14})(1 - y_{23})];$$
$$N_{24} = -2[y_{13}(y_{12} - y_{13} - y_{24} + y_{34}) + (y_{12}y_{34} - y_{13}y_{24}) +$$
$$+ (1 - y_{14})(1 - y_{23})];$$
$$N_{34} = 2[y_{12}(y_{12} - y_{13} - y_{24} + y_{34}) - (y_{12}y_{34} - y_{13}y_{24}) +$$
$$+ (1 - y_{14})(1 - y_{23})];$$
$$Q_{12} = -Q_{13} = [y_{12} - y_{13} - y_{14}(y_{24} - y_{34})];$$
$$Q_{24} = -Q_{34} = [y_{24} - y_{34} - y_{14}(y_{12} - y_{13})];$$
$$D_{12} = -2[y_{12} + y_{34}(y_{12}y_{34} - y_{14}y_{24}) - y_{14}y_{24} - y_{13}y_{23} + y_{14}y_{23}y_{34})];$$
$$D_{13} = -2[y_{13} + y_2(y_{12}y_{34} - y_{13}y_{24}) - y_{14}y_{34} - y_{12}y_{23} + y_{14}y_{23}y_{24})];$$
$$D_{24} = -2[y_{24} + y_{13}(y_{12}y_{34} - y_{13}y_{24}) - y_{12}y_{14} - y_{23}y_{34} + y_{13}y_{14}y_{23})];$$
$$D_{34} = -2[y_{34} + y_{12}(y_{12}y_{34} - y_{13}y_{24}) - y_{13}y_{14} - y_{23}y_{24} + y_{12}y_{14}y_{23})];$$
$$M_{12,12} = M_{13,13} = M_{34,34} = Q_{12,12} = Q_{13,13} = Q_{24,24} = Q_{34,34} = -2;$$
$$N_{12,12} = 4y_{34}; \quad N_{13,13} = 4y_{24}; \quad N_{24,24} = 4y_{13}; \quad N_{34,34} = 4y_{12};$$
$$D_{12,12} = -2(1 - y_{34}^2); \quad D_{13,13} = -2(1 - y_{24}^2);$$
$$D_{24,24} = -2(1 - y_{13}^2); \quad D_{34,34} = -2(1 - y_{12}^2);$$
$$M_{12,13} = 2y_{14}; \quad M_{12,24} = 2;$$
$$N_{12,13} = -2(y_{24} + y_{34}); \quad N_{12,24} = -(y_{13} + y_{34});$$
$$Q_{12,13} = 2; \quad Q_{12,24} = 2y_{23};$$
$$D_{12,13} = 2(y_{14} - y_{24}y_{34}); \quad D_{12,24} = 2(y_{23} - y_{13}y_{34});$$
$$M_{12,34} = -2y_{24}; \quad M_{13,24} = -2y_{14};$$
$$N_{12,34} = 2[2(y_{12} + y_{34}) - y_{13} - y_{24}];$$
$$N_{13,24} = 2[2(y_{13} + y_{24}) - y_{12}y_{34}];$$
$$Q_{12,34} = 2y_{23}; \quad Q_{13,24} = -2y_{23};$$

$$D_{12,34} = 2(2y_{12}y_{34} - y_{13}y_{24} - y_{14}y_{23});$$

$$D_{13,24} = 2(2(y_{13}y_{24} - y_{12}y_{34} - y_{14}y_{23});$$

$$M_{13,34} = 2; \quad M_{24,34} = 2y_{14};$$

$$N_{13,34} = -2(y_{12} + y_{24}); \quad N_{24,34} = -2(y_{12} + y_{13});$$

$$Q_{13,34} = 2y_{23}; \quad Q_{24,34} = 2;$$

$$D_{13,34} = 2(y_{23} - y_{12}y_{24}); \quad D_{24,34} = 2(y_{14} - y_{12}y_{13});$$

$$y_{14} = y_{23} = \text{const}; \quad y_{im} = \frac{1}{5k}[(1 + k'^2)\mathbf{K}(k) - 2\mathbf{E}(k)];$$

$$\frac{\partial y_{im}}{\partial x_1} = \frac{1}{5b}[2k'^2\mathbf{K}(k) - (2 - k^2)\mathbf{E}(k)];$$

$$\frac{\partial^2 y_{im}}{\partial x_1^2} = -2\frac{\partial^2 y_{im}}{\partial x_3^2} = \frac{k^3}{5k'^2}[1];$$

$$\frac{\partial^2 y_{im}}{\partial x_2^2} = -\frac{1}{10k'^2}(k\rho_1 + kd + b)\{b[2] + k^2(\rho_1 + d)[1]\};$$

$$\frac{\partial^2 y_{im}}{\partial x_4^2} = \frac{\partial^2 y_{im}}{\partial x_5^2} = \frac{k}{40k'k}\{k'^2[(2 - 3k^2 + 2k^4)\mathbf{E}(k) -$$

$$- k'^2(2 - k^2)\mathbf{K}(k)] - 4kd(d + b)[1]\};$$

$$\frac{\partial^2 y_{im}}{\partial x_3 \partial x_4} = \frac{1}{20k'^2}\{b[2] + 2k^3d[1]\};$$

$$\frac{\partial^2 y_{im}}{\partial x_2 \partial x_5} = \frac{1}{20k'^2k}(k\rho_1 + kd + b)\{b[2] + 2k^3d[1]\};$$

$$[1] = [k'^2\mathbf{K}(k) - (1 - 2k^2)\mathbf{E}(k)];$$

$$[2] = [k'^2(4 + k^2)\mathbf{K}(k) - (4 - k^2 - 2k^4)\mathbf{E}(k)];$$

$$\text{at } im = 12 \quad b = k', \quad d = e, \quad \rho_1 = 0;$$

$$\text{at } im = 13 \quad b = k', \quad d = -e, \quad \rho_1 = 0;$$

$$\text{at } im = 24 \quad b = -k', \quad d = e, \quad \rho_1 = 2h;$$

$$\text{at } im = 34 \quad b = -k', \quad d = -e, \quad \rho_1 = 2h;$$

$\mathbf{K}(k)$, $\mathbf{E}(k)$ are complete elliptical integrals of the 1st and 2nd kind of the absolute value of k_{im}, and $k_{im}^2 = [1 + (x_1 - \rho_1 - d)^2]^{-1}$; $k'^2 = 1 - k^2$.

The zero term $U_m(q_0)$ is omitted in decomposition (16) since potential energy is determined accurately to a constant, and equality to zero of the coefficient is the necessary condition of equilibrium of $(x_1 - x_{10})$ the system with a gravitational force. Using the following condition

$$\frac{\partial U_m}{\partial y_{12}}\frac{\partial x_{12}}{\partial x_1}\bigg|_0 + \frac{\partial U_m}{\partial y_{13}}\frac{\partial y_{13}}{\partial y_1}\bigg|_0 + \frac{\partial U_m}{\partial y_{25}}\frac{\partial x_{24}}{\partial y_1}\bigg|_0 + \frac{\partial U_m}{\partial y_{34}}\frac{\partial y_{34}}{\partial x_1}\bigg|_0 = mg \qquad (19)$$

we can find the value of x_{10} at which gravitational force of SE is balanced by magnetic interaction forces, and which is placed into expression (16).

In order to obtain the dynamic equations of the sensor, we use the results discussed above and formula for kinetic energy of a free body in the form (16):

$$T = \frac{1}{2}m(\dot{\rho}^2 + \rho^2\dot{\alpha}^2 + \dot{\zeta}^2) + \frac{1}{2}A(\dot{v}\sin\varphi - \dot{\psi}\sin v\cos\varphi)^2 +$$

$$+ \frac{1}{2}B(\dot{v}\cos\varphi + \dot{\psi}\sin v\sin\varphi)^2 + \frac{1}{2}C(\dot{\varphi} + \dot{\psi}\cos v])^2. \qquad (20)$$

Let us suppose that the principal moments of inertia of the sensor with respect to axes rigidly bound to coordinate system A, B, C are equal to $4ma^2$ and let us go over to dimensionless coordinates $x_1, \ldots, x_6$ and dimensionless time $\tau = t\omega$ by introducing characteristic frequency of sensor oscillations ω. Then dimensionless kinetic energy $\tilde{T}$ will be

$$2\tilde{T} = 2T(4ma^2\omega^2)^{-1} = \dot{x}_1^2 + x_1^2\dot{x}_2^2 + \dot{x}_4^2 + \dot{x}_5^2 + \dot{x}_6^2 + 2\dot{x}_5\dot{x}_6\cos x_4. \qquad (21)$$

Applying Lagrange equations of the first kind

$$\frac{d}{dt}\frac{\partial L}{\partial q_i} = \frac{\partial L}{\partial q_i}, \qquad (22)$$

where $L = T - U$ is Lagrange function, we obtain the required dynamic equations of the sensor:

$$\ddot{x}_1 = x_1\dot{x}_2^2 - \gamma\frac{\partial U}{\partial x_1} - x_1\dot{x}_2^2 - \gamma\frac{\partial^2 U}{\partial x_1^2}\bigg|_0 (x_1 - x_{10});$$

$$\ddot{x}_2 = -x_1^{-2}\left[2x_1\dot{x}_1\dot{x}_2\;\gamma\frac{\partial^2 U}{\partial x_2^2}\bigg|_0\left(x_2 - \frac{\pi}{2}\right) + \gamma\frac{\partial^2 U}{\partial x_2\partial x_3}\bigg|_0 x_5\right];$$

$$\dot{x}_3 = -\gamma\frac{\partial^2 U}{\partial x_3^2}\bigg|_0 x_3 - \gamma\frac{\partial^2 U}{\partial x_3\partial x_4}\bigg|_0\left(x_4 - \frac{\pi}{2}\right);$$

$$\ddot{x}_4 = -\dot{x}_5\dot{x}_6\sin x_4 - \gamma\frac{\partial^2 U}{\partial x_4^2}\bigg|_0\left(x_4 - \frac{\pi}{2}\right) - \gamma\frac{\partial^2 U}{\partial x_3\partial x_4}\bigg|_0 x_3;$$

$$\ddot{x}_5 = -\sin^{-2}x_4\left[\dot{x}_4\dot{x}_5\sin x_4\cos x_4 - \dot{x}_4\dot{x}_6\sin x_4 + \gamma\frac{\partial^2 U}{\partial x_5^2}\bigg|_0 x_5 + \right.$$

$$\left. + \gamma\frac{\partial^2 U}{\partial x_2\partial x_5}\bigg|_0\left(x_2 - \frac{\pi}{2}\right)\right];$$

$$\ddot{x}_6 = -\sin^{-2}x_4\left[\dot{x}_4\dot{x}_6\sin x_4\cos x_4 - \dot{x}_4\dot{x}_5\sin x_4 + \gamma\frac{\partial^2 U}{\partial x_5^2}\bigg|_0 x_5\cos x_4 + \right.$$

$$\left. + \gamma\frac{\partial^2 U}{\partial x_2\partial x_5}\bigg|_0\left(x_2 - \frac{\pi}{2}\right)\cos x_4\right]. \qquad (23)$$

Here the numeric value $\gamma = \Psi_1^2(8Lma^2\omega^2)^{-1}$ is determined from condition

$$\gamma \sum_{\substack{i,m=1 \\ i \neq m}} \frac{\partial U_m}{\partial y_{im}} \frac{\partial y_{im}}{\partial x_1}\bigg|_0 = \frac{g}{2a\omega^2}. \tag{24}$$

Thus, the expressions for potential and kinetic energies and differential equations of motion of the sensor in the form convenient for numerical analysis of stability and dynamics of gravity-inertial devices are obtained.

4 Optimal Synthesis of Chaotic Dynamics

In the optimization methods described in [12] sensor dynamics throughout the state space are represented either with a single set of coupled maps

$$y_i(n+1) = f_i[y(n), \gamma, a], \quad i = 1, \ldots, N, \quad y \in \mathbb{R}^N \tag{25}$$

or a set of ordinary differential equations

$$\dot{y}_i(t) = f_i[y(t), \gamma, a], \quad i = 1, \ldots, N, \quad y \in \mathbb{R}^N, \tag{26}$$

where $y \in \Gamma \subset \mathbb{R}^n$ is a state vector; $a \in \mathbb{R}^n$ is a parameter vector; γ represents a noise term. If $\gamma(t) = 0$ and $a = $ const, the equation (26) defines a deterministic dynamical system. The time series of sensor measurement is then a sequence of observations $\{s_n\}, M = 1$, where $s_n = h[y(t = n \triangle t)]$, with a measurement function h and a sampling $\triangle t$. The number of observed variables is assumed to be sufficient to embed the dynamics. The functions $\{f_i\}$ may be of any form, but are usually taken to be a series expansion. This method has been successfully tested with Taylor- and Fourier-series expansion. In this manner, the modeling is done by finding the best expansion coefficients to reproduce the experimental data. Often, the form of the functions $\{f_i\}$ is known, but the coefficients are unknown. For example, this situation occurs frequently with rate equations for measurement processes. The added information greatly reduces the number of undetermined parameters, thus making the modeling computationally more efficient.

The modeling procedure begins with the step of choosing some trial coefficients. The error in these parameters can be computed by taking each data point $x(t_n)$ as an initial condition for the model equations. The predicted value $y(t_{n+1})$ can then be calculated for CM's as

$$y_i(n+1) = f_i[x(n), a], \quad i = 1, \ldots, N,$$

or for ODE's as

$$y_i(t_{n+1}) = x_i(t_n) + \int_{t_n}^{t_{n+1}} f_i[y(t'), a]dt, \quad i = 1, \ldots, N \tag{27}$$

and compared to the experimentally determined value. It is well known that more stable models can often be obtained by comparing the prediction and the experimental data several time steps into the future. For the present analysis, we will predict the value only to the time of the first unused experimental data point. The error in the model is thus obtained by summing these differences

$$F = \frac{1}{N(M-1) - N_c} \sum_{i=1}^{M} \sum_{j=1}^{N} \frac{1}{\sigma_{ij}^2} [y_j(t_i) - s_j(t_i)]^2, \qquad (28)$$

where N_c is the number of free coefficients a_i, M is the number of data points, and σ_{ij} is the error in the jth vector component of the ith calibration measurement. The task of finding the optimal model parameters has now been reduced to a minimization problem. Thus, the best parameters are determined by

$$\min_a F(a, y), \quad \alpha_i^{\min} \leq \alpha_i \leq \alpha_i^{\max}, \quad i = 1, \ldots, r, \qquad (29)$$

where α_i are the system characteristics of the sensor (fractal dimension, point-wise dimension, information dimension, generalized dimension, embedding dimension, Lyapunov dimension, metric entropy et al.)

A minimal embedding of dimension N is determined by means of Hausdorff dimension d or any other generalized dimension. The essence of the present method is as follows. We consider relatively slow parameter a. As a result, we should solve the corresponding *constrained optimization problem*.

Therefore the ability to determine these coefficients rests upon the strength of the algorithm employed to search through the space of parameters. Since this has been formulated as a standard F_ν^2 identification problem, the normal statistical tests can be applied. Typically, $F_\nu \simeq 1$ implies that the modeling was successful; however, more sophisticated tests can be applied as well, e.g., the F test. If the experimental errors σ_{ij} are unavailable, the normalization factor can simply be removed from equation (28). This means that the F_ν tests cannot be applied, but the best possible model can still be determined by locating the global minimum of F_ν in the parameter space.

4.1 Construction of the Sensor

Mechanically, the sensor represents a free body designed as a rigid pack of two coaxial short-circuited superconducting coils, and suspended in magnetic field of two stationary superconducting current coils whose axes in nonperturbed state coincide with the axis of the sensor. The probe is positioned between stationary coils. The distance between each stationary coils and nearest to it a free coil is much less than the distance them. The stationary coils are powered by currents with the same direction and strength. Then, they are shorted out, and magnetic fluxes induce into free coils from stationary coils, such that each coil is attracted to the next stationary coil, i.e. the sensor is stretched by magnetic forces. An acceleration component is registered by a superconducting magnetometer.

Stationary suspension coils are connected in such a way that after their energizing with currents they form two independent loops with a common section which incorporates a measuring coil. Currents flowing in the formed loops are subtracted at the measuring coil. Since loops are powered by the same currents, at the initial instant of time a current in measuring coil is equal to zero. When the probe is displaced along its axis under the effect of acceleration, a current in one stationary coil increases by the value ΔI_2 whereas in the other it decreases by ΔI_2. In this case the current $\Delta I_1 + \Delta I_2$ will flow through the measuring coil .

However, increments of currents of opposite signs arise only in case of displacement along the axis. But, if perturbations arise in the direction perpendicular to the axis or along the angle of inclination of sensor axis, currents in both loops either increase or decrease. Current in the measuring coil will not vary. Due to the symmetry of the magnetic system, the sensor is invariant to the mentioned perturbations.

5 Chaotic Dynamics of the Levitated Probe

The connection of the shift y_1 of the probe in the magnetic field by means of measured signal u_3 and an output signal of the sensor can be described by an equation for state variables $y_3, \ldots, y_6$ and some functional z (a model of the quantum interferometer S). This model admits [9],[11] the following bilinear model (BM)

$$\dot{y} = Ay + (Bu_1 + Cu_1^2)y + Du_1 + Eu_1^2 + Fu_3 + Gu_4, \quad z = Ly, \qquad (30)$$

where A, B, C, D, E, F, G, L are matrix; $y \in \widehat{Y} \subset \mathbb{R}^2$; $z \in \mathbb{R}^1$.

Then there exists some possibilities for optimization of information characteristics of the measurement using the parameter matrix a and control $u(\widehat{n})$. On the base of these characteristics we can provide a matrix and topological behaviours of discrete approximation of the BM $\{T, \widehat{Y}, S, \Psi\}$ using symbolic dynamic methods. Here $\{\widehat{T}^n; \ n \in \mathbb{Z}\}$ is *cascade*; $T : \widehat{Y} \to \widehat{Y}$, $\Psi : \widehat{Y} \to \mathcal{L}$ is the map "input-output" of the system S, $\mathcal{L}$ is a finite alphabet. A further optimization of the sensor can be reached near a *Smale's horseshoe* of additional Lebesgue measure of the dynamical system $\{T, \widehat{Y}, S, \Psi\}$.

The requirement of the equilibrium of the probe is provided by a feedback

$$\widehat{u}_1 + \widehat{\alpha}\widehat{u}_1 = \widehat{\alpha}r(y - u_0)$$

in the simplified model

$$\ddot{y} - \widehat{u}_4 = \widehat{u}_1 + \widehat{u}_3.$$

Here $u_0(t)$ is a fixed relation of the time of a probe; $\widehat{u}_1 = d_2 u_1$, $\widehat{u}_3 = f_2 u_3$, $\widehat{u}_4(y) = g_2 u_4 = \delta y + \widehat{K}(y)y$, δ, r, $\widehat{\alpha}$ are constants, $\widehat{K}(y) = (1/\widehat{B})(y^2 - 1)(y^2 - B)$, $\widehat{B} > 1$. Under the parameter $r = 0$, the feedback realizes the three stable states $y = 0, \pm\sqrt{\widehat{B}}$ and two saddle points $y = \pm 1$. Under some values of the parameters $(\delta, \widehat{a}, r, \widehat{B})$ the probe u_0 will be move from one point to another.

However under different values of the parameters we will have a *limit cycles* and a chaotic mode. If $u_0 = \widehat{u}_3 = 0$ and $r = r_0$ then the origin of the coordinate system will be unstable saddle point of the spiral type. A numerical model of a measurement has chaotic properties (*strange attractor*) which can be used for constructing a better sensitivity measurement.

Using the linear model near a stable point of the probe

$$\dot{x} = Ax + Du_1 + Fu_3 + Gu_4, \quad z = Lx, \quad x \in \mathbb{R}^2, \tag{31}$$

a normalized polynomial $\Theta(\lambda) = \alpha_1\lambda^2 + \alpha_2\lambda + \alpha_3$ with a negative real part of roots and the method of synthesis of a nonperturbed motion under control $y = 0$, $u_1 = 0$, we can find a stabilizing control

$$u_1 = -(\alpha_3 + a_{11})a_{17}^{-1}\alpha^{-1}z. \tag{32}$$

6 Nonlinear Dynamics and Chaos

Consider a coil element of length, β, carrying constant current I moving over a continuous-sheet guideway. In the high speed limit, the force on the coil will be given by the field due to an image coil below the guideway of opposite current direction.

Under these very generous assumptions we can derive an equation for the vertical motion (called heave) of the form

$$m\ddot{z} + \delta\dot{z} - \frac{\mu_0 I^2 \beta}{2\pi z} = -mg + m\frac{V_0^2 4\pi^2 A_0}{\Lambda^2}\cos\omega t, \tag{33}$$

where $\omega = 2\pi V_0/\Lambda$ and an arbitrary damping term has been added. This system can be written in the form of a third-order autonomous system of first-order differential equations:

$$\dot{z} = v,$$
$$\dot{v} = -cv + \frac{b}{z} - a + f\cos\phi,$$
$$\dot{\phi} = \omega. \tag{34}$$

This system of equations can easily be numerically integrated in time using a Runge–Kutta or other suitable algorithm. The trajectory is easily projected onto the phase plane of z versus v. As the amplitude of the guideway waviness is increased, one can see a change in the geometry of the motion from elliptical to a distorted ellipse to chaotic motion. The chaotic motion is better viewed by looking at a strobescopic view of the dynamics by plotting (z_n, v_n) at discrete values of the phase $\phi = \omega t$ or $t_n = 2\pi n/\omega$. This picture is called a *Poincaré map* (Moon, 1992). In contrast to the unordered continuous time the Poincaré map shows a fractal-like structure. This type of chaotic motion with fractal structure is called a *strange attractor*. It indicates that the dynamics are very sensitive to the initial conditions.

Chaotic-like dynamics in a levitated model moving over a rotating-wheel guideway have been observed by Moon [8].

The magnetic force between a permanent magnet and a temperature superconductor such as YBCO is hysteretic near the critical temperature. Hystcretic forces are both nonlinear and dissipative and can produce complex nonlinear dynamics. A permanent magnet is restrained to move laterally over a YBCO superconductor. As the gap between the magnet and the superconductor is decreased, the dynamics of the magnet become increasingly complex in a pattern called *period doubling*. Subharmonic frequencies appear in the spectrum of the form $m\omega/n$, where $n = 2, 4, \ldots, 2k$. This bifurcation behavior is shown in the Poincaré map as a function of the magnet-YBCO gap. At a critical gap, the motion becomes chaotic. Another tool for observing chaotic motions is to plot a *return map* on one of the state variables, say X_{n+1}, versus X_n, where X_n is the displacement of the magnet at discrete times synchronous with the driving amplitude – that is, $t_n = 2\pi n/\omega$. The return map shows a simple parabolic shape. This map is similar to a very famous equations of chaos known as the *logistic map* (Moon, 1992):

$$X_{n+1} = aX_n(1 - X_n). \tag{35}$$

For $a > 3.57$ the dynamics may become chaotic, and this equation generates a probability density function.

This simple experiment again indicates that although magnetically levitated bodies are governed by deterministic forces, the nonlinear nature of the forces can generate complex and sometimes unpredictable dynamics which are sensitive to initial conditions and changes in other system parameters. Thus, care in design of such systems should include exploration of the possible nonlinear behavior of levitation devices.

7 Conclusions

In this chapter, we describe a superconducting gravity meter, its mathematical model, and a nonlinear controller that stabilize a probe at the equilibrium state. We have also presented a mathematical model of the superconducting suspension which is based on a magnetic levitation. A nonlinear control algorithm has been implemented for the purpose of maintaining chaotic behavior in the sensor.

References

1. E. Brandt. Levitation in physics. *Science*, 243:349–355, 1989.
2. E. Brandt. La Lévitation. *Recherche*, 224–229, 1990.
3. W. Braunbeck, W. Freishwebende Körper in electrischen and magnetischen Feld. *Z. Phys.*, 112(11–12):753–763, 1939.
4. R. Horst, P. Pardalos. *Hadbook of Global Optimization.* Kluwer Academic Publishers, Dordrecht, 1995.
5. J. Jensns. *The Mathematical Theory of Electricity and Magnetism.* Cambridge, CUP, 1925.
6. V. Kozorez, O. Cheborin. On stability of equilibrium in a system of two ideal current rings. *Dokl. Akad. Nauk UkrSSR, Ser. A*, 4: 80–81, 1988.
7. F. Moon. Chaotic vibrations of a magnet near a superconductor. *Phys. Lett. A*, 132(5):249–251, 1988.
8. F. Moon. *Chaotic and Fractal Dynamics.* John Wiley & Sons, New York, 1992.
9. P. Pardalos, P. Knopov, S. Urysev, V. Yatsenko. Optimal estimation of signal parameters using bilinear observation. In Rubinov, A. and Gloveredited, B., editors, *Optimization and Related Topics.* Kluwer Academic Publishers, Dordrecht–Boston–London, 103-116, 2001.
10. D. White, G. Woodson, G. *Electromechanical Energy Conversion.* John Wilev and Sons. Inc., New York,1959.
11. V. Yatsenko. Estimating the signal acting on macroscopic body in a controlled potential well. *Kibernetika*, 2:81–85, 1989.
12. V. Yatsenko and P. Pardallos. Global optimization of cryogenic-optical sensor. In Sensors, Systems, and Next-Generation Satellites, K. W.H. Fujisada, J. Lirie (Eds.), Proc. SPIE 4550, 2001.-P. 433 - 441.
13. V. Yatsenko, P. Pardallos, and J. Principe. Cryogenic-optical sensor for the higly ensitive gravity meters. *Advance Sensor, Systems, and Next-Generation Satellites V*, Proc. SPIE, 4881:549-557, 2002.
14. V. Yatsenko. Functional structure of the cryogenic optical sensor and mathematical modeling of signal. *Cryogenic Optical Systems and Instruments*, Proc. of SPIE, 5172:97-107, 2003.

Stochastic Optimization and Worst–case Decisions

Nalan Gülpinar, Berç Rustem, Stanislav Žaković

Imperial College London,
Department of Computing, South Kensington Campus,
London, SW7 2AZ, UK.
E-mail: {ng2,br,zakovic}@doc.ic.ac.uk

Summary. In this chapter, we are concerned with decision making methods for dynamic systems under uncertainty. We consider expected value optimization of stochastic systems and worst-case robust strategies. Stochastic decision-making involves uncertainty and consequently risk. An important tool to address the inherent error for forecasting uncertainty is worst-case analysis. From the risk management point of view, minimax yields the best strategy determined simultaneously with the worst state of the underlying system. Worst-case analysis is a robust framework for decisions under uncertainty as the actual performance of the decision has a non-inferiority property. The significance of robust strategies is increasingly recognized as attitudes towards risk evolve in diverse areas. We present worst-case approach to macroeconomics policy making and financial portfolio management.

1 Introduction

Worst-case analysis (minimax) is a robust framework for decision making under uncertainty. The starting point for the worst-case optimization is based on Rustem & Howe [12]. Minimax solution is determined simultaneously with the worst state of the underlying system. The actual performance of the decision has a non-inferiority property. The significance of worst-case robust strategies is increasingly recognized as attitudes toward risk evolve in economics, financial markets and engineering.

Model-based policy design entails a reasonable specification of the underlying model and an appropriate characterization of the uncertainties. The latter can be an exogenous effect, parameter uncertainty or uncertainty regarding the structure of the model (which requires a setting that admits rival structures). In this chapter we consider methods that address those types of uncertainty and present two applications of minimax to decision making, namely macroeconomics policy making and financial portfolio management. The uncertainties are characterized either in terms of a number of rival scenarios or ranges in which the uncertain parameters or exogenous effects may vary.

Such characterization of uncertainty leads to discrete and continuous minimax models. The discrete minimax approach determines the optimal strategy in view of all specified rival scenarios simultaneously, rather than any single scenario. The continuous minimax strategy provides a guaranteed optimal performance in view of continuum of scenarios varying between upper and lower bounds. Thus, there are an infinite number of future scenarios in the continuous minimax framework.

Financial portfolio management problem can be modeled using single-stage and multi-stage stochastic programming [4], [9]. The future is seen in terms of scenarios that are essentially a discrete set of realization of uncertainties. In the multi-stage case, each scenario evolves into a set of scenarios in the next stage, with associated probabilities [2]. Instead of risk and return forecasts over a single-stage, the forecast takes the shape of a scenario tree, which divides the investment horizon into discrete time intervals, at which the portfolio may be rebalanced. Classical measure of risk in the Markowitz mean-variance approach requires the knowledge of the first and second moments of the distribution of returns and capture errors in the data such as mean and covariance matrix of the returns. Therefore, the imprecise nature of the moment forecasts needs to be tackled to reduce the risk of decision-making on the wrong scenario.

An alternative worst-case approach is the H^∞ formulation [1]. The H^∞ approach transforms the original minimax problem with box constraints, which may be convex with respect to the uncertain variables, to a concave maximization problem by an appropriate choice of a penalty parameter γ. This requires the solution of a minimax problem, convex in the minimization (i.e. policy) variables and concave in the maximization variables (i.e. uncertainties). This is a saddle point. If the original intention is to design a robust policy for a given range of uncertainties, H^∞ is not really an appropriate tool. The formulation is also extremely sensitive to the choice of γ which may result in policies which are either more optimistic or more pessimistic than intended. However, it does provide some degree of robustness cover.

The rest of this chapter is organized as follows. In section 2, we describe minimax approach and discuss some issues that arise from the approach. Sections 3 and 4 focus on applications of worst-case analysis to macroeconomics policy making and financial portfolio management, respectively. Section 5 summarizes our conclusions.

2 The Minimax Approach

The general minimax optimization problem can be defined as:

$$\min_{x \in X} \max_{v \in V} \ F(x, v),$$
$$s.t. \ \ X \subset R^n, \tag{1}$$
$$V \subset R^m,$$

where $F : R^{n+m} \to R$. The aim of the worst-case approach is to minimize the objective function with respect to the worst possible outcome of the uncertain variables v. Therefore, x is chosen to minimize the objective function, where nature chooses v to maximize it.

If V is a finite set, (1) becomes the discrete minimax problem:

$$\min_x \ \max_{v \in V} \ F(x, v),$$
$$s.t. \ \ V = \{v_1, v_2, ..., v_k\}. \tag{2}$$

Introducing a more familiar notation,

$$F(x, v_j) = f^j(x), \quad j = 1, 2, ..., k$$

the discrete minimax problem (2) can be reformulated as

$$\min_x \ \max_{j=1,2,...,k} \ f^j(x). \tag{3}$$

It can be shown that minimax problem (3) is equivalent to the following nonlinear programming problem:

$$\min_{x,z} \ z,$$
$$s.t. \ f^j(x) \le z, \quad \forall j = 1, 2, ..., k, \tag{4}$$

Using the fact that the maximum over a set of scalars is equal to the maximum over their convex hull, (3) can be equivalently expressed as the continuous minimax problem

$$\min_x \max_\beta \ \sum_{i=1}^{k} \beta_i f^i(x)$$
$$s.t. \ \sum_{i=1}^{k} \beta_i = 1, \tag{5}$$
$$\beta_i \ge 0, \quad \forall i.$$

If V has an infinite number of elements, then (1) is called continuous minimax which can be stated as the following semi-infinite optimization problem

$$\min_{x,z} \ z,$$
$$s.t. \ F(x, v) \le z, \quad \forall v \in V, \tag{6}$$

Notice that there is an infinite number of constraints corresponding to the elements in V.

Let

$$\Phi(x) = \max_{v \in V} F(x, v), \tag{7}$$

for all x. We call $\Phi(x)$ the max function. Therefore, (1) can be written as

$$\min_{x \in X} \Phi(x). \qquad (8)$$

To solve (8) a quasi-Newton algorithm is used. The algorithm generates a descent direction based on a subgradient of $F(x,.)$ and uses an approximate Hessian in the presence of possible multiple maximizers of (7) as well as a step size strategy that ensures sufficient decrease in $\Phi(x)$ at each iteration. Problem (8) poses several difficulties:

- $\Phi(x)$ is in general continuous but may have kinks, so it might not be differentiable. At a kink the maximizer is not unique and the choice of subgradient to generate a search direction is not simple;
- $\Phi(x)$ may not be computed accurately as it would require infinitely many iterations of an algorithm to maximize $F(x, y)$;
- a global maximum is required in view of possible multiple solutions. The use of a local maximum cannot guarantee a monotonic decrease in $\Phi(x)$.

When the cost or objective function is convex with respect to the uncertain variables the maximum will correspond to one or more vertices of the hypercube defined by the upper and lower bounds on the uncertain variables. If the objective function is concave with respect to the uncertainties the maximum may lie anywhere within the hypercube.

The minimax algorithms and applications to a number of problems in engineering, finance and macroeconomics are presented in [12], [15], [16]. In this chapter we present both discrete and continuous minimax models arising in macroeconomics and finance.

3 Macroeconomics Policy Making

In recent years, a number of central banks have announced inflation targets and have adopted an explicit inflation targeting framework for optimal macroeconomics policy making. Orphanides and Wieland [10] use a simple macroeconomics model of inflation (π_t), output gap (y_t) and interest rates (r_t) to investigate different motives for inflation point versus inflation zone targeting. In the first case, the policymaker varies short-term nominal interest rates in order to stabilize inflation around a point target whereas in the second case, the emphasis is on containing inflation within a target range. Inflation point targeting arises naturally in linear models of the economy with a quadratic loss function for the policymaker (the L-Q model in [10]). Orphanides and Wieland show that inflation zone targeting may be motivated by a non-linear, or more precisely, zone-linear Phillips curve relationship between the change in inflation and the output gap (the ZL-Q model in [10]).

In the minimalist macroeconomics model of [10], the two key variables for the policy decision process are inflation and output. The policy instrument is the short term nominal interest rate. The dynamic structure of the model is represented by a single lag of inflation in the Phillips curve and a single lag of the output gap in the aggregate demand equation. It is appropriate, therefore, to interpret the length of a period to be rather long, say half a year to a year.

3.1 Worst-case Inflation Targeting

In every period t, the policymaker sets the interest rate, r_t, with the objective to maintain inflation π_t, close to a desired target and output gap y_t close to the economy's natural level. To describe the policymaker's welfare loss during a period t, a per-period loss function is specified as

$$l_t = l(\pi_t, y_t).$$

The per-period loss facing the policymaker in period $t + 1$, l_{t+1}, can be expressed as a weighted average of the deviation of inflation π from its desired target π^* and the output deviation from the economy's natural level y:

$$l_{t+1} = \omega(\pi_{t+1} - \pi^*)^2 + (1 - \omega)y_{t+1}^2, \quad \omega \in (0, 1). \tag{9}$$

The following two equations describe the evolution of the economy:

$$\begin{bmatrix} \pi_{t+1} \\ y_{t+1} \end{bmatrix} = \begin{bmatrix} 1 & \alpha\rho \\ 0 & \rho \end{bmatrix} \begin{bmatrix} \pi_t \\ y_t \end{bmatrix} + \begin{bmatrix} -\alpha\xi \\ -\xi \end{bmatrix} r_t + \begin{bmatrix} \alpha\delta + \alpha u_{t+1} + e_{t+1} \\ \delta + u_{t+1} \end{bmatrix}, \tag{10}$$

where r_t represents the real interest rate, e_{t+1} and u_{t+1} are random, zero-mean shocks:

$$-\sigma \leq u_t, e_t \leq \sigma, \quad \forall t, \tag{11}$$

and $\alpha, \rho, \delta, \xi$ are given model parameters. The objective function is defined in terms of a sum of discounted per-period losses as follows:

$$F(r, v) = \sum_{t=0}^{\infty} \beta^t l_{t+1}, \tag{12}$$

with the uncertainty $v = (u_t, e_t)'$ and the discount factor $\beta < 1$.

An alternative approach, which could be used in this framework, is considered by Tetlow and von zur Muehlen in [14] and Hansen and Sargent in [5]. In these approaches, the policymaker chooses the parameters x_1 and x_2 of the feedback law

$$r_t = x_1\pi_t + x_2 y_t \tag{13}$$

to minimize welfare losses that are maximized over v. This rule is referred to as feedback rule. We can then formulate the minimax problem as

$$\min_{x_1, x_2} \max_{v \in V} F(x, v), \tag{14}$$

where the objective function F is given by (12), the constraints on the systems are given by the model (10) and the feedback law is represented by (13).

3.2 The H^∞ Approach

Another approach to robust design, with minimax origin, is the H^∞ framework (Basar and Bernhard [1]). Consider the following problem

$$\min_x \max_v \ F(x, v)$$

$$s.t. \ \|v\|_2^2 \leq C. \tag{15}$$

If F is convex in x and convex in v, the solution of this problem lies on the boundary of the constraint $\|v\|_2^2 \leq C$. Furthermore, generally, there are multiple maximizers. The constraint is implicitly imposed using a penalty formulation that discourages the transgression of the constraint. Thus, the H^∞ formulation is given by

$$\min_x \max_v F(x, v) - \gamma^2 \|v\|_2^2, \tag{16}$$

This problem is convex in x and, for smaller values of γ, convex in v. Hence, it can have multiple maxima. As γ increases, the augmented objective function (16) becomes concave in v at some value of γ. It is this value of γ that H^∞ seeks. Since the transformed objective is now convex in x and concave in v, it yields a robust solution that is also a saddle point.

H^∞ is not concerned with the original robust decision problem. H_∞ was developed for the specific purpose of generating feedback controls that will ensure a robust policy if the underlying uncertainty (noise) deviates from its base. It is designed to have a unique worst-case, corresponding to the saddle-point solution. It transforms the original problem to the determination of a saddle–point that provides some degree of robustness. Furthermore, the computation is highly sensitive to the choice of γ. The results for two different choices of γ illustrate this point and are presented in Figures 1 and 2. In the former, for $\gamma = 1$, we observe two maxima (i.e. two worst-case realizations of the shocks (u, e)) for our macroeconomics model, consisting of worst-case paths for interest rate, inflation and output gap.

In Figure 2, with $\gamma = 3$, the uncertainties are forced towards zero and we have a saddle point. Therefore, only one maximum is observed for interest rate, inflation and output gap.

Robustness and Optimality of Minimax

Let x^*, v^* solve (1). Then, the following inequality is valid for all feasible v

$$F(x^*, v^*) \geq F(x^*, v).$$

Let x^{**}, v^{**} be the optimal solution of (16). Then

$$F(x^{**}, v^{**}) - \gamma^2 \|v^{**}\|_2^2 \geq F(x^{**}, v) - \gamma^2 \|v\|_2^2, \text{ for all feasible } v.$$

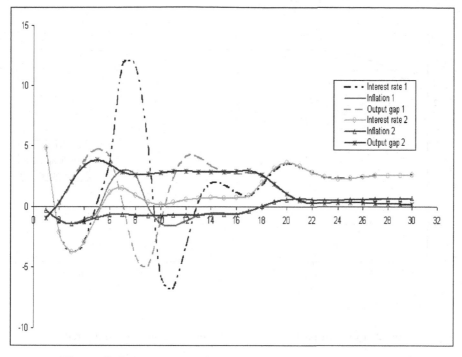

Fig. 1. Inflation patterns for $\gamma = 1$. Two maxima encountered.

The above inequalities simply state the optimality of v^*, v^{**} for the corresponding problems (1) and (16). They also signify the robustness of minimax in that performance is assured to improve if the worst-case v^* or v^{**} does not happen. Similarly, under the same assumptions, for all feasible x, we have

$$F(x^*, v^*) \le F(x, v^*),$$

and

$$F(x^{**}, v^{**}) - \gamma^2 \|v^{**}\|_2^2 \le F(x, v^{**}) - \gamma^2 \|v^{**}\|_2^2.$$

3.3 Expected Value Optimization

We also consider expected value optimization of nonlinear function $F(x, v)$ and compare the results obtained with the minimax approach. Random shocks u_t, e_t are now assumed to be normally distributed, zero-mean variables:

$$u_t, e_t \sim N(0, \sigma^2), \quad \forall t.$$

We formulate the problem as minimization of expected value of $F(x, v)$ as

$$\min_{x_1, x_2} E(F(x, v)),$$

$$s.t. \quad v = (u_t, e_t)' \sim N(0, \sigma). \tag{17}$$

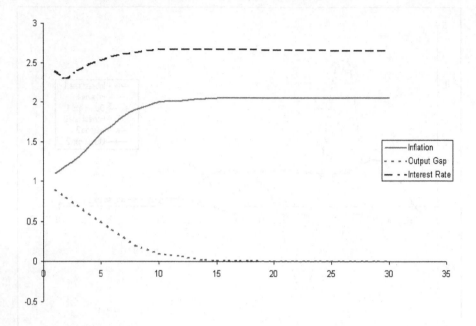

Fig. 2. Inflation patterns for $\gamma = 3$. One maximum encountered.

Let x^e be

$$x^e = \arg\min E(F(x, v)).$$

Given the minimax solution (x^*, v^*), the following inequality is valid

$$F(x^*, v^*) \leq F(x^e, v^*).$$

In the computational experiments, we set $\pi^* = 0, t = 20, \beta = 0.9$ and the same weight $w = \frac{1}{2}$ for both inflation and output gap. The model parameters are estimated by Orphanides and Wieland [10]. We present the results of our computational experiments obtained with different bounds on the uncertainties in Table 1. As the shocks (uncertainties u, e) are additive in the model, the feedback rules $x^* = (x_1^*, x_2^*)'$ are the same for all cases. Notice that the objective function values increase as the boundaries on uncertainties are increased. The results of minimizing expected value show a similar conclusion - the control $x^e = (x_1^e, x_2^e)'$ is the same due to the shocks appearing linearly in the model, and also the expected loss increases as the uncertainty increases. It can be observed from the results that expectation of the loss is always lower than the worst-case

$$E_v(F(x^e, v)) \leq F(x^*, v^*).$$

We evaluate the consequence of applying the expected value optima x^e (corresponding to different levels of uncertainty) when the worst-case scenario v^* is realized; in other words, $F(x^e, v^*)$ is computed. We also investigate the consequence of adopting the worst-case optima x^* (corresponding to different bounds) in view of stochastic uncertainty, that is $E_v(F(x^*, v))$. The results of cross evaluation are presented in Table 1.

Table 1. Minimax and expected value optima and their cross evaluation.

Minimax Optimum				
bounds	x_1^*	x_2^*	$F(x^*, v^*)$	$E_v(F(x^*, v))$
$\frac{1}{2}\sigma_u, \frac{1}{2}\sigma_e$	5.217	1.873	10.943	6.583
σ_u, σ_e	5.217	1.873	43.772	26.334
$\frac{3}{2}\sigma_u, \frac{3}{2}\sigma_e,$	5.217	1.873	101.252	72.638
Minimized Expectation				
distribution	x_1^e	x_2^e	$E_v(F(x^e, v))$	$F(x^e, v^*)$
$N(0, \frac{\sigma_u^2}{4}), N(0, \frac{\sigma_e}{4}^2)$	1.857	1.930	4.013	12.564
$N(0, \sigma_u^2), N(0, \sigma_e^2)$	1.857	1.930	16.053	66.543
$N(0, (\frac{3\sigma_u}{2})^2), N(0, (\frac{3\sigma_e}{2})^2)$	1.857	1.930	36.119	129.733

We compare adopting the worst-case feedback rule in the stochastic framework and the expected value optimization feedback rule when the worst-case is realized. The expected performance of the former (completed using Monte Carlo simulation) is observed to be much better than the performance of the latter (for example 6.583 is the expectation, while the worst-case value is 10.943):

$$E_v(F(x^*, v)) \le F(x^*, v^*).$$

The situation rapidly changes when x^e, the feedback rules obtained from minimizing expectation are used. In case when such rules are used and the worst-case scenario happens, the loss could increase up to 60%

$$F(x^*, v^*) \le F(x^e, v^*).$$

Therefore, although the expected value optimization performs better on the average, minimax approach guards against the worst possible scenarios and provides the upper bound for the loss function. Should any other scenario be realized, then better performance and lower loss are guaranteed.

4 Financial Portfolio Management

In financial portfolio management, the maximization of return for a level of risk is the accepted approach to decision making. A fundamental example is the single-stage Markowitz [9] model in which expected portfolio return is maximized and risk measured by the variance of portfolio return is minimized, [9]. The single-stage asset allocation problem can be extended to a multi-stage framework using stochastic programming. In the multi-stage case, after the initial investment, one can rebalance the portfolio (subject to any desired bounds) to maximize profit at the investment horizon and minimize the risk at discrete stages and redeem at the end of the stage. In this section we are concerned with both single-stage and multi-stage mean-variance financial portfolio management problems.

4.1 Single-stage Mean-Variance Optimization

Markowitz Model

The single-stage model of Markowitz considers a portfolio of n assets defined in terms of a set of weights w^i for $i = 1, \cdots, n$, which sum to unity. In other words, the initial budget is normalized to 1 as $\mathbf{1}'\mathbf{w} = 1$. Given an expected rate of return $\bar{r}$, the optimal portfolio is defined in terms of the solution of the following quadratic programming problem:

$$\min_{\mathbf{w}} \{< \mathbf{w}, \Lambda\mathbf{w} > | \mathbf{w}'\mathbf{r} = \bar{r}, \ \mathbf{1}'\mathbf{w} = 1, \ \mathbf{w} \geq 0\}$$

where Λ is the covariance matrix of asset returns. The quadratic program yields the minimum variance portfolio. Many traditional portfolio analysis models seek only to maximize expected return. This can be achieved with a classical stochastic linear programming formulation which incorporates the mean term. This is a risk-neutral approach which does not take risk-attitudes into account.

If the investor currently has holdings of assets $1, \ldots, n$, then vector $\mathbf{p}$ (scaled so that $\mathbf{1}'\mathbf{p} = 1$) represents the current position. If the investor currently has no holdings (wishing to buy), then $\mathbf{p} = \mathbf{0}$. Let $\mathbf{b}$ and $\mathbf{s}$ define decision variables for transactions of buying and selling, respectively. Then the allocation of the initial budget is represented with the following constraints:

$$\mathbf{p} + \mathbf{b} - \mathbf{s} = \mathbf{w} \tag{18}$$

Transaction costs, τ, incurred by moving to strategy $\mathbf{w}$ from current position $\mathbf{p}$, subject to costs $\mathbf{c_b}$, $\mathbf{c_s}$ can be incorporated into the mean-variance optimization model. The transaction cost of the purchase or sale is formulated as

$$\mathbf{c}_b'\mathbf{b} + \mathbf{c}_s'\mathbf{s} = \tau \tag{19}$$

The trade off between expected return and risk is achieved by solving the following quadratic programming problem;

$$\min \ \alpha(\mathbf{w} - \overline{\mathbf{w}})'\Lambda(\mathbf{w} - \overline{\mathbf{w}}) - \left[(\mathbf{w} - \overline{\mathbf{w}})'\mathbf{r} - \tau\right]$$

subject to

$$1'\mathbf{w} = 1$$
$$\mathbf{p} + \mathbf{b} - \mathbf{s} = \mathbf{w}$$
$$\mathbf{c}_b'\mathbf{b} + \mathbf{c}_s'\mathbf{s} = \tau$$
$$\mathbf{w}, \mathbf{b}, \mathbf{s} \geq 0$$

where the scaling constant α determines the level of risk-aversion optimized for. By sliding from $\alpha = 0$ (total risk-seeking) to $\alpha = \infty$ (total risk aversion), the entire range of efficient investment strategies is obtained. These investment strategies basically defines the efficient frontier.

Single-stage Minimax Portfolios

The mean-variance framework is based on a single forecast of return and risk. In reality, however, it is often difficult or impossible to rely on a single forecast. There are different rival risk and return estimates, or scenarios. The inaccuracy in forecasting can be addressed through the specification of rival scenarios. These are used with forecast pooling using stochastic programming; for example see [6], [7], [8]. Robust pooling using minimax has been introduced in [11] and [12]. Minimax optimization is more robust to the realization of worst-case scenarios than considering a single scenario or an arbitrary pooling of scenarios. It is suitable for situations which need protection against risk of adopting the investment strategy based on the wrong scenario. There are two minimax models; discrete and continuous. The discrete minimax approach determines the optimal investment strategy in view of all specified discrete rival scenarios simultaneously, rather than any single scenario. Its disadvantage is that it requires the specification of a number of discrete scenarios. An alternative approach that addresses the specification of the return forecast in terms of a range given by upper and lower bounds is the continuous minimax. The continuous minimax strategy provides a guaranteed optimal performance in view of continuum of scenarios varying between upper and lower bounds. Thus, there are an infinite number of future scenarios in the continuous minimax framework. The reader is referred to [11], [13] for single-stage minimax problem and to [3] for the corresponding continuous minimax model, but a short summary is as follows.

Discrete Minimax

Let J and I be a number of rival return and risk scenarios, respectively. Let K denote the number of benchmarks provided. A compact representation of the minimax portfolio allocation problem is as follows:

$$\min_{\mathbf{w}} \left\{ \alpha \cdot \max_{i,k}\{(\mathbf{w} - \overline{\mathbf{w}}_k)' \Lambda_i (\mathbf{w} - \overline{\mathbf{w}}_k)\} - \min_{j,k}\{(\mathbf{w} - \overline{\mathbf{w}}_k)' \mathbf{r}_j - t(\mathbf{w})\} \right\} \quad (20)$$

where $i = 1, \cdots, I$, $j = 1, \cdots, J$ and $k = 1, \cdots, K$. Function $t(\mathbf{w})$ represents the transaction costs incurred by moving to strategy $\mathbf{w}$ from current position $\mathbf{p}$, subject to costs $\mathbf{c_b}$, $\mathbf{c_s}$. In order to solve (20), we reformulate the problem as a quadratically constrained mathematical program:

$$\min \ \alpha\nu - \mu + \gamma\mathbf{b}'\mathbf{s}$$

subject to

$$\mathbf{1}'\mathbf{w} = 1$$
$$\mathbf{p} + \mathbf{b} - \mathbf{s} = \mathbf{w}$$
$$\mathbf{c_b}'\mathbf{b} + \mathbf{c_s}'\mathbf{s} = \tau \qquad (21)$$
$$(\mathbf{w} - \overline{\mathbf{w}}_k)'\Lambda_i(\mathbf{w} - \overline{\mathbf{w}}_k) \leq \nu, \quad i = 1, \cdots, I, \ k = 1, \cdots, K$$
$$(\mathbf{w} - \overline{\mathbf{w}}_k)'\mathbf{r}_j - \tau \geq \mu, \quad j = 1, \cdots, J, \ k = 1, \cdots, K$$
$$\mathbf{w}, \mathbf{b}, \mathbf{s} \geq \mathbf{0}$$

If $I = 0$ (no risk scenarios provided), then quadratic constraints are omitted and the objective function becomes $\min_\mathbf{w} -\mu \equiv \max_\mathbf{w} \mu$, which is a purely linear problem. If $J = 0$ (no return scenarios provided), then linear performance constraints are omitted and the objective function becomes simply $\min_\mathbf{w} \nu$. When $K = 1$, then only one benchmark portfolio is considered.

Continuous Minimax

The classical Markowitz framework can be extended to the continuous minimax with upper and lower bounds on the return scenarios and various rival risk scenarios. The minimax model integrates benchmark relative computations in view of scalable (not fixed) transaction costs. Assume that the return forecast of assets is defined by the bounds $\mathbf{r}^l \leq \mathbf{r} \leq \mathbf{r}^u$. In view of I rival risk scenarios and a range of return forecasts, the mean-variance optimization problem can be formulated as the following minimax problem;

$$\min_\mathbf{w} \left\{ \max_{\mathbf{r}^l \leq \mathbf{r} \leq \mathbf{r}^u} -\{(\mathbf{w} - \overline{\mathbf{w}})'\mathbf{r} - \tau\} + \alpha \cdot \max_i \{(\mathbf{w} - \overline{\mathbf{w}})'\Lambda_i(\mathbf{w} - \overline{\mathbf{w}})\} \right\} \qquad (22)$$

where τ represents the transaction costs. We define

$$\mathbf{x}^+ - \mathbf{x}^- = \mathbf{w} - \overline{\mathbf{w}}$$

so that $\mathbf{x}^+, \mathbf{x}^- \geq 0$ and $\mathbf{x}^+ \cdot \mathbf{x}^- = 0$. Notice that if $\mathbf{w} > \overline{\mathbf{w}}$, then $\mathbf{x}^+ > 0$ and $\mathbf{x}^- = 0$; if $\mathbf{w} < \overline{\mathbf{w}}$, then $\mathbf{x}^+ = 0$ and $\mathbf{x}^- > 0$. Since the expected return of portfolio is a linear function of $\mathbf{r}$, there are worst-case scenarios which are at the lower or at the upper bounds of given range. In other words, either $x_i^+ r_i^l$ or $x_i^- r_i^u$ is realized. Thus, the minimax problem (22) becomes

$$\min_{\mathbf{x}^+, \mathbf{x}^-} \left\{ -\{(\mathbf{x}^-)'\mathbf{r}^u + (\mathbf{x}^+)'\mathbf{r}^l - \tau\} + \alpha \cdot \max_i \{(\mathbf{x}^+ - \mathbf{x}^-)'\Lambda_i(\mathbf{x}^+ - \mathbf{x}^-)\} \right\}$$

This is equivalent to the following quadratically constrained mathematical program whose optimal solution provides the worst-case investment strategy,

$$\min_{\mathbf{x}^+,\mathbf{x}^-} \ -\{(\mathbf{x}^-)'\mathbf{r}^u + (\mathbf{x}^+)'\mathbf{r}^l - \tau\} + \alpha\nu + \gamma\mathbf{b}'\mathbf{s} + \beta(\mathbf{x}^+)'\mathbf{x}^-$$

subject to

$$\mathbf{1}'\mathbf{w} = 1$$
$$\mathbf{p} + \mathbf{b} - \mathbf{s} = \mathbf{w}$$
$$\mathbf{c}_b'\mathbf{b} + \mathbf{c}_s'\mathbf{s} = \tau \qquad\qquad (23)$$
$$\mathbf{x}^+ - \mathbf{x}^- = \mathbf{w} - \overline{\mathbf{w}}$$
$$(\mathbf{w} - \overline{\mathbf{w}})'\Lambda_i(\mathbf{w} - \overline{\mathbf{w}}) \leq \nu \qquad i = 1, \cdots, I$$
$$\mathbf{w}, \mathbf{b}, \mathbf{s}, \mathbf{x}^+, \mathbf{x}^- \geq \mathbf{0}$$

where γ and β are penalty terms.

We evaluate the performance of the discrete and continuous minimax investment strategies (obtained by solving the minimax optimization problems (21) and (23), respectively) in terms of the worst-case risk-return frontiers. For the computational experiments, three covariance matrices are estimated using the historical data which consists of monthly prices of 10 FTSE100 stocks through the 1990's. The forecast bounds are determined around the historical mean and the bounds are chosen as (10%± standard deviation) for each asset. In order to use in the discrete minimax model, we selected three rival return scenarios within this range; namely at the lower bound, upper bound and central mean. The minimax mean-variance optimization model constructs an optimal portfolio simultaneously with the worst-case scenario.

Figure 3 illustrates the performance of continuous minimax (at the top) and discrete minimax (at the bottom) approaches and noninferiority of continuous and discrete worst-case optimization models over 3 selected rival return scenarios within the range of return forecast. These results show that the worst-case investment strategy has the best lower bound performance which can only be improved when any scenario other than the worst-case is realized. Therefore, non-inferiority of minimax optimization ensures the robustness of the strategy. Note that in Figure 3 (at the bottom), the discrete worst-case investment strategy coincides with the efficient frontier obtained by evaluating the minimax strategy on lower-bound based scenario. In this case, benchmark relative worst-case strategy is adjusted to cover lower-bound based scenario as the rival worst-case.

4.2 Multi-stage Mean-variance Optimization

In this section, we first briefly summarize multi-stage mean-variance optimization model; see [4] for more details. Then we extend it to mean-variance minimax model with multiple rival risk and return scenarios. The minimax model

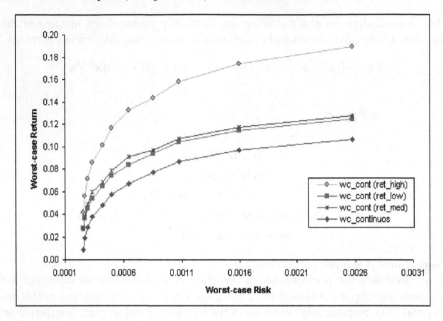

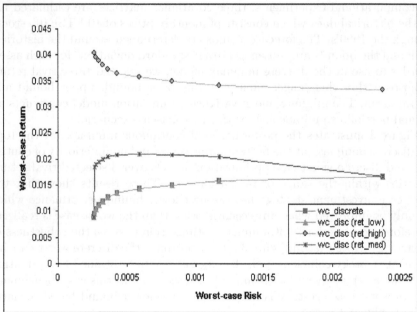

Fig. 3. Robustness of continuous minimax and discrete minimax models.

integrates benchmark relative computations in view of scalable (not fixed) transaction costs. Minimax mean-variance optimization constructs an optimal portfolio simultaneously with the worst-case scenario. Robustness arises from the non-inferiority of the worst-case optimal strategy.

Multi-stage Mean-Variance Model

We consider n risky assets and construct a portfolio over an investment horizon T. After the initial investment ($t = 0$), the portfolio may be restructured at discrete times $t = 1, \ldots, T - 1$ in terms of both return and risk, and redeemed at the end of the investment horizon ($t = T$). Multi-stage portfolio optimization entails the construction of a scenario tree representing a discretized estimate of uncertainties and associated probabilities in future stages. The multi-stage stochastic mean-variance approach takes account of the approximate nature of the discrete set of scenarios by considering a variance term around the return scenarios. Hence, uncertainty on return values of instruments is represented by a discrete approximation of a multivariate continuous distribution as well as the variability due to the discrete approximation.

A scenario is defined as a possible realization of the stochastic variables $\{\rho_1, \ldots, \rho_T\}$. Hence, the set of scenarios corresponds to the set of leaves of the scenario tree, N_T, and nodes of the tree at level $t \geq 1$ (the set N_t) correspond to possible realizations of ρ^t. A set of interior nodes of the scenario tree, excluding the root node and leaves, is denoted by N_I. A node of the tree (or event) is represented by $\mathbf{e} = (s, t)$, where s is a scenario (path from root to leaf), and stage t specifies a particular node on that path. The root of the tree is $\mathbf{0} = (s, 0)$ and the ancestor (parent) of event $\mathbf{e} = (s, t)$ is $a(\mathbf{e}) = (s, t - 1)$. $p_{\mathbf{e}}$ is the conditional probability of event $\mathbf{e}$, given its parent event $a(\mathbf{e})$. Figure 4 displays an example of scenario tree. The approaches to generate scenario tree are described in [2].

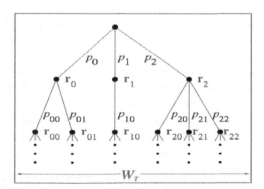

Fig. 4. Scenario tree for multi-stage mean-variance.

Let $\mathbf{w_e}, \mathbf{b_e}, \mathbf{s_e}$ define decision variables for asset allocations and transactions of buying and selling at node $\mathbf{e}$, respectively. The dynamic structure of the model is captured by investment strategy which is defined by asset weights at each interior node of the scenario tree as follows:

$$r_e \circ \mathbf{w}_{a(e)} + (1 - \mathbf{c_b}) \circ \mathbf{b_e} - (1 + \mathbf{c_s}) \circ \mathbf{s_e} = \mathbf{w_e}, \qquad \mathbf{e} \in N_I \qquad (24)$$

Since transactions of buying and selling at the last stage are not allowed, asset weights at $t = T$ are computed as

$$r_e \circ \mathbf{w}_{a(e)} = \mathbf{w_e}, \qquad \mathbf{e} \in N_T \qquad (25)$$

We require subsequent transactions (buy $= \mathbf{b}_t$, sell $= \mathbf{s}_t$) not to alter the wealth within the stage t. Hence, we have the condition

$$\mathbf{1}'\mathbf{b_e} - \mathbf{1}'\mathbf{s_e} = 0, \qquad \mathbf{e} \in N_I \qquad (26)$$

The allocation of the initial budget of 1 can be represented with the following constraints:

$$\mathbf{p} + (1 - \mathbf{c_b})\mathbf{b_0} - (1 + \mathbf{c_s})\mathbf{s_0} = \mathbf{w_0} \qquad (27)$$

$$\mathbf{1}'\mathbf{b_0} - \mathbf{1}'\mathbf{s_0} = 1 - \mathbf{1}'\mathbf{p} \qquad (28)$$

The expected wealth at the last stage W_T is the expected portfolio return and computed as

$$W_T = \sum_{e \in N_T^k} P_e(\mathbf{w}_{a(e)} - \overline{\mathbf{w}}_{a(e)})' r_e$$

The risk is measured as the variance of the portfolio return relative to the benchmark $\overline{\mathbf{w}}$ and formulated as

$$\sum_{t=1}^{T} \alpha_t \sum_{e \in N_t} P_e \left[(\mathbf{w}_{a(e)} - \overline{\mathbf{w}}_{a(e)})'(\Lambda_e + \hat{\mathbf{r}}_e\hat{\mathbf{r}}_e')(\mathbf{w}_{a(e)} - \overline{\mathbf{w}}_{a(e)}) \right]$$

where P_e is probability of event $\mathbf{e}$ and computed as $P_e = \prod_{i=1,\ldots,t} P_{(s,i)}$. α_t weights the risk of period t and the portfolio risk is measured as a weighted sum of risk at $t = 1, \ldots, T$.

The multi-stage portfolio reallocation problem can be expressed as the minimization of the trade off between risk and expected wealth subject to constraints which describe the growth of the portfolio along all the various scenarios and bounds on the decision variables. Box constraints are defined on $\mathbf{w_e}$, $\mathbf{b_e}$, $\mathbf{s_e}$ for each event $\mathbf{e} \in N_I$ to prevent the short sale and enforce any restriction imposed by the investor.

$$\min_{\mathbf{w},\mathbf{b},\mathbf{s}} \; \gamma \sum_{t=1}^{T} \alpha_t \sum_{e \in N_t} P_e \left[(\mathbf{w}_{a(e)} - \overline{\mathbf{w}}_{a(e)})'(\Lambda_e + \hat{\mathbf{r}}_e\hat{\mathbf{r}}_e')(\mathbf{w}_{a(e)} - \overline{\mathbf{w}}_{a(e)}) \right]$$

$$- \sum_{e \in N_T^k} P_e(\mathbf{w}_{a(e)} - \overline{\mathbf{w}}_{a(e)})' r_e$$

subject to

$$\mathbf{p} + (1 - \mathbf{c_b})\mathbf{b_0} - (1 + \mathbf{c_s})\mathbf{s_0} = \mathbf{w_0}$$

$$\mathbf{1'b_0} - \mathbf{1's_0} = 1 - \mathbf{1'p}$$

$$r_\mathbf{e} \circ \mathbf{w}_{a(\mathbf{e})} + (1 - \mathbf{c_b}) \circ \mathbf{b_e} - (1 + \mathbf{c_s}) \circ \mathbf{s_e} = \mathbf{w_e}, \qquad \mathbf{e} \in N_I$$

$$\mathbf{1'b_e} - \mathbf{1's_e} = 0, \qquad \mathbf{e} \in N_I$$

$$r_\mathbf{e} \circ \mathbf{w}_{a(\mathbf{e})} = \mathbf{w_e}, \qquad \mathbf{e} \in N_T$$

$$\mathbf{w_e^L} \le \mathbf{w_e} \le \mathbf{w_e^U}, \qquad \mathbf{e} \in N_I$$

$$0 \le \mathbf{b_e} \le \mathbf{b_e^U}, \qquad \mathbf{e} \in N_I$$

$$0 \le \mathbf{s_e} \le \mathbf{s_e^U}, \qquad \mathbf{e} \in N_I$$

The level of risk aversion optimized for is determined by the scaling constant γ. When $\gamma = 0$, the pure risk-seeking investment strategy (at the highest end of the efficient frontier) is obtained by solving a linear programming problem. When $\gamma = \infty$, completely risk-averse strategy (at the lowest end of the efficient frontier) is obtained by solving the quadratic programming problem (by ignoring the expected portfolio return).

Multi-stage Minimax Portfolios

Multi-stage mean-variance optimization framework can be extended to worst-case design with multiple rival return and risk scenarios. The optimal portfolio is constructed (relative to benchmark) simultaneously with the worst-case to take account of all rival scenarios. The portfolio is balanced at each stage incorporating scalable (not fixed) transaction cost and its relative performance is measured in terms of returns and the volatility of returns.

Assume that the risk scenarios are considered at each event of the future realizations. Therefore, we have the same number of covariance matrices, $I_\mathbf{e}$, at each node of the scenario tree, $\mathbf{e} \in N_t$ for $t = 1, \cdots, T$. The covariance matrices and scenario tree are an input to the minimax model. Rival return scenarios are determined by the scenario tree as number of events at the first stage since investors wish to survive at the first stage. For instance, the scenario tree in Figure 5 consists of three rival return scenarios. Let K denote the number of rival return scenarios. Given the rival risk and return scenarios (or scenario tree), the general minimax model for multi-stage asset allocation problem is formulated as follows;

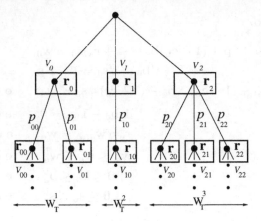

Fig. 5. Multi-stage worst-case mean-variance on a scenario tree.

$$\min_{\mathbf{w,b,s}} \left\{ \gamma \sum_{t=1}^{T} \alpha_t \sum_{\mathbf{e} \in N_t} \max_i \left[P_{\mathbf{e}}(\mathbf{w}_{a(\mathbf{e})} - \overline{\mathbf{w}}_{a(\mathbf{e})})'(\Lambda_i + r'_{\mathbf{e}}r_{\mathbf{e}})(\mathbf{w}_{a(\mathbf{e})} - \overline{\mathbf{w}}_{a(\mathbf{e})}) \right] \right\}$$

$$+ \min_{\mathbf{w,b,s}} \left\{ -\min_k \left[\sum_{\mathbf{e} \in N_T^k} P_{\mathbf{e}}(\mathbf{w}_{a(\mathbf{e})} - \overline{\mathbf{w}}_{a(\mathbf{e})})' r_{\mathbf{e}} \right] \right\}$$

$$\equiv \min_{\mathbf{w,b,s}} \left\{ \gamma \sum_{t=1}^{T} \alpha_t \sum_{\mathbf{e} \in N_t} \max_i \left[J_{\mathbf{e}}^i(\mathbf{w}) \right] - \min_k \left[W_T^k(\mathbf{w}) \right] \right\}$$

where

$$J_{\mathbf{e}}^i(\mathbf{w}) = P_{\mathbf{e}}(\mathbf{w}_{a(\mathbf{e})} - \overline{\mathbf{w}}_{a(\mathbf{e})})'(\Lambda_i + r'_{\mathbf{e}}r_{\mathbf{e}})(\mathbf{w}_{a(\mathbf{e})} - \overline{\mathbf{w}}_{a(\mathbf{e})})$$

$$W_T^k(\mathbf{w}) = \sum_{\mathbf{e} \in N_T^k} P_{\mathbf{e}}(\mathbf{w}_{a(\mathbf{e})} - \overline{\mathbf{w}}_{a(\mathbf{e})})' r_{\mathbf{e}}$$

for $i \in I_{\mathbf{e}}$, $k = 1, \cdots, K$, $t = 1, \cdots, T$ and $\mathbf{e} \in N_t$. Let $\nu_{\mathbf{e}}$ and μ be the worst-case risk at node $\mathbf{e} \in N_t$ and the worst-case return, respectively. In order to solve the minimax problem above, we reformulate it as a quadratically constrained mathematical program

$$\min_{\mathbf{w,b,s}} \gamma \sum_{t=1}^{T} \alpha_t \sum_{\mathbf{e} \in N_t} \nu_{\mathbf{e}} - \mu$$

subject to

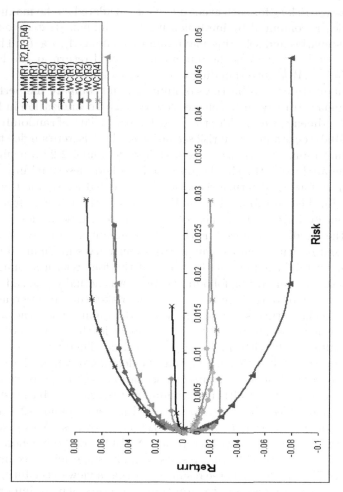

Fig. 6. Minimax versus single scenario optimization and the worst-case analysis.

$$\mathbf{p} + (1 - \mathbf{c_b})\mathbf{b}_0 - (1 + \mathbf{c_s})\mathbf{s}_0 = \mathbf{w}_0$$

$$\mathbf{1}'\mathbf{b}_0 - \mathbf{1}'\mathbf{s}_0 = 1 - \mathbf{1}'\mathbf{p}$$

$$r_{\mathbf{e}} \circ \mathbf{w}_{a(\mathbf{e})} + (1 - \mathbf{c_b}) \circ \mathbf{b_e} - (1 + \mathbf{c_s}) \circ \mathbf{s_e} = \mathbf{w_e}, \qquad \mathbf{e} \in N_I$$

$$\mathbf{1}'\mathbf{b_e} - \mathbf{1}'\mathbf{s_e} = 0, \qquad \mathbf{e} \in N_I$$

$$r_{\mathbf{e}} \circ \mathbf{w}_{a(\mathbf{e})} = \mathbf{w_e}, \qquad \mathbf{e} \in N_T$$

$$\sum_{\mathbf{e} \in N_T^k} P_{\mathbf{e}}(\mathbf{w}_{a(\mathbf{e})} - \overline{\mathbf{w}}_{a(\mathbf{e})})' r_{\mathbf{e}} \geq \mu, \quad k = 1, \cdots, K$$

$$P_{\mathbf{e}}(\mathbf{w}_{a(\mathbf{e})} - \overline{\mathbf{w}}_{a(\mathbf{e})})'(\Lambda_i + r_{\mathbf{e}}'r_{\mathbf{e}})(\mathbf{w}_{a(\mathbf{e})} - \overline{\mathbf{w}}_{a(\mathbf{e})}) \leq \nu_{\mathbf{e}}, \quad i \in I_{\mathbf{e}}, \ \mathbf{e} \in N_t,$$

$$t = 1, \cdots, T$$

$$\mathbf{w_e}^L \leq \mathbf{w_e} \leq \mathbf{w_e}^U, \qquad \mathbf{e} \in N_I$$

$$0 \leq \mathbf{b_e} \leq \mathbf{b_e}^U, \qquad \mathbf{e} \in N_I$$

$$0 \leq \mathbf{s_e} \leq \mathbf{s_e}^U, \qquad \mathbf{e} \in N_I.$$

The worst-case risk $\nu_\mathbf{e}$ for each event $\mathbf{e} \in N_t$ is calculated as the maximum risk value, which is computed by implementing the minimax strategy on specific rival risk scenarios and selecting maximum one among Λ_i, $i \in I_\mathbf{e}$. The worst-case return μ is obtained as the minimum expected return at the last stage among $W_T^1, \cdots, W_T^K$ corresponding to each sub-scenario tree.

We demonstrate the robustness of minimax and compare the performance of the minimax strategy with a single scenario-based optimization in Figure 6. For the experiments, we consider the same historical data of randomly selected ten FTSE100 stocks and 3 rival risk scenarios used in discrete single-stage minimax optimization. The scenario tree with 3 stages and 4-2-2 branching topology is generated using the simulation based method, described in [2]. Therefore, we have four rival return scenarios (represented by R_i for $i = 1, \cdots, 4$ in Figure 6). The top four plots in Figure 6 are the efficient frontiers obtained from optimizing with each of the rival return scenarios alone. The curve in the middle of Figure 6 is the robust minimax strategy based on all rival risk and return scenarios. The bottom four curves in Figure 6 represent what actually would happen if the worst of the four return scenarios (with respect to the optimized portfolio in question) was actually realized. Not surprisingly, the more optimistic the worst-case risk-return frontier associated with a particular return scenario, the worse the failure when the worst-case occurs. A worst-case optimal strategy would yield the best decision determined simultaneously with the worst-case scenario. Therefore, the worst-case strategy protects against risk of adopting the decision based on the wrong scenario. In order to illustrate the effect of scenario trees on the performance of worst-case analysis, the multi-stage minimax strategies are backtested using different scenario trees. The scenario trees with 3 stages and 2-2-2 and 4-2-2 branching topologies are generated by antithetic variates (ANT), sequential method (SEQ) and Sobol low-discrepancy (SOBOL) random generator.

Figure 7 presents backtesting results of multi-stage minimax investment strategies at 75% risk level. These results show that a view of the future defined in terms of a scenario tree and its branching structure play an important role on the performance of multi-stage stochastic programming models.

5 Conclusions

In this chapter, we address issues arising in optimization under uncertainty. Special emphasis is given to robustness of minimax and algorithmic issues, such as multiple solutions and finding global worst-cases. In particular discrete and continuous models are discussed and contrasted with stochastic optimization. A number of examples illustrates applications of worst-case analysis. Expected value optimization is compared with worst-case robust strategies. The cost of implementing a worst-case robust decision needs to be evaluated in view of its expected performance in a stochastic setting. A decision based on

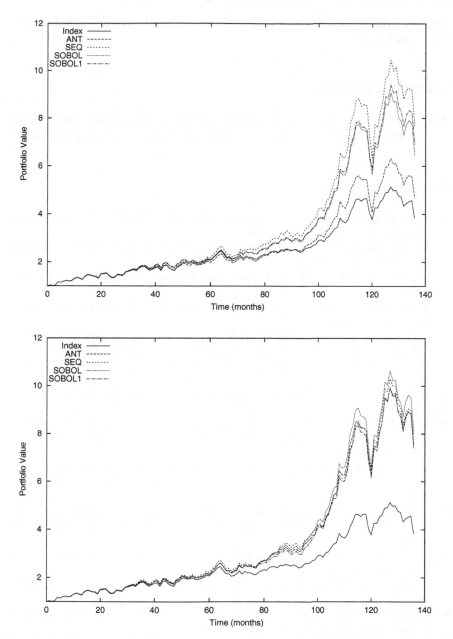

Fig. 7. Backtesting with different scenario trees generated with 3 stages 2-2-2 and 4-2-2 branching, at 75% risk level.

expected value optimization, on the other hand, needs to be justified in view of the worst-case.

Acknowledgments

This research was supported by EPSRC grant GR/T02560/01. The authors are grateful to anonymous referee for helpful comments and suggestions.

References

1. T. Basar and P. Bernhard, H^∞-Optimal Control and Related Minimax Design Problems, Birkhauser, Boston, 1991.
2. N. Gulpinar, B. Rustem, and R. Settergren, *Journal of Economics Dynamics and Control* **28**, 1291 (2004).
3. N. Gulpinar and B. Rustem, Forthcoming in *Numerical Methods in Finance*, 2005.
4. N., Gulpinar, B. Rustem, R. Settergren, *Innovations in Financial and Economic Networks*, 3, (2003), p. 46-63.
5. L. Hansen, L. and T. J. Sargent, *Review of Economic Studies* **66**, 873 (1999).
6. P. Kall, "Stochastic Linear Programming", Springer, Berlin, (1976).
7. M. J. Lawrence, R. H. Edmunson and M.J., O'Connor, *Management Science* **32**, 1521 (1986).
8. S. Makridakis and R. Winkler, *Management Science* **29**, 987 (1983).
9. H., Markowitz, *Journal of Finance* **7**, 77 (1952).
10. A. Orphanides and V. Wieland, *European Economic Review* **44**, 1351 (2000).
11. B. Rustem, R. Becker and W. Marty, *Journal of Economic Dynamics and Control* **24**, 1591 (2000).
12. B. Rustem and M. Howe, "Algorithms for Worst-Case Design and Applications to Risk Management", Princeton University Press, London and New Jersey, (2002).
13. B. Rustem and R. Settergren, in *Computational Methods in Decision Making, Economics and Finance: Optimization Models*, 75 , Kluwer Academic Publishers (2002).
14. R. J. Tetlow and P. von zur Muehlen, *Journal of Economic Dynamics and Control* **25**,911 (2001).
15. S. Zakovic, C. C. Pantelides and B. Rustem, *Annals of Operations Research* **99**, 59 (2000).
16. S. Zakovic, B. Rustem and V. Wieland, in *Decision and Control in Management Science*, Kluwer Academic Publishers, (2002).

Decentralized Estimation for Cooperative Phantom Track Generation

Tal Shima[1], Phillip Chandler[1] and Meir Pachter[2]

[1] Air Vehicles Directorate, Air Force Research Laboratory
   Wright-Patterson AFB, OH
   E-mail: shima_tal@yahoo.com, phillip.chandler@wpafb.af.mil
[2] Department of Electrical Engineering
   Air Force Institute of Technology
   Wright-Patterson AFB, OH
   E-mail:meir.pachter@afit.edu

Summary. A decentralized estimation-decision strategy is derived for a team of electronic combat air vehicles (ECAVs) deceiving a network of radars. For the deception, a phantom target track is cooperatively generated by each ECAV applying range delay on the individual radar pulses. To continuously obtain a feasible phantom track, the team must tightly coordinate the phantom's trajectory so as not to violate any of the system constraints. The coordination is performed by a decentralized decision process for which each ECAV periodically transmits its constraints on feasible tracks. To perform the decision process with minimal communication between the ECAVs, a decentralized estimation algorithm is proposed where each ECAV continuously estimates the states of its teammates and their respective radar position based on their individual transmitted constraints. Thus, all the information obtained in the individual messages is extracted and group coordination is obtained. Moreover, if there are gaps in communication the team coordination can be maintained. Simulation results confirm the viability of the proposed decentralized estimation-decision team strategy.

1 Introduction

Radio detection and ranging (radar) is achieved by transmitting radio waves and listening to the returning echoes [3]. Detection is achieved if the returning echoes are strong enough to be distinguished from the background noise. Ranging can be determined by measuring the radio waves round trip time to the target. Radar system electronic counter measures were developed early on [3]. The earliest and simplest is chaff, consisting of metal-coated dielectric fibers; dispensed in large numbers, they can produce strong radar echoes. Another simple method is noise jamming by raising the level of the background against which target returns must be detected. A more sophisticated method involves creating a false phantom target by delaying a received pulse, from a threat radar, for a period corresponding to the desired additional range of the false target. This method is often termed range deception.

Electronic combat air vehicles (ECAVs) are designed so as to have low detectability, making them essentially invisible to conventional radars. In order to spoof an enemy radar, an ECAV can employ various electronic counter measures, including range deception as discussed above. In [2] such a problem was analyzed. Different feasible phantom tracks were investigated and closed form solutions were obtained for the ECAV's trajectory given specific phantom tracks. As a counter measure, a radar network can simultaneously track a target and by correlating the tracks distinguish between feasible and phantom targets. To counter this ability, a team of ECAVs, consisting of the same number of vehicles as radars, can cooperatively engage a radar network and create a coherent phantom target track without being detected [2]. In [1] a cooperative control algorithm for such an ECAV team deceiving a radar network has been proposed and different phantom track trajectories have been studied.

Cooperative control algorithms can be implemented in a redundant centralized manner in which the decision algorithm is replicated over the multiple agents in the team. Assuming perfect communication all vehicles will have the same information set and thus come up with the same team plan. This will result in synchronized action. However, in a realistic scenario communication constraints are expected. Information flow constraints, such as communication delays, may produce different information sets for the different UAVs in the team leading to multiple strategies. For the team phantom track deception of a radar network, this will result in multiple uncorrelated tracks, defeating the deception process.

In this chapter a cooperative decision-estimation algorithm is proposed for cooperatively creating a radar phantom track with communication constraints. The remainder of this chapter is organized as follows. In the next section the one-on-one radar deception problem is posed. Then, the group decision process and its requirements are discussed. The estimation process of team members' states and the related decision process is then presented. Concluding remarks are offered in the last section.

2 One-On-One Problem

In this section we present, based on [2], the problem of radar deception by one ECAV. The deception is performed using a range delay of the radar signal and thus the ECAV is positioned on the line of sight (LOS) between the stationary radar and the phantom target. The range delay and the ECAV's motion determine the phantom target track. A schematic view of the two dimensional deception geometry is shown in Fig. 1.

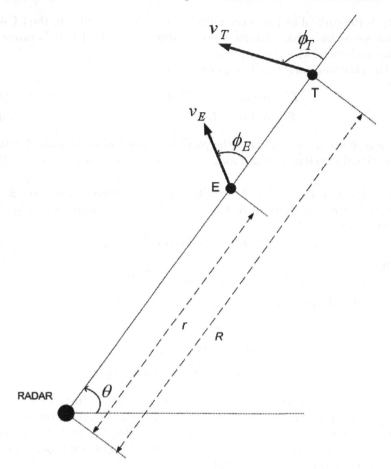

Fig. 1. One-on-one deception engagement

For simplicity we assume that the ECAV and phantom target have constant speeds, denoted v_E and v_T respectively. We define the normalized parameter

$$\alpha \equiv v_T/v_E > 0 \tag{1}$$

Without loss of generality we normalize the ranges from the radar to the ECAV and to the phantom target, r and R respectively, by the initial range to the phantom target R_0; and normalize the speeds by v_E. Thus, the normalized equations of motion for the ECAV, expressed in a polar coordinate system (r, θ) attached to the stationary radar, are

$$\dot{r} = \cos \phi_E \quad ; \quad r(0) = r_0/R_0 \tag{2a}$$
$$\dot{\theta} = \sin \phi_E/r \quad ; \quad \theta(0) = \theta_0 \tag{2b}$$

where ϕ_E is the control and r_0 is the initial range from the radar to the ECAV. Note that we assume that, due to antenna limitations, the ECAV's range to the radar is confined to $r \in [r_{min}, r_{max}]$.

For the phantom target these equations are

$$\dot{R} = \alpha \cos \phi_T \quad ; \quad R(0) = 1 \tag{3a}$$
$$\dot{\theta} = \alpha \sin \phi_T / R \quad ; \quad \theta(0) = \theta_0 \tag{3b}$$

where ϕ_T is the control. Note that the phantom target must be placed within the operational envelope of the radar and thus its range is confined to $R \in [0, R_{max}]$

By construction the angle θ is identical in both systems; and thus, given ϕ_T, the angle ϕ_E is selected such that the relationship from Eqs. (2b) and (3b) holds

$$\sin \phi_E / r = \alpha \sin \phi_T / R \tag{4}$$

Consequently

$$\cos \phi_E = \pm \sqrt{1 - (\alpha r \sin \phi_T / R)^2} \tag{5}$$

Using Eqs. (2), (3), and (5) we obtain

$$\dot{R} = \alpha \cos \phi_T \quad ; \quad R(0) = 1 \tag{6a}$$
$$\dot{\theta} = \alpha \sin \phi_T / R \quad ; \quad \theta(0) = \theta_0 \tag{6b}$$
$$\dot{r} = \beta \sqrt{1 - (\alpha r \sin \phi_T / R)^2}; \quad r(0) = r_0 / R_0 \tag{6c}$$

where $\beta = 1$ or -1 and is selected such that the ECAV's constraints are not reached and to maintain that $r \leq R \ \forall \ t > 0$ (i.e. the ECAV is between the radar and the phantom target, on the LOS). Note that the system has three states (R, θ, r), two outputs (R, θ), and one input (ϕ_T). For a solution to exist the following condition must be satisfied

$$|\sin \phi_T| \leq \frac{R(t)}{\alpha r(t)} \quad \forall \quad t > 0 \tag{7}$$

meaning that the phantom target angular rate does not exceed the maximum angular rate achievable by the ECAV. For a detailed analysis of feasible trajectories under different constraints see [2].

3 Team Decision Problem

Let $V = \{1, 2, ..., n\}$ be a set of ECAVs employing range deception on a network of n radars; and let $T = \{1, 2, ..., n\}$ be the set of these radars. The team must cooperatively create a phantom target while not exceeding any of the ECAV kinematic restrictions.

A schematic view of the cooperative deception problem is shown in Fig. 2 where (R, θ) is a globally known polar coordinate system and each ECAV $i \in V$

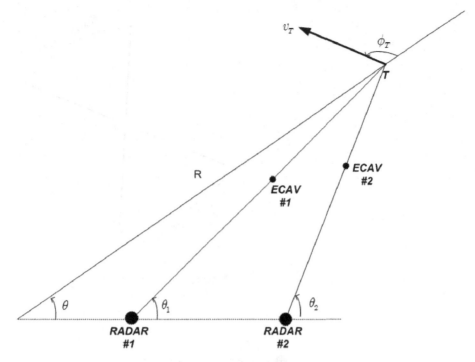

Fig. 2. Multiple radars deception engagement

also has a local polar coordinate system (R_i, θ_i). Without loss of generality we assume that the radars are positioned on a line.

The ECAVs coordinate by exchanging information regarding feasible phantom trajectories. Thus, we employ a decision process in which each ECAV communicates to his teammates two variables, ϕ_{Ti}^u and ϕ_{Ti}^l, representing the sector for a feasible phantom target heading ϕ_T (see Fig. 3). This sector, computed based on the maximum angular rate the ECAV can follow, transformed to the global coordinate system, is

$$S_i = \left[-\phi_{Ti}^l, \phi_{Ti}^u\right] \cup \left[\pi - \phi_{Ti}^l, \pi + \phi_{Ti}^u\right] \quad ; \quad i = 1, ..., n \tag{8}$$

where

$$\phi_{Ti}^u = \phi_{Timax} + \theta_i - \theta \quad ; \quad i = 1, 2, ..., n \tag{9a}$$
$$\phi_{Ti}^l = \phi_{Timax} - \theta_i + \theta \quad ; \quad i = 1, 2, ..., n \tag{9b}$$

and ϕ_{Timax} is computed from Eq. (7) such that

$$\phi_{Timax} = \sin^{-1} \frac{R_i}{\alpha r_i} \leq \pi/2 \, , \quad \frac{R_i}{\alpha r_i} \leq 1 \quad ; \quad i = 1, 2, ..., n \tag{10}$$

Note that if $\frac{R_i}{\alpha r_i} > 1$ then $\phi_{Timax} = \pi/2$, meaning that the ECAV can follow any phantom target heading. Using Eq. (8) the sector for a feasible target

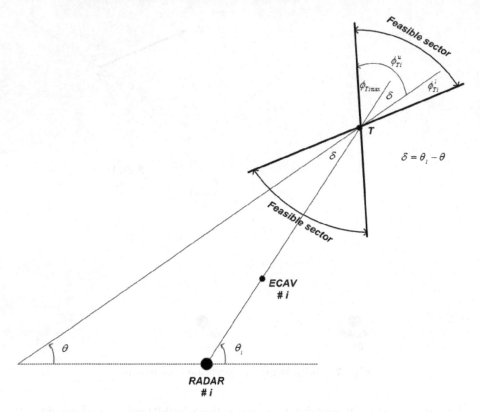

Fig. 3. Feasible phantom target heading

heading is

$$S = S_1 \cap S_2 ... \cap S_n \tag{11}$$

The required phantom target heading ϕ_T^r is an input to the system. It is computed based on the required phantom path trajectory, chosen based on operational considerations. The actual target heading is selected such that

$$\phi_T = \underset{\phi \in S}{argmin} |\phi - \phi_T^r| \tag{12}$$

Note that only if $\phi_T^r \in S$ then $\phi_T = \phi_T^r$; and if Eq. (12) has two solutions then we choose the one with the smaller value.

An example of the cooperative deception engagement between 3 ECAVs and 3 radars is presented in Figs. 4, 5. The required and actual phantom target trajectories, as well as those of the ECAVs, are plotted in Fig. 4. The instantaneous LOS at $t = 0, 1, ..., 5$ from the phantom target to the 3 deceived radars, are also plotted. The required and actual heading as well as the relevant constraints from each ECAV (ϕ_{Ti}^u) are plotted in Fig. 5. From both figures it is evident that up to about 3.2 sec from the beginning of the scenario the actual phantom trajectory exactly follows the required one. At that moment the angular rate limit of ECAV 2 is reached. From then on this constraint is active and the actual phantom trajectory can no longer follow the required one. Note that since the coordinate system of ECAV 1 coincides with the global one and it can follow any phantom target heading, its constraint ϕ_{T1}^u is constant and equal to $\pi/2$.

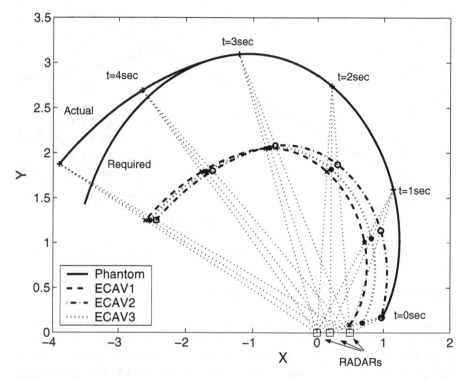

Fig. 4. Trajectories for the deception engagement between 3 ECAVs and radars

The decision process discussed above may be implemented in a centralized manner, performed on one of the ECAVs. The selected phantom heading ϕ_T may then be transmitted to the rest of the group. The process can also be implemented in a redundant centralized manner in which the algorithm is replicated across the UAV group. Provided there exists perfect communication then all ECAVs will come up with the same phantom target heading, resulting in a synchronized action generating a phantom target.

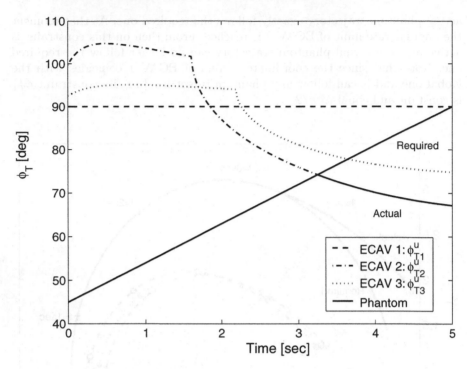

Fig. 5. Constraints in a deception engagement between 3 ECAVs and radars

4 Estimation-Decision Process

The decision process presented in the previous section assumes perfect communication within the ECAV team. For coordination under limited communication bandwidth, each ECAV employs an estimator on its teammates states. This process enables the continuous estimation of each ECAV's constraints, which are required for the cooperative decision process.

Consider the two variables ϕ^u_{Ti} and ϕ^l_{Ti} that each ECAV transmits to the group. Since, based on Eqs. (9)-(10), these variables contain information on the states of vehicle $i \in V$ they may be viewed as noiseless measurements. Thus, we define the measurements of each system as

$$z_{i1} = \phi^u_{Ti} + \theta + \omega_1 \tag{13a}$$
$$z_{i2} = \phi^l_{Ti} - \theta + \omega_2 \tag{13b}$$

where ω_1 and ω_2 are fictitious white noises and θ is the angle from the globally known reference system.

From Eqs. (6) the equations of motion are

$$\dot{R}_i = \alpha \cos \phi_{Ti} \quad ; \quad R_i(0) = 1 \tag{14a}$$

$$\dot{\theta}_i = \alpha \sin \phi_{Ti}/R \quad ; \quad \theta_i(0) = \theta_{i0} \tag{14b}$$

$$\dot{r}_i = \beta \sqrt{1 - (\alpha r_i \sin \phi_{Ti}/R_i)^2} \quad ; \quad r_i(0) = r_{i0}/R_{i0} \tag{14c}$$

where

$$\phi_{Ti} = \phi_T - \theta_i + \theta \tag{15}$$

Note that the heading ϕ_T is the input to the system, known to the entire team. Using Eqs. (13)-(15) an EKF can be constructed for all $i \in V$. Thus, each ECAV can employ $n - 1$ EKFs on its teammates. An example of an estimation process is given in Figs. 6, 7 for $\alpha = 1.5$, $\phi_T^r(t) = \pi/4 + t\pi/20$, and a communication rate of $1Hz$. From Fig. 6 it is evident that the trajectory estimation is quite accurate, and is converging. From Fig. 7 it is evident that the angle θ is estimated quite accurately after only one observation update, while the estimation of the range r converges more slowly. This is expected since the measurements include direct information only on the angle θ (by subtraction of the two measurements).

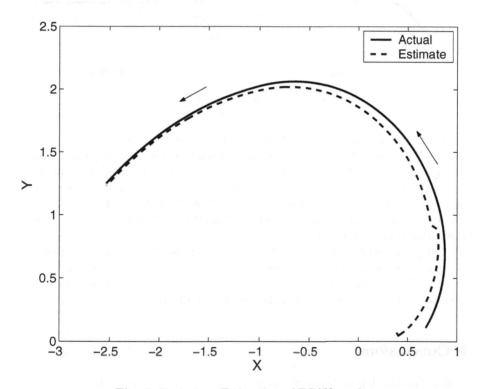

Fig. 6. Trajectory Estimation of ECAV no. 3.

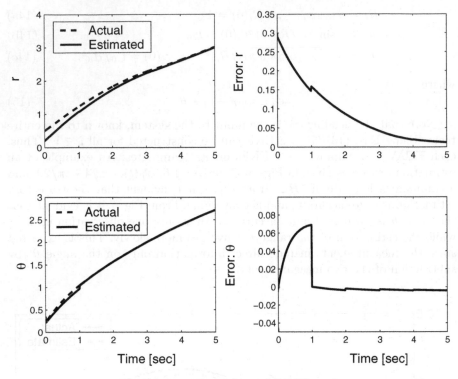

Fig. 7. State Estimation of ECAV no. 3.

Each ECAV knows the phantom's position in the global coordinate system (R, θ). Thus, using the estimates of the teammates states (θ_i, R_i) the position of each radar $i \in T$ being deceived can also be estimated. An example of such an estimation process for the same scenario described above is plotted in Fig. 8 in a Cartesian reference frame. Note that the initial estimate is $(\hat{X}, \hat{Y}) = (0.05, -0.17)$ and it is evident that after the first observation update the estimation error is very small.

Using the estimation process discussed above each ECAV can continuously estimate the constraints of its teammates on the feasible phantom target trajectory. Thus, team coordination can be retained with a slow communication rate or even when there are gaps in the communication.

5 Conclusions

A novel estimation-decision strategy, for the cooperative deception of a radar network by a team of ECAVs, has been proposed. For the deception, a phantom target track is cooperatively generated by each ECAV applying range

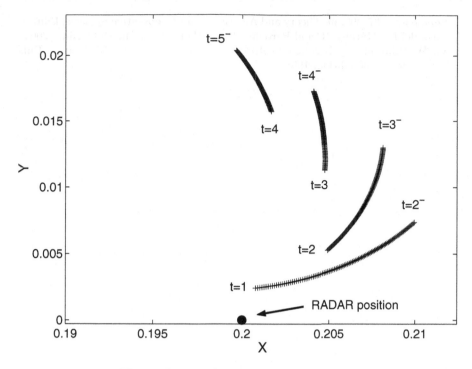

Fig. 8. Position Estimation of radar no. 3.

delay on the individual radar pulses. The team must tightly coordinate the phantom's trajectory so as not to violate any of the system constraints.

The coordination is performed by a decentralized decision process for which each ECAV periodically transmits his constraints on feasible tracks. It has been shown that for the continuous generation of a phantom target track by a team of ECAVs there is no need for high bandwidth communication. For team coordination, each ECAV estimates its teammates states enabling it to continuously estimate each ECAVs constraints, which are in turn required for the cooperative decision process. The respective radar position of each ECAV can also be estimated based on the communicated constraints on feasible target tracks.

References

1. D.H.A. Maithripala and S. Jayasuriya. Radar deception through phantom track generation. In *Proceedings of the American Control Conference*, Arlington, Virginia, 2005. American Automatic Control Council.
2. M. Pachter, P.R. Chandler, K.B. Purvis, S.D. Waun, and R.A. Larson. *Multiple RADAR Phantom Tracks from Cooperating Vehicles using Range-Delay Decep-*

tion, pages 367–390. in: Theory and Algorithms for Cooperative Systems, Editors: Grundel D., Murphy, R. and Pardalos, P.M., World Scientific Publishing, 2004.

3. G. W. Stimson. *Introduction to Airborne Radar*, pages 439–455. Scitech Publishing, Inc., 2nd edition, 1998.

Information Flow Requirements for the Stability of Motion of Vehicles in a Rigid Formation

Sai Krishna Yadlapalli, Swaroop Darbha and Kumbakonam R. Rajagopal

Department of Mechanical Engineering,
Texas A&M University,
College Station,
TX -77843-3123, USA.
E-mail: kris5372@tamu.edu, {dswaroop,krajagopal}@mengr.tamu.edu

Summary. It is known in the literature on Automated Highway Systems that information flow can significantly affect the propagation of errors in spacing in a collection of vehicles. This chapter investigates this issue further for a homogeneous collection of vehicles, where in the motion of each vehicle is modeled as a point mass and is digitally controlled. The structure of the controller employed by the vehicles is as follows: $U_i(z) = C(z) \sum_{j \in S_i} (X_i - X_j - \frac{L_{ij} z}{z-1})$, where $U_i(z)$ is the (z- transformation of) control action for the ith vehicle, X_i is the position of the ith vehicle, L_{ij} is the desired distance between the ith and the jth vehicles in the collection, $C(z)$ is the discrete transfer function of the controller and S_i is the set of vehicles that the ith vehicle can communicate with directly. This chapter further assumes that the information flow is undirected, i.e., $i \in S_j \iff j \in S_i$ and the information flow graph is *connected*. We consider information flow in the collection, where each vehicle can communicate with a maximum of $q(n)$ vehicles. We allow $q(n)$ to vary with the size n of the collection. We first show that $C(z)$ cannot have any zeroes at $z = 1$ to ensure that relative spacing is maintained in response to a reference vehicle making a maneuver where its velocity experiences a steady state offset. We then show that if the control transfer function $C(z)$ has one or more poles located at $z = 1$, then the motion of the collection of vehicles will become unstable if the size of the collection is sufficiently large. These two results imply that $C(1) \neq 0$ and $C(1)$ must be well defined. We further show that if $\frac{q(n)}{n} \to 0$ as $n \to \infty$, then there is a low frequency sinusoidal disturbance of at most unit amplitude acting on each vehicle such that the maximum error in spacing response increase at least as $\Omega\left(\sqrt{\frac{n^3}{q^3(n)}}\right)$. A consequence of the results presented in this chapter is that the maximum of the error in spacing and velocity of any vehicle can be made insensitive to the size of the collection only if there is at least one vehicle in the collection that communicates with at least $\Omega(n)$ other vehicles in the collection. We also show that there can be at most one vehicle that communicates with $\Omega(n)$ vehicles and that any other vehicle in the collection can only communicate with at most p vehicles, where p depends only on the chosen controller and the its sampling time.

1 Introduction

Recent advances in a variety of technologies such as communication, computation, sensing and actuation have enabled the development and increased the possibility of deployment of collections of Unmanned Vehicles (UVs) (or simply vehicles) for a wide variety of tasks. UVs are central to automating driving tasks in an Automated Highway System (AHS) [1], the dynamic positioning of mobile offshore bases for creating a runway for large aircrafts [2] and for information gathering in dangerous environments [13]. There seem to be potentially many advantages to deploying UVs in collections for certain tasks: flexibility, ease of reconfiguration and lower cost of deploying collections of smaller UVs as compared to deploying a larger UV being some of them. In order to realize these potential advantages, the problem of coordinating the motion of the collection of vehicles must be addressed and this chapter is devoted to an analysis of this problem.

It is conceivable that a collection of vehicles will be required to maintain (or remain close to) specified discernible geometric patterns during its motion. We call such a collection of vehicles as a formation if every vehicle aids in the maintenance of the specified geometric pattern by coordinating its motion through communication with or sensing other vehicles in the collection. Of recent interest in the research community is the rigid formation of vehicles, where it is desired that the distance between any two vehicles remain constant throughout the motion. In such a case, the desired motion of every vehicle in a formation is determined by the desired motion of a few vehicles in the collection. Since vehicles in a formation are coupled dynamically by feedback, errors in spacing and velocity (defined as the deviation in the position and velocity from their respective desired values) of a vehicle propagate from one vehicle in the formation to the other.

In an AHS, such rigid formations (referred to as a platoons) are desired from the viewpoint of maintaining safety and enhancing the throughput of vehicles on a section of a congested highway. A rigid formation is helpful for localization in partially known environments in the case of mobile robots [7], and in drag reduction via close formation flight [4, 5].

An issue with the design of controllers for vehicles in a collection is that of *collective stability* of the controlled motion of vehicles. This issue arises because errors in spacing and velocity of a vehicle propagate to others in the collection. Intuitively, the collective stability requires the following: With a specified identical controller on every vehicle and with the vehicles starting at their desired positions and velocities, for any given bound, ϵ, is there a bound, δ, *independent of the size of the collection*, on the magnitude of *any* disturbing force that can act on *any* vehicle, so that as the errors in spacing and velocity propagate with the choice of controllers, they always remain smaller than ϵ ? The requirement of the independence of δ from the size of the collection captures the scalability of the stability of motion with the specified controllers. We will mean that controller is scalable if the above requirement of collective stability of controlled motion is met. Since no formation can ever be rigid, we will say that an "approximately rigid formation" can be synthesized if one can synthesize a scalable controller.

In this chapter, we are interested in the synthesis of scalable controllers, which take into account an additional consideration - that of spatial shift-invariance (i.e. controller is not dependent on the index of the vehicle or the size of the collection). From a practical viewpoint, such a controller is easy to develop and implement on every vehicle. This is important for applications such as the Adaptive Cruise Control (ACC) System for ground vehicles, because one does not know *a priori* how many vehicles with an ACC System are placed in succession in traffic. In [9, 6], controllers that used the information about the index of the vehicle in the collection were synthesized; however, for them to achieve an approximately rigid linear formation, the control gains had to increase with the index of the vehicle in a geometric manner and from a practical viewpoint, this is unrealistic since it will lead to saturation of control effort even with small errors in spacing and velocity. For this reason and for the simplicity of treatment, we only consider the restricted class of controllers for further investigation.

The synthesis of an "approximate rigid formation" is strongly influenced by the communication pattern between the vehicles. If the formation has the knowledge of the information of a reference vehicle in the collection, then errors in the spacing and velocity resulting from a disturbance acting on a vehicle, can be made to attenuate as it propagates from one vehicle to another [3, 8]. To date, it is believed that the information of one vehicle must be available to $\Omega(n)$ [1], where n is the number of vehicles in the formation, if one were to construct approximate rigid formations. The results in [3, 8] and even in this chapter point in this direction. The following question naturally arises and is the focus of investigation in this chapter: How does a pattern of communication amongst vehicles affect the propagation of errors? Specifically, with a specified pattern of communication amongst them, can an approximately rigid formation be synthesized? If the answer to the latter question is in the affirmative, one can employ the same controller in each of the vehicles irrespective of the size of the collection, i.e., one can design a "scalable" control system with the given information flow.

The main results of this chapter concern the *necessary conditions* on the information structure for the synthesis of approximately rigid formations and are as follows: If the motion of each vehicle can be represented as the motion of a unit mass under the action of a control force and a disturbance and that the information flow graph is undirected, we show that there is no "scalable" control system if every vehicle can only communicate with at most $q(n)$ vehicles, where n is the size of the collection and $q(n)$ satisfies $\lim_{n \to \infty} \frac{q(n)}{n} = 0$. We show this result by constructing a sinusoidal disturbance of at most unit

---

[1] A function $p(n)$ is $\Omega(q(n))$ if there exists a non-zero constant $c \neq 0$ and a $N^* > 0$ such that $p(n) \geq cq(n)$ for all $n > N^*$

magnitude acting on each vehicle at an appropriately chosen low frequency that results in a maximum error in spacing is of $\Omega(\sqrt{\frac{n^3}{q^3(n)}})$. A consequence of this result is that at least one vehicle in the collection must communicate with at least $\Omega(n)$ other vehicles in the collection for a "scalable" controller to exist. We also show that if the controller incorporates an integral action, the motion of the collection is necessarily unstable for all sizes of the collection greater than a critical value.

We also show that there can be only one vehicle communicating with $\Omega(n)$ vehicles. All other vehicles can only communicate with p other vehicles, where p is dependent only on the plant and controller transfer functions and the sampling time. The chapter is organized as follows: In Section II, we formulate the problem precisely for one-dimensional formations and prove the results stated above. In Section III, we provide corroborating simulations. In Section IV, we summarize the results of this chapter.

2 Problem Formulation for a String of Vehicles Traveling Along a Straight Line

In this section, we consider a string of vehicles moving along a straight line. The first vehicle, which we call reference vehicle, executes maneuvers with bounded velocity and acceleration. The reference vehicle is referred to as lead vehicle in the Automated Highway System (AHS) literature. For each $i \geq 2$, the ith vehicle desires to maintain a fixed following distance $L_{i,i-1}$ from its predecessor. Initially, all vehicles are assumed to be at their desired position and the velocity of all the vehicles are identical. Since, most of the applications require implementation of control law digitally, it would be helpful to consider z-transformations. A detailed treatise of the z-transformations and digitally controlled systems can be found in [12].

2.1 Model of a Vehicle

Let $x(t)$ denote the position of a vehicle measured from the origin of an inertial reference frame at time t. Let T be the sampling time of the controller. We assume that the position of the vehicle remains constant in the time period $[kT, kT + T)$ for $k \geq 0$ and is $x(kT)$. For the sake of brevity, we represent $x(kT)$, $u(kT)$ and $d(kT)$ by $x(k)$, $u(k)$ and $d(k)$ respectively. Then one may express the Z transformation, $X(z)$, of $x(k)$ in terms of the z transformations, $U(z)$ and $D(z)$ of $u(k)$ and $d(k)$ respectively as follows:

$$X(z) = \frac{T^2(z+1)}{2(z-1)^2}[U(z) + D(z)] + \frac{z}{z-1}x(0) + \frac{Tzv(0)}{(z-1)^2}. \tag{1}$$

where $v(0)$ is the initial velocity of the vehicle.

2.2 Further Assumptions and Formulation of the Problem

We make the assumption that the information flow graph is undirected; if a vehicle A transmits the information concerning its state directly to a vehicle B, then vehicle B transmits the information concerning its state directly to vehicle A. Therefore, if S_i is the set of vehicles the ith vehicle in the collection can communicate directly with, this assumption implies that $j \in S_i \Rightarrow i \in S_j$. If the ith vehicle, V_i, and the jth vehicle, V_j, are in direct communication with each other, we refer to the ordered pair (i, j) as a communication link. We particularly assume that the information available to the ith vehicle in the collection is $x_i(k) - x_j(k) - L_{ij}$, where $j \in S_i$ and L_{ij} is the desired distance to be maintained between the ith and the jth vehicles. We restrict the size of S_i (given by $|S_i|$) to be at most q, which may be a function of the size, n of the collection.

We also assume that the information flow graph representing the communication pattern is *connected*. By connectedness, we mean that every vehicle in the collection *should be able* to communicate with every other vehicle in the collection, even if they are not communicating directly, through a sequence of already existing communication links. We further assume that the structure of the control law used by each vehicle, other than the reference vehicle, is the same. Specifically, we consider the following structure

$$U_i(z) = -C(z) \sum_{j \in S_i} (X_i(z) - X_j(z) - \frac{L_{ij}z}{z - 1}), \qquad (2)$$

where $C(z)$ is a rational discrete transfer function. Let $x_{ref}(k) \in \Re$ be the position of the reference vehicle at instant k. The desired position $x_{i,des}(k)$ is related to the position of the reference vehicle x_{ref} through a constant offset L_i, i.e., $x_{i,des}(k) - x_{ref}(k) - L_i \equiv 0$. We define the error in spacing, $e_i(k)$ of the ith vehicle to be the deviation of its position from the desired position, i.e.,

$$e_i(k) := x_i(k) - x_{i,des}(k) = x_i(k) - x_{ref}(k) - L_i.$$

Since the desired formation corresponds to the vehicles moving as a rigid body in a pure translational maneuver, the desired deviation $L_{ij} := x_{i,des}(k) - x_{j,des}(k)$ is constant throughout the motion and equals $L_i - L_j$.

Let $E_i(z)$ be the Z transformation of the error in spacing, $e_i(k)$ of the ith vehicle. Let $x_{ref}(k)$ be the position of the reference vehicle at instant k and let $\bar{x}(k) := x_{ref}(k) - x_{ref}(0)$ be the displacement of the reference vehicle from its initial position in the time period $[kT, (k + 1)T)$. Then $X_{ref}(z) = \frac{x_{ref}(0)z}{z-1} + \bar{X}(z)$. If all the initial positions of the vehicles were chosen to correspond to the rigid formation, then $x_i(0) - x_{ref}(0) - L_i \equiv 0$. With such a choice of initial conditions and the choice of control law given in equation (2) for the plant described by equation (1), evolution equations for the errors in spacing can be expressed compactly as:

$$[I_{n-1} + \frac{C(z)T^2(z+1)}{2(z-1)^2}K_1]E(z) = \frac{T^2(z+1)}{2(z-1)^2}D(z) + \tilde{X}(z), \qquad (3)$$

where $E(z)$ and $D(z)$ are the respective Z transformations of the vector of errors of the following vehicles and the disturbances acting on them. The term $\tilde{X}(z)$ is a vector of dimension $n-1$ and every element of this vector is $\bar{X}(z)$. The term I_{n-1} is an identity matrix of dimension $n-1$ and K_1 is the principal minor obtained by removing the first row and column of the Laplacian K of the information flow graph defined as follows: For $j \neq i$, $K_{ij} = -1$ if vehicles i and j communicate directly; otherwise $K_{ij} = 0$. The ith diagonal element is defined as $K_{ii} = -\sum_{j \neq i} K_{ij}$. The Laplacian K is essentially the stiffness matrix obtained by connecting springs of unit spring constant between vehicles that communicate directly.

2.3 Problem Formulation

The following are the objectives of the control law given by equation (2):

1. In the absence of any disturbance on every vehicle in the formation, it is desired that for every $i \geq 2$, $\lim_{k \to \infty} e_i(k) = 0$, when the reference vehicle executes a maneuver where its speed asymptotically reaches a constant value.
2. In the presence of disturbances of atmost unit in magnitude, it is desirable that there exist a constant $M_R > 0$ such that $\max\{|e_i(k)|, |\dot{e}_i(k)|\} \leq M_R$ for every size of the collection.

The second objective ensures that the control law given by equation (2) is scalable. The problem is to determine conditions on the information flow graph (through constraints on K_1) and on the controller (through constraints on $C(s)$) so that these two objectives are met.

2.4 Analysis

Let us analyze the first requirement. Since the speed of the reference vehicle reaches a constant value, v_f, asymptotically, we have: $\lim_{k \to \infty} \bar{v}(k) = v_f = \lim_{z \to 1} \frac{(1-z^{-1})^2}{T} \bar{X}(z)$. Therefore, $\lim_{z \to 1}(1 - z^{-1})^3 \bar{X}(z) = 0$. Since, the first requirement must be satisfied in the absence of disturbing forces we have $D(z) \equiv 0$. If $\det[I_{n-1} + \frac{C(z)T^2(z+1)}{2(z-1)^2}K_1]$ is Schur, we have:

$$\lim_{z \to 1}(1 - z^{-1})E(z)$$

$$= \lim_{z \to 1}[I_{n-1} + \frac{C(z)(z+1)T^2}{2(z-1)^2}K_1]^{-1}(1 - z^{-1})\tilde{X}(z),$$

$$= \lim_{z \to 1}[(z-1)^2 I_{n-1} + \frac{T^2 C(z)(z+1)}{2}K_1]^{-1}z^2 \lim_{z \to 1}(1 - z^{-1})^3 \tilde{X}(z) = 0.$$

Therefore, the steady state error requirement is readily met if $\det[I_{n-1} + \frac{C(z)T^2(z+1)}{2(z-1)^2}K_1]$ is Hurwitz, i.e., if the controlled motion of formations is stable.

The second condition, in fact, concerns the stability of the controlled motion of formations.

Below we prove the main result concerning the stability of the controlled motion by using the mechanical analogy between the Laplacian of the information flow graph and the stiffness matrix, which essentially provides a way to address the propagation of errors. A route to instability in structural mechanics, for systems that do not have a rigid body mode, is that the smallest eigenvalue of the stiffness matrix converges to zero. In the context of vehicles, the smallest eigenvalue of the Laplacian K is zero, which corresponds to the rigid body mode, i.e., all vehicles have the same non-trivial displacement. A way to get a system without a rigid body mode is to ground one of the vehicles; in our case, for the sake of analysis of propagation of errors, there is no loss of generality in attaching the reference vehicle to the ground, that is, we set $\bar{X}(z) = 0$ from Equation 3.

The mechanical analogy indicates the following line of proof:

1. The smallest eigenvalue, λ, of K_1 goes to zero as $n \to 0$.
2. Let v be the corresponding eigenvector scaled in such a way that $||v||_\infty = 1$. The analogy indicates the examination of $e(k)$ when $d(k) = sin(\omega kT)v$ where ω is the first natural frequency or close to the first natural frequency.

Convergence of the Smallest Eigenvalue of K_1

Since K_1 is symmetric, we use Rayleigh's inequality to construct an upper bound for the smallest eigenvalue, λ. For that we construct an assumed mode, v_a, in the following way: We keep the reference vehicle grounded and displace other by one unit. Since the assumed mode shape indicates the amount by which every mass is displaced, all the elements of v_a, are equal. From the use of Rayleigh's inequality, it follows that:

$$\lambda \le \frac{<v, K_1 v>}{<v, Lv>} \le \frac{q_r}{n-1} \le \frac{q(n)}{n-1}$$

The above result holds for information flow graphs which are only subject to the constraint that each vehicle may only communicate with a specified number of vehicles. In certain types of regular formations such as a square formation or a cubic formation, where each vehicle can only communicate with vehicles within a certain distance from it, more structure can be imposed on the graphs such as the one dealt in the following proposition:

Proposition 1. *Consider information flow graphs that are connected. Suppose each vehicle in the collection may only communicate with m other vehicles in the collection, m being a constant. Further, suppose that the distribution of vehicles is such that the number of vehicles p(k), with k as the length of the* *communication path* [2] *to the reference vehicle be bounded by:*

$$\alpha k^r \le p(k) \le \beta k^r, \quad k = 1, \ldots, l_0$$

for some positive constants α , β and r. The term l_0 is the diameter [3] *of the graph considered. Then, the smallest eigenvalue λ of K_1 converges to zero in the following manner: There exists a $N^* > 0$ such that for all $n > N^*$ for any such information flow graph considered,*

$$\lambda \le \frac{m}{\alpha}(\frac{\beta}{\alpha})^{\frac{r+3}{r+1}} \frac{1}{n^{\frac{2}{r+1}}}. \tag{4}$$

Proof: We will use Rayleigh's inequality again to construct an upper bound for the smallest eigenvalue λ, of K_1. The assumed mode v_a, is constructed as follows: Using Bellman-Ford or a similar algorithm, we find the length of the communication path, l_i of the ith vehicle from the reference vehicle and assign it to the ith entry of v_a. Then, entries in v_a corresponding to any two communicating vehicles, will differ atmost by unity. Then,

$$< v_a, K_1 v_a > = \frac{1}{2} \sum_{i=1}^{n} \sum_{j \in S_i} (v_i - v_j)^2 \le \frac{1}{2} \sum_{i=1}^{n} |S_i| = \frac{mn}{2}$$

Let us now consider $< v_a, v_a > = 1^2 p(1) + 2^2 p(2) + \cdots + l_0^2 p(l_0)$.

$$\alpha(1^{2+r} + \cdots + l_0^{2+r}) \le < v_a, v_a > \le \beta 1^{2+r} + \cdots + l_0^{2+r} \tag{5}$$

Therefore using Rayleigh's inequality,

$$\lambda \le \frac{v_a, K_1 v_a}{v_a, v_a}$$

$$\le \frac{mn}{2} \frac{1}{1^2 p(1) + \ldots + (l_0 - 1)^2 p(l_0 - 1) + l_0^2 p(l_0)}$$

$$\le \frac{mn}{2\alpha} \frac{1}{1^{2+r} + 2^{2+r} + \ldots + (l_0 - 1)^{2+r}}$$

$$\le \frac{mn}{2\alpha} \frac{1}{\int_0^{l_0-1} x^{2+r} dx}$$

$$= \frac{mn}{2\alpha} \frac{r+3}{(l_0 - 1)^{r+3}}.$$

---

[2] For vehicles A and B that do not communicate directly, the length, l, of the communication path between A and B is the minimum number of intermediate vehicles $V_1, V_2, \ldots, V_l$ such that (1) A and V_1 communicate directly, (2) V_l and B communicate directly and (3) for all $1 \le i \le l - 1$, V_i and V_{i+1} communicate directly.

[3] The diameter of a graph, l_0, is the maximum value of the length between all possible pairs of vehicles that do not communicate directly.

Next, we formulate an upper bound on l_0. Since the total number of vehicles, excluding the reference vehicle, in the collection is $n-1$, it follows that $p(1) + \ldots + p(l_0) = n - 1$ and hence $n - 1 \leq \beta \sum_{k=0}^{l_0} k^r$. Using the fact that:

$$\sum_{k=0}^{l_0-1} k^r \leq \int_0^{l_0} x^r \, dx = \frac{l_0^{r+1}}{r+1} \leq \sum_{k=0}^{l_0} k^r.$$

it follows that

$$n \leq 1 + \beta \frac{(l_0 + 1)^{r+1}}{r+1}$$

$$\Rightarrow l_0 + 1 \geq \left(\frac{(n-1)(r+1)}{\alpha}\right)^{\frac{1}{r+1}}.$$

From the above inequality, we are guaranteed that $l_0 \to \infty$ as $n \to \infty$ for all information graphs considered. It follows that there exists a $N^* > 0$ such that for all $n > N^*$ and for any information flow graph considered in this corollary, the following inequality holds:

$$\lambda \leq \frac{mn}{2\alpha} \frac{r+3}{(l_0 - 1)^{r+3}}$$

$$\leq \frac{mn}{\alpha} \left(\frac{\beta}{l_0 + 1}\right)^{r+3} (2(l_0 - 1)^{r+3} \geq (l_0 + 1)^{r+3} \forall l_0 \geq l^*)$$

$$\leq \frac{m}{\alpha n^{\frac{2}{r+1}}} \left(\frac{\beta}{r+1}\right)^{\frac{r+3}{r+1}}$$

∎

Remark 1. If $r < 1$, the bound in the corollary is a tighter one than the one given by Lemma 1.

In the latter part of the chapter, various information flow graphs are considered, where the vehicle communication pattern is randomly assigned subject to the constraint that every vehicle can at most communicate directly with a pre-specified number of vehicles. The numerical results obtained for them corroborate with the above bound. Next, we use the upper bound on the convergence of λ of K_1 to 0 for analyzing the propagation of errors due to disturbances acting on the vehicles.

Analysis of the Propagation of Errors

Since $\lambda \to 0$ as $n \to \infty$,

1. If $C(z)$ does not have a pole at $z = 1$, there exists a sinusoidal disturbance acting on each vehicle of at most unit amplitude and of frequency proportional to $\sqrt{\lambda}$ that results in amplitudes of errors in spacing of the order of $\Omega\left(\sqrt{\frac{(n-1)^3}{q(n)^3}}\right)$.

2. If $C(z)$ has a pole at z $= 1$, then there is a critical size N^* of the collection such that for all $n > N^*$, at least one root of the equation

$$1 + \lambda \frac{T^2 C(z)(z+1)}{2(z-1)^2} = 0$$

is outside the unit circle; in other words, the controlled motion of the collection is unstable.

Lemma 1. *If $C(z)$ has a pole at unity and if $\lambda \to 0$ as the size of the collection, n, approaches ∞, then there exists a critical size N^* of the formation, such that for any size $n > N^*$ of the formation, the motion of the formation becomes unstable.*

Proof: For the problem considered in this section, if $C(z)$ has l poles at unity, it can be factored as $C(z) = \frac{L(z)}{(z-1)^l}$, $(l > 0)$ for some $L(z)$ that does not have any poles at the origin. We can write the closed loop characteristic equation $\Delta(z)$ as,

$$\delta(z) := (z-1)^{l+2} + \lambda \frac{T^2 L(z)(z+1)}{2} = 0.$$

Consider the following bilinear transformation which maps the z-plane to w-plane:

$$w = \frac{z-1}{z+1}$$

Because of the above mapping, $\delta(z)$ is Schur (i.e., all roots of $\delta(z)$ lie inside the unit cricle) if and only if the roots of $\Delta(w) = 1 + \lambda \frac{\tilde{C}(w)T^2(1-w)^{l+1}}{2^{l+2}w^{l+2}}$ have negative real parts. We note that $\Delta(w)$ is Hurwitz only if $L(0) \neq 0$. We further note that $\Delta(w)$ is Hurwitz iff $w^m \Delta(1/w)$ is Hurwitz, where m is the degree of the polynomial $\Delta(w)$. Next, we analyze the root locus of $\Phi(w) = 1 + K \frac{\tilde{C}(w)}{w^{l+2}}$, $K := \frac{2^{l+2}T^2}{\lambda}$ and $\tilde{C}(w) = \frac{1}{\tilde{C}(1/w)(1-\frac{1}{w})^{l-1}}$. Since $\tilde{C}(w)$ is always proper, it is clear that the root locus of $\Phi(w)$ has at least $l+2$ asymptotes. Thus, as $K \to \infty$, $(l+2)$ root loci move along lines that make the following angles with the positive real axis.

$$\phi_j = \frac{180° + 360°(j-1)}{l+2}, \qquad j = 1, 2, ..., l+2$$

Since $l \geq 1$, it is clear that at least one asymptote, along which one encounters a RHP pole, resulting in the instability of the closed loop as K increases. In other words, for arbitrarily small λ, if $C(z)$ has a pole at a unity, the motion of the formation becomes unstable. Hence, if $C(z)$ has more than a pole at unity, it is evident that there exists a critical size N^* of the formation, such that for any size $n > N^*$ of the formation, the motion of the formation becomes unstable. ∎

Remark 2. From $\Phi(w)$ and the subsequent analysis, it is apparent that if $C(1) < 0$ and $l = 0$, then one root is outside the unit circle, resulting in instability of motion.

The following theorem addresses the main result for platoons and it relates the propagation of errors in a platoon due to a disturbance of at most unit magnitude acting on each vehicle.

Theorem 1. *If $C(z)$ does not have a pole at unity and $C(1)$ is positive, then errors in spacing grow at least as $O\left(\sqrt{\frac{n^3}{q^3(n)}}\right)$; in other words, no control law of the type considered in this chapter is scalable to arbitrarily large collections if $\frac{q(n)}{(n)} \to 0$ as $n \to \infty$.*

Proof:
    Consider the error propagation equation (3) with $\tilde{X}(z) = 0$.

$$[I_{n-1} + \frac{C(z)T^2(z+1)}{2(z-1)^2}K_1]E(z) = \frac{T^2(z+1)}{2(z-1)^2}D(z). \tag{6}$$

We consider sinusoidal disturbances acting on all the vehicles, given by $d(k) = sin(\omega kT)v$. Using the assumed form of disturbance vector, equation 6 can be simplified as:

$$E(z) = \frac{T^2(z+1)}{2(z-1)^2 + \lambda T^2(z+1)C(z)}D(z). \tag{7}$$

Since $C(z)$ does not have a pole at unity, $C(1) \neq 0$. The frequency response of the discrete transfer function is obtained by moving along the unit circle, i.e., $z = e^{jwT}$. We use the following parametrization of e^{jwT}, to simplify the expression:

$$e^{j\bar{w}T} = \frac{1 + jwT/2}{1 - jwT/2}.$$

It should be noted that every point on the unit circle can be uniquely parameterized. The frequency of the sinusoidal disturbance, w_o, is assumed to be $\sqrt{\lambda C(1)}rad/s$. Then, the maximum amplitude of the error response is $||e(k)||_\infty$ is given by the magnitude of the following complex number:

$$\frac{T^2(1 - jw_oT/2)}{w_o^2T^2\underbrace{(1 - \frac{(1 - jw_oT/2)\bar{C}(jw_o)}{\bar{C}(0)})}_{\theta(w_o)}},$$

where $\bar{C}(jw_o) = C(\frac{1+jw_oT/2}{1-jw_oT/2})$. Since $\theta(w_o)$ has a root at zero, let $|\theta(w_o)| = w_o{}^p|\tilde{\theta}(w_o)|$, where $\tilde{\theta}(0) \neq 0$ and $p \geq 1$. Therefore, the amplitude ratio is

$$\frac{1}{(w_o)^{p+2}} \left| \frac{(1 - jw_o T/2)}{\tilde{\theta}(w_o)} \right|.$$

As $\lambda \to 0$, it is evident that $w_o \to 0$ and hence, the amplitude ratio grows to infinity as

$$\frac{\alpha}{|\tilde{\theta}(0)|} \frac{1}{(\sqrt{\lambda})^{p+2}},$$

where $p \geq 1$ and $\alpha = (\frac{1}{\sqrt{\tilde{C}(0)}})^{p+2}$. Since $p \geq 1$ as $\lambda \to 0$, $e(k)$ grows at least as

$$\frac{\alpha}{|\tilde{\theta}(0)|} \frac{1}{(\sqrt{\lambda})^3}.$$

Therefore, the maximum amplitude of the errors in spacing over all the vehicles for sufficiently large size of the formation is of $\Omega(\frac{1}{(\sqrt{\lambda})^3})$. Hence, a scalable control algorithm requires an information flow graph, where at least one vehicle in the collection communicates directly with at least $\Omega(n)$ vehicles. ∎

The following result shows the limitations due the digital control of motion of vehicles.

Theorem 2. *There is a limit m^* on the maximum number of vehicles connected to each vehicle other than the reference vehicle, which depends on the controller, $C(z)$ and its sampling time T.*

Proof: We consider sinusoidal disturbances acting on the vehicle of the form $d(k) = sin(\omega kT)v_m$, where v_m is the eigenvector of K_1 corresponding to λ_m, the largest eigenvalue of K_1. The equation 6 can then be simplified as

$$E(z) = \frac{T^2(z+1)}{2(z-1)^2 + \lambda_m T^2(z+1)C(z)} D(z). \qquad (8)$$

Next, we analyze the root locus of $\Delta(z) = 1 + \frac{\mu C(z)(z+1)}{(z-1)^2}$, where $\mu = \frac{\lambda_m T^2}{2}$. Since $C(z)$ is proper, it is clear that as $\mu \to \infty$, one root of $\Delta(z)$, goes to ∞. Let μ^* be the critical value where this branch of root locus crosses the unit circle. Hence for all $\mu > \mu^*$, at least one root of $\Delta(z)$ is outside the unit circle, resulting in instability of controlled motion. Hence for the stability of closed loop characteristic equation we require:

$$\lambda_m \leq \frac{2\mu^*}{T^2}. \qquad (9)$$

Let m is the maximum number of vehicles in the collection each vehicle can communicate with, apart from the reference vehicle. Let p be the index of the vehicle with m as its degree [4]. We construct the assumed mode shape,

---

[4] There could be more than one such vehicle. As per the remark that follows, it does not matter which one is picked.

v_m as follows: We displace the pth vehicle by one unit, and let the rest of the vehicles grounded. We have from the Rayleigh's inequality:

$$\lambda_m \geq \frac{< v_m, K_1 v_m >}{< v_m, v_m >} \geq m$$

Using the above derived bound in 9, we get $m < \frac{2\mu^*}{T^2} = m^*$ for stable motion of digitally controlled vehicles. Hence, every vehicle other than the reference vehicle can only connect to a maximum of m^* vehicles. ∎

Remark 3. For a given size of vehicles, it can be seen that as the sampling time increases, the closed loop characteristic equation becomes unstable. The critical sampling time is, $T^* = \sqrt{\frac{2\mu^*}{\lambda_m}}$.

3 Simulations

For the purposes of numerical simulation, we consider the motion of collection of vehicles moving in a straight line. Each vehicle is assumed to be a point mass. The structure of the control law used is as mentioned in equation (2). We consider a string of vehicles moving in a straight line trying to maintain constant distance amongst them. We describe the corresponding results below. We consider six vehicles, indexed from 1 to n. The set of vehicles that the first vehicle communicates with directly is the second vehicle, i.e. $S_1 = \{2\}$. For $i = 2, \ldots, n - 1$, the set S_i of vehicles the ith vehicle communicates with directly is $\{i-1, i+1\}$ and $S_n = \{n-1\}$. A PD (Proportional-Derivative) controller is used for feeding back the error in spacing and is given by $C(z) = 2 + \frac{5(1-z^{-1})}{T}$. Figure 1 shows the convergence of λ to 0 as the length of the string increases. Figure 3 shows the propagation of errors in spacing in a string of six vehicles. It shows how errors amplify in response to a sinusoidal disturbance acting on the last vehicle along the string, as we move away from the reference vehicle (vehicle indexed 1).

The above simulations are repeated with randomly generated information flow architectures. The convergence of λ to 0 for various random graphs with a maximal degree constraint of 4 is shown in Figure 4. It can be observed that though the information flow graphs are random, the upper bound on λ of K_1 seems to hold good for all the cases even for a small size of the collection. Figure 5 shows the migration of dominant pole towards $z = 1$ as the size of the collection of vehicles increases. In Figure 6 illustrates how the critical sampling time varies with the size of the collection. From Remark 2.4 we infer that for a fixed size of collection and the sampling time, as we increase m, at some point the digitally controlled motion of the vehicles becomes unstable. Figure 7 shows the stable motion when $m = 8$ with a sample time, $T = 100ms$ and $C(z)$ as specified earlier. It should be noted that the condition posed in Remark 2.4 is a necessary condition, but not sufficient. Hence, it is possible that controlled motion can become unstable even when $m < m^*$.

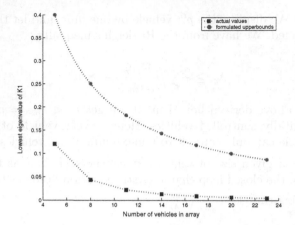

Fig. 1. The variation of λ (lowest eigenvalue of K_1) with n, for a string of n vehicles with each vehicle connected to the vehicles directly behind and ahead of it.

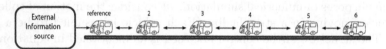

Fig. 2. Predecessor and follower based information flow pattern in the string

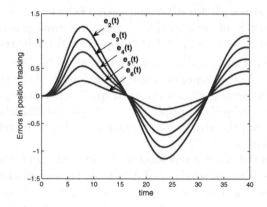

Fig. 3. Propagation of the errors along the string

4 Conclusions

In this chapter, we have considered information flow graphs for a collection of vehicles, where there is a constraint on the maximum number of vehicles in the collection every vehicle can communicate with directly. We have related how the smallest eigenvalue λ of a principal minor of the Laplacian of information flow graph goes to zero. We then showed that the motion of vehicles is unstable

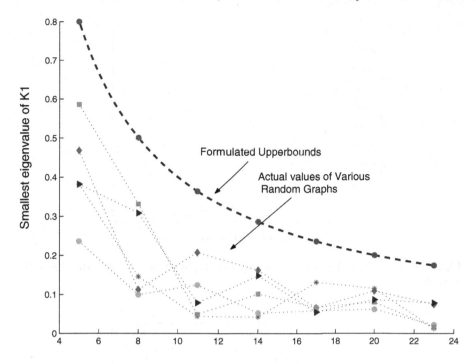

Fig. 4. Variation of λ with n , for a string of n vehicles, connected in a random fashion to a maximum of 4 other vehicles

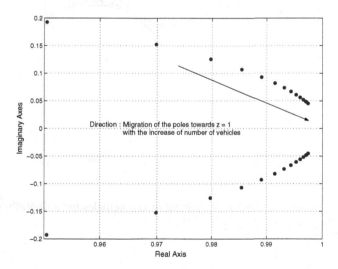

Fig. 5. Plot showing the migration of the dominant pole to outside of unit circle with increase in the number of vehicles

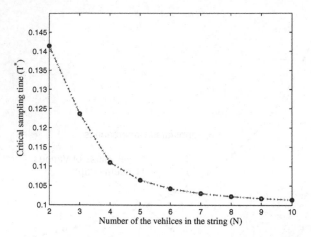

Fig. 6. Plot showing the variation of critical sampling time, T^* with the size of the collection.

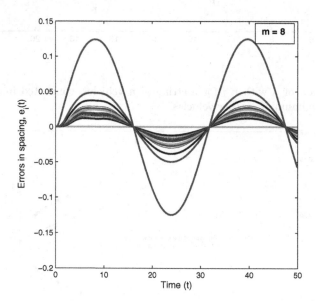

Fig. 7. Plot showing the errors in spacing of the vehicles for a digitally controlled stable motion

if the controller transfer function $C(z)$ had one or more poles at $z = 1$ and that it must have no zeroes at $z = 1$ to track ramp inputs resulting from the reference vehicle moving at a constant velocity. We further showed that if $\lambda \to 0$, there is a disturbance of sufficiently low frequency acting on each vehicle of at most unit magnitude which results in errors in spacing of $\Omega\left(\sqrt{\frac{n^3}{q(n)^3}}\right)$. The error propagation and the stability results for the continuous systems are analogous and are discussed in [14].

References

1. D. Swaroop and J. K. Hedrick, *String Stability of interconnected systems IEEE Transactions on Automatic Control*, vol 41 , pp. 349-357, March 1996.
2. Aniruddha G. Pant, *Mesh stability of formations of Unmanned Aerial vehicles*, Ph.D Thesis, University of California, Berkeley, 2002.
3. D. Swaroop, *String Stability of interconnected systems: An application to platooning in automated highway systems*, Ph.D. Thesis, University of California, Berkeley, 1994.
4. Singh, Sahjendra N., Rong Zhang, Phil Chandler, and Siva Banda, *Decentralized Adaptive Close Formation Control of UAVs*, AIAA 2001-0106, 39th AIAA Aerospace Sciences Meeting & Exhibit, Reno, NV, Jan. 2001.
5. Pachter, Meir, John J. DAzzo, and Andrew W. Proud, *Tight Formation Flight Control*, AIAA Journal of Guidance, Control and Dynamics, Vol. 24, No. 2, MarchApril 2001.
6. Khatir M., and E.J. Davison, *Bounded Stability and Eventual String Stability of a Large Platoon of Vehicles using Non-Identical Controllers*, 2004 IEEE Control and Decision Conference, Paradise Island, Dec. 2004, to appear.
7. Tolga Eren, Brian D. O. Anderson, A. Stephen Morse, Walter Whiteley, and Peter N. Belhumeur. *Operations on rigid formations of autonomous agents*. Communications in Information and Systems, 2004. to appear.
8. Peter Joseph Seiler, *Coordinated control of Unmanned Aerial Vehicles*, Ph.D Thesis, Department of Mechanical Engineering, University of California, Berkeley, 2001.
9. Swaroop, D., Hedrick, J.K., Chien, C.C. and Ioannou, P.A., "*A Comparison of Spacing and Headway Control Laws for Automatically Controlled Vehicles,* " Vehicle System Dynamics Journal, Vol. 23, No. 8, pp. 597-625, 1994.
10. J. A. Fax and R.M.Murray, *Information Flow and cooperative control of vehicle formations*, Proceedings of the IFAC World Congress, Barcelona, Spain, pp. 2360-2365, July, 2002.
11. P. J. Seiler, Aniruddha Pant and J. K. Hedrick, *Preliminary Investigation of Mesh stability for Linear Systems*, IMECE99/DSC-7B-1, 1999.
12. G. F. Franklin, J. D. Powell and M. L. Workman, *Digital Control of Dynamic Systems*, Prentice Hall, 1997.
13. Aniruddha Pant, Pete Seiler, T. John Koo, Karl Hedrick, *Mesh Stability of Unmanned Aerial Vehicle Clusters*, Proceedings of American Control Conference 2001, pp. 62-68.
14. Sai Krishna Y, Swaroop Darbha and K.R. Rajagopal *Information flow and its relation to the stability of motion of vehicles in a rigid formation*, Proceedings of American Control Conference 2005, pp. 1853-1858.

Formation Control of Nonholonomic Mobile Robots Using Graph Theoretical Methods

Wenjie Dong[1] and Yi Guo[2]

[1] Department of Electrical & Computer Engineering
University of Central Florida
Orlando, FL 32816
[2] Department of Electrical & Computer Engineering
Stevens Institute of Technology
Hoboken, NJ 07030
E-mail: yguo1@stevens.edu

Summary. In this chapter, formation control of mobile robots with nonlinear models is considered. Two controllers are proposed with the aid of the dynamic feedback linearization technique, the time-scaling technique and properties of Laplacian matrix. The proposed controllers ensure the group of mobile robots to move in a desired formation. Existing results in formation control using graph theoretical methods are extended to nonlinear systems of high dimensions. Simulation results show the effectiveness of the proposed controllers.

1 Introduction

Cooperative control of multiple systems has received a lot of attention recently due to its challenging features and application importance. To name a few, multiple vehicles are used in rescue mission, moving a large object, troop hunting, formation control, and cluster of satellites [37, 22, 13, 36, 2, 12, 13, 31, 16, 38, 23]. Different control strategies have been proposed, which include behavior based approach, virtual structure approach, leader following approach, and graph theoretical approach.

Arkin [1] studied cooperation without communication for multiple robots foraging and retrieving objects in a hostile environment. Later, this behavior-based approach was extended to formation control of multiple robots [2]. In the behavioral approach [2, 15, 27, 34, 32, 30], the control action for each robot is defined to be a weighted average of the control associated with each desired behavior. Possible behaviors include collision avoidance, obstacle avoidance, and formation keeping. Using this approach, it is easy to derive control strategies even when vehicles have multiple competing objectives. Generally, it is hard to analyze the behavioral performance analytically.

The virtual structure approach was proposed for formation control of mobile robots by Lewis [19]. It is used to formation control of spacecrafts and multi-satellite systems [29, 16]. In the virtual structure approach, the entire desired formation is treated as a single entity. Desired states for each vehicle in the formation can be specified by place-holders in the virtual structure. Using this approach, it is easy to prescribe the coordinated behavior for the group and the virtual structure can maintain formation during maneuvers. However, the requirement of the formation to act as a virtual structure limits the class of potential applications.

The leader-following approach is another important method [35, 8, 9, 22, 33]. In this approach, some mobile robots are designated as leaders while others as followers. The leaders track desired trajectories, and the followers track desired trajectories with respect to the leaders. The advantage of this approach is its simplicity in that the reference trajectories of the leaders are pre-defined and the internal stability of the formation are guaranteed by individual robot's control laws. A disadvantage is that the whole system fails if the leader fails.

The graph theoretical approach was proposed for cooperative control of multiple linear systems by Fax and Murray [11]. Different control laws were designed with the aid of graph theory [24, 10, 7, 17]. Communication links among systems are described by Laplacian matrices. Each vehicle is treated as a vertex and the communication links between vehicles are treated as edges. Stability of the whole system is guaranteed by stability of each modified individual linear system. However, the methods are limited to linear models of vehicle systems.

In this chapter, we discuss formation control of nonholonomic mobile robots using graph theoretical methods. We propose two formation control approaches to achieve formation stability. In the first approach, the robot's model is transformed to a linear system by dynamic feedback linearization. The controller is then designed based on graph theory. In the second approach, with the aid of the time-scaling technique, a time-varying parameter is introduced in the control law. The two controllers ensure the group of mobile robots to move in a desired formation. While most existing results use linear vehicle models, we discuss cooperative control of nonholonomic mobile robots and design global controllers to achieve formation. We extended earlier results using graph theoretical methods to nonlinear systems with high dimensions.

The rest of the chapter is organized as follows. In Section 2, we introduce terminologies in graph theory which will be used in the chapter. In Section 3, the formation control problem is defined. In Section 4, we design two controllers for the above defined problem. In Section 5, simulation results are presented. Finally, we conclude the chapter in Section 6.

2 Graph Theory

In this section, some terminologies and basic properties used in this chapter are listed. Interested readers please refer to literature [6, 21, 5].

A directed graph $\mathbb{G}$ consists of a set of vertices $\mathbb{V}$ and a set of edges $\mathbb{E} \subset \mathbb{V}^2$, where $e = (\alpha, \beta) \in \mathbb{E}$ and $\alpha \in \mathbb{V}$, $\beta \in \mathbb{V}$. The first element of e is denoted $tail(e)$, and the second of element of e is denoted $head(e)$. For all e, we assume that $tail(e) \neq head(e)$, which means that the graph has no self-loop. We also assume that each element of $\mathbb{E}$ is unique. A graph is called un-directed if $(\beta, \alpha) \in \mathbb{E}$ for any $(\alpha, \beta) \in \mathbb{E}$. The in-degree of a vertex α, denoted $d_i(\alpha)$, is the number of edges with α as its head. The out-degree of a vertex α, denoted $d_o(\alpha)$, is the number of edges with α as its tail. A vertex is called balanced if its in-degree is equal to its out-degree. A graph is called balanced if all of its vertex is balanced. Let $\tilde{\mathbb{E}}$ be the set of reverse edges of $\mathbb{G}$ obtained by reversing the order of nodes of all the edges. The mirror of $\mathbb{G}$ denoted by $\tilde{\mathbb{G}}$ is an un-directed graph with vertices $\mathbb{V}$ and edges $\mathbb{E} \cup \tilde{\mathbb{E}}$.

A path on $\mathbb{G}$ of length N from α_0 to α_N is an ordered set of distinct vertices $\{\alpha_0, \ldots, \alpha_N\}$ such that $(\alpha_{i-1}, \alpha_i) \in \mathbb{E}$ for all $i \in [1, N]$. A graph is called strongly connected if any two different vertices α and β in $\mathbb{V}$ there exists at least one path from α to β. A graph is called disconnected if there exists a disjoint subsets of vertices which cannot be joined by any path. The diameter $\mathbb{D}$ of a strongly connected graph $\mathbb{G}$ is the maximum distance between any two vertices of $\mathbb{G}$. For any vertex β in $\mathbb{V}$, the neighbor of β, denoted by $\mathbb{J}_\beta$, is the set of all vertices α such that $(\alpha, \beta) \in \mathbb{E}$.

Assume that the vertices of $\mathbb{G}$ are enumerated and each is denoted as α_i. The adjacency matrix of a graph, denoted as $G(\mathbb{G})$, is a square matrix of size $|\mathbb{V}|$ and defined by

$$G_{i,j} = \begin{cases} 1, & \text{if } (\alpha_i, \alpha_j) \in \mathbb{E} \\ 0, & \text{otherwise.} \end{cases}$$

The degree matrix of a graph, denoted $D(\mathbb{G})$, is also a square matrix of size $|\mathbb{V}|$ and defined by $D_{i,i} = d_o(\alpha_i)$ and $D_{i,j} = 0 (i \neq j)$. Assume the graph $\mathbb{G}$ is strongly connected, D is nonsingular. The Laplacian matrix of the graph is defined as

$$L = I - D^{-1}G.$$

Some properties of Laplacian matrix are very useful [17].

Property 1. Zero is an eigenvalue of L. The associated eigenvector is $\mathbf{1}$.

Property 2. If $\mathbb{G}$ is strongly connected, the multiplicity of the zero eigenvalue is one.

Property 3. All eigenvalues of L lie in a disk of radius 1 centered at the point $1 + 0j$ in the complex plane.

Property 4. If $\mathbb{G}$ is strongly connected, then each nonzero eigenvalues λ of L satisfies

$$\lambda \geq \frac{1}{\mathbb{D}\sum_{i\in V} d_{i,i}}.$$

Let $A = \{a_{ij}\}_{m\times n}$ and $B = \{b_{ij}\}_{p\times q}$ are two matrices, the Kronecker product of A and B is defined as

$$A \otimes B = \begin{pmatrix} a_{11}B & \cdots & a_{1n}B \\ \vdots & \ddots & \vdots \\ a_{m1}B & \cdots & a_{mn}B \end{pmatrix}.$$

By the definition, the following properties are easily derived.

1. given three matrices A, B, and C,

$$A \otimes (B + C) = A \otimes B + A \otimes C,$$

$$(B + C) \otimes A = B \otimes A + C \otimes A,$$

2. If AC and BD exists, then

$$(A \otimes B)(C \otimes D) = AC \otimes BD.$$

3 Problem Statement

Assume there are m mobile robots moving on the plane. For simplicity, all robots are assumed to have the same structure. Noting the results in literature [4, 20], we assume that there exist suitable transformations such that the kinematics of robot $j(1 \leq j \leq m)$ can be transformed globally or locally into the following chained system

$$\begin{cases} \dot{z}_{1,j}(t) & = u_{1,j}(t), \\ \dot{z}_{2,j}(t) & = z_{3,j}(t)u_{1,j}(t), \\ \quad\vdots \\ \dot{z}_{n-1,j}(t) = z_{n,j}(t)u_{1,j}(t), \\ \dot{z}_{n,j}(t) & = u_{2,j}(t), \end{cases} \tag{1}$$

with $(z_{1,j}(t), z_{2,j}(t))$ being the X and Y positions of robot j in the $X - Y$ plane.

Remark 1. For many types of wheeled mobile robots, we can always find global or local transformations such that the kinematics of the robots are transformed into chained form (1) with the first two states being the position of the robot in the plane [20]. See Section 5 for one example.

Let graph $\mathbb{G}$ describe the communication links between the group of mobile robots and L_G be the Laplacian matrix corresponding to the graph $\mathbb{G}$. Each vertex represents a robot and each edge represents a communication link between two robots. For each robot i, $\mathbb{J}_i$ denotes its neighbors. In this chapter, we assume that each robot only knows its own state and the relative positions to its neighbors. That is, in the controller design, $u_i(t)$ is assumed to be a function of $z_i(t)$ and $z_i(t) - z_j(t)$ for each $j \in \mathbb{J}_i$.

Given constant vectors $[h_{j,x}, h_{j,y}] \in \mathbb{R}^2 (1 \leq j \leq m)$ which form a desired form $\mathbb{F}$ in the plane. m robots are said to be in formation if

$$z_{1,j}(t) - h_{j,x} = z_{1,i}(t) - h_{i,x}(i \neq j) \tag{2}$$
$$z_{2,j}(t) - h_{j,y} = z_{2,i}(t) - h_{i,y}(i \neq j). \tag{3}$$

In this chapter, the formation control problem is defined as designing control laws $u_i(t)(1 \leq i \leq m)$ such that m robots are in formation as time tends to infinity, i.e., design control laws $u_i(t)(1 \leq i \leq m)$ such that

$$\lim_{t \to \infty} (z_{1,i}(t) - z_{1,j}(t)) = h_{i,x} - h_{j,x}(i \neq j) \tag{4}$$
$$\lim_{t \to \infty} (z_{2,i}(t) - z_{2,j}(t)) = h_{i,y} - h_{j,y}(i \neq j). \tag{5}$$

Remark 2. In the formation control, (2)-(3) means that points $(z_{1,j}(t), z_{2,j}(t))(1 \leq j \leq m)$ form the same form as $(h_{j,x}, h_{j,y})(1 \leq j \leq m)$. Since $(z_{1,j}(t), z_{2,j}(t))(1 \leq j \leq m)$ are positions of mobile robots in the plane, the m robots form the desired formation $\mathbb{F}$.

In the following, we omit the argument t for state variables, for example, $z_{1,i}(t)$ is simply written as $z_{1,i}$.

4 Controller Design

4.1 Approach I

To design the controller, we first transform (1) into a linear system by the dynamic feedback linearization technique.

Lemma 1. *If* $\zeta_{1,j}(t) \neq 0(\forall t)$, *by the dynamic controller*

$$\begin{cases} \dot{\zeta}_{1,j} &= \zeta_{2,j}, \\ &\vdots \\ \dot{\zeta}_{n-3,j} &= \zeta_{n-2,j}, \\ \dot{\zeta}_{n-2,j} &= r_{1,j}, \\ u_{1,j} &= \zeta_{1,j}, \\ u_{2,j} &= (r_{2,j} - g_{n-1,j}(z_{4,j}, \ldots, z_{n,j}, \zeta_{1,j}, \ldots, \zeta_{n-3,j}))/\zeta_{1,j}^{n-2}, \end{cases} \tag{6}$$

where $g_{n-1,j}$ is calculated by following recursion

$$g_{2,j} = z_{3,j}\zeta_{2,j},$$

$$g_{k,j} = (k-1)z_{k+1,j}\zeta_{1,j}^{k-2}\zeta_{2,j} + \sum_{l=3}^{k}\frac{\partial g_{k-1,j}}{\partial z_{l,j}}z_{l+1,j}\zeta_{1,j}$$

$$+\sum_{l=1}^{k-1}\frac{\partial g_{k-1,j}}{\partial \zeta_{l,j}}\zeta_{l+1,j}, (k = 3,\ldots, n-3),$$

system (1) is transformed into

$$\dot{q}_j = A_v q_j + B_v r_j \tag{7}$$

where $A_v = I_2 \otimes A_0$, $B_v = I_2 \otimes B_0$, $q_j = [q_{1,j}^T, q_{2,j}^T]^T$, $r_j = [r_{1,j}, r_{2,j}]^T$, and

$$A_0 = \begin{bmatrix} 0 & 1 & 0 & \cdots & 0 & 0 \\ 0 & 0 & 1 & \cdots & 0 & 0 \\ \vdots & \vdots & \vdots & \ddots & \vdots & \vdots \\ 0 & 0 & 0 & \cdots & 0 & 1 \\ 0 & 0 & 0 & \cdots & 0 & 0 \end{bmatrix}, B_0 = \begin{bmatrix} 0 \\ 0 \\ \vdots \\ 0 \\ 1 \end{bmatrix}, q_{1,j} = \begin{bmatrix} z_{1,j} \\ \dot{z}_{1,j} \\ \vdots \\ \dfrac{d^{n-2}z_{1,j}}{dt^{n-2}} \end{bmatrix}, q_{2,j} = \begin{bmatrix} z_{2,j} \\ \dot{z}_{2,j} \\ \vdots \\ \dfrac{d^{n-2}z_{2,j}}{dt^{n-2}} \end{bmatrix}.$$

Proof: The result can be obtained by direct calculation. ∎

Let

$$q = [q_1^T, \ldots, q_m^T]^T, r = [r_1^T, \ldots, r_m^T]^T,$$

$$A = I_m \otimes A_v, B = I_m \otimes B_v,$$

we have

$$\dot{q} = Aq + Br. \tag{8}$$

Let

$$\hbar_{j,1} = [h_{j,x}, 0, \ldots, 0]^T \in \mathbb{R}^{n-1},$$
$$\hbar_{j,2} = [h_{j,y}, 0, \ldots, 0]^T \in \mathbb{R}^{n-1},$$
$$\hbar_j = [\hbar_{j,1}^T, \hbar_{j,2}^T]^T,$$
$$\hbar = [\hbar_1^T, \ldots, \hbar_m^T]^T,$$

and

$$\rho = L_1(q - \hbar),$$

where $L_1 = L_G \otimes I_{2(n-1)}$, we have the following result.

Lemma 2. *If the communication graph $\mathbb{G}$ is directed and strongly connected and $\rho = 0$, the robots are in formation $\mathbb{F}$.*

Proof: Since $L_1(q - \hbar) = 0$, $q = \hbar + a\mathbf{1}$. Therefore, $z_{1,j} - h_{j,x} = z_{n,j} - h_{j,y}(1 \leq j \leq m)$, which means the robots are in formation. ∎

Next, we design r such that ρ converges to zero.

Lemma 3. *For system (8), if the communication graph $\mathbb{G}$ is directed and strongly connected, the control law*

$$r = \Gamma L_1(q - \hbar)$$

makes ρ converge to zero, where $\Gamma = I_{2m} \otimes \Gamma_1$, $\Gamma_1 = [g_1, g_2, \ldots, g_{n-1}]$ is chosen such that the matrix $A_0 + \lambda B_0 \Gamma_1$ is stable (Hurwitz) for each nonzero eigenvalue λ of the communication Laplacian matrix L_G.

Proof: We follow the idea in the proof of proposition 4.3 [17]. First, it is easy to prove that the eigenvalues of $A + B\Gamma L_1$ are those of $A_v + \lambda B_v \Gamma_v$ for λ an eigenvalue of L_G. Next, we consider the following system

$$\begin{bmatrix} \dot{q} \\ \dot{\hbar}_p \end{bmatrix} = \begin{bmatrix} A + BFL_1 & -BFL_1 \\ 0 & 0 \end{bmatrix} \begin{bmatrix} q \\ \hbar_p \end{bmatrix} := \Pi \begin{bmatrix} q \\ \hbar_p \end{bmatrix} \tag{9}$$

where $\hbar_p = [h_{1,x}, h_{1,y}, \ldots, h_{m,x}, h_{m,y}]^T$. Define the subspace $\mathbb{S}$ by

$$\mathbb{S} = \left\{ \begin{bmatrix} q \\ \hbar_p \end{bmatrix} : L_1(q - \hbar) = 0 \right\}.$$

For any vector $\sigma \in \mathbb{S}$,

$$\sigma = \begin{bmatrix} \mathbf{1} \otimes a + b \otimes \begin{bmatrix} 1 \\ 0 \end{bmatrix} \\ b \end{bmatrix}$$

where a and b are vectors. Therefore,

$$\Pi\sigma = \begin{bmatrix} \mathbf{1} \otimes A_0 a \\ 0 \end{bmatrix} \in \mathbb{S},$$

which means $\mathbb{S}$ is Π-invariant and the restriction of the transformation induced by Π on $\mathbb{S}$ is $diag[A_0, 0]$. The eigenvalues of the restriction of the transformation induced by Π on the quotient space $\mathbb{R}^{2mn}/\mathbb{S}$ are those of $A_0 + \lambda B_0 \Gamma_1$ for λ a nonzero eigenvalue of L_G. By the assumption that these eigenvalues have negative real parts, the quotient space is stable. Therefore, for any solution $[q, \hbar_p]^T$ of (9), it tends to the subspace $\mathbb{S}$ as time tends to infinity. From the definition of $\mathbb{S}$, $L_1(x - \hbar)$ tends to zero. ∎

In Lemma 3, λ may be several different values. Γ_1 is designed to simultaneously stabilize $A_0 + \lambda B_0 \Gamma_1$ for all non-zero eigenvalues λ. Noting the special structures of A_0 and B_0, we have the following lemma.

Lemma 4. *If the communication graph L_G is directed and strongly connected and $n \leq 4$, there always exists $\Gamma_1 = [g_1, \ldots, g_{n-1}]$ such that $A_0 + \lambda B_0 \Gamma_1$ is stable for each $\lambda(\neq 0)$ of L_G.*

Proof: Noting the special structures of A_0 and B_0, we have the following characteristic polynomial

$$det(\alpha I_{n-1} - A_0 - \lambda B_0 \Gamma_1) = \alpha^{n-1} - \lambda g_1 \alpha^{n-2} - \cdots - \lambda g_{n-2} \alpha - \lambda g_{n-1}.$$

By Kharitonov theorem [3], it can be proved that the characteristic polynomial is stable if Γ_1 is chosen as follows.

$$g_1 < 0, g_2 < 0 \qquad\qquad\qquad\text{if } n = 3;$$
$$g_1 < 0, g_2 < 0, -\frac{g_1 g_2}{2\mathbb{D}\sum_{i\in\mathbb{V}} d_{ii}} < g_3 < 0 \ \text{if } n = 4.$$

∎

In Lemma 4, we assume $n \leq 4$ because the dimensions of many wheeled mobiles are less than four. For high dimension systems, we can use Kharitonov theorem [3] to find Γ_1 if it exists.

Theorem 1. *For system (1),if the communication graph $\mathbb{G}$ is directed and strongly connected, controller (6) with*

$$r = \Gamma L_1 (q - \hbar)$$

make (4) and (5) satisfied, where $\Gamma = I_{2m} \otimes \Gamma_1$, $\Gamma_1 = [g_1, g_2, \ldots, g_{n-1}]$ is chosen such that the matrix $A_0 + \lambda B_0 \Gamma_1$ is stable (Hurwitz) for each nonzero eigenvalue λ of the communication Laplacian matrix L_G. Moreover, Γ_1 always exists if $n \leq 4$.

Controller (6) solve the formation control of nonholonomic mobile robots with the aid of graph theory. In the formation control, $u_{1,j}$ is required to be non-zero all the time, which means that each robot is required to move along X-axis. By the proof, this can be guaranteed by suitably choosing the control parameters and the initial conditions of the dynamic controller.

4.2 Approach II

To overcome the singularity of the controller, we propose another controller in two steps under the condition that the communication graph is directed, balanced, and strongly connected. In the first step, $u_{1,j}(1 \leq j \leq m)$ are designed such that (4) is satisfied based on the properties of Laplacian matrix. In the second step, $u_{2,j}(1 \leq j \leq m)$ are designed such that (5) is satisfied with the aid of the time-scaling technique and the properties of Laplacian matrix. In the first step, we have the following lemma.

Lemma 5. *For system (1), if the communication graph is directed and strongly connected, control law*

$$u_{1,j} = \dot{\alpha} - \sum_{i\in\mathbb{J}_j}(z_{1,j} - z_{1,i} - h_{j,x} + h_{i,x}), 1 \leq j \leq m \qquad\qquad (10)$$

make (4) satisfied, where $\alpha(t)$ is any differentiable function.

Proof: Let
$$y_1 = [y_{1,1}, y_{1,2}, \ldots, y_{1,m}^T$$
where
$$y_{1,j} = z_{1,j} - \alpha - h_{j,x},$$
with the control law (10), we have
$$\dot{y}_{1,j} = -\sum_{i \in \mathbb{J}_j} (y_{1,i} - y_{1,j}).$$
So
$$\dot{y}_1 = -L_G y_1. \tag{11}$$

Since the graph $\mathbb{G}$ is directed and strongly connected, L_G is a symmetric nonnegative definite matrix. There exists an non-singular matrix S such that
$$S^{-1} L_G S = \Lambda = \mathrm{diag}[0, J_2, \ldots, J_m]$$
where J_j is the Jordan form corresponding to eigenvalue $\lambda_j > 0(2 \leq j \leq m)$. Let the state transformation $\xi = S^T y_1$, we have
$$\dot{\xi} = \Lambda \xi.$$

Therefore,
$$\xi(t) = \begin{bmatrix} \xi_1(0) \\ \xi_2(t), \\ \vdots, \\ \xi_(t) \end{bmatrix}$$
where $\xi_j (2 \leq j \leq m)$ exponentially converge to zero. Noting $y_1 = S\xi$, so
$$y_{1,j}(t) = \sum_{i=1}^{m} s_{j,i} \xi_i(t) = s_{j,1} \xi_1(0) + \sum_{i=2}^{m} s_{j,i} \xi_i(t).$$

Since $L_G S = S\Lambda$,
$$L_G \begin{bmatrix} s_{1,1} \\ s_{2,1} \\ \vdots \\ s_{m,1} \end{bmatrix} = 0$$

Noting L_G is a Laplacian matrix corresponding to a directed strongly connected matrix,
$$s_{1,1} = s_{2,1} = \cdots = s_{m,1}.$$
So,
$$\lim_{t \to \infty} y_{1,i} = \lim_{t \to \infty} y_{1,j} (i \neq j).$$
Therefore, $\lim_{t \to \infty}(z_{1,i} - z_{1,j}) = h_{i,x} - h_{j,x} (i \neq j).$ ■

In control law (10), α is a control parameter which plays an important role in designing $u_{2,j}$ in the next step with suitable assumption.

In the second step, we first have the following lemma.

Lemma 6. *For system (1), if the communication graph is directed, balanced and strongly connected, with the control law (10), $u_{1,j}(1 \leq j \leq m)$ exponentially converge to $\dot{\alpha}$ with the least rate $\lambda_2(L_G)$ (i.e., the least non-zero eigenvalue of L_G). If $\dot{\alpha}(t) \geq \epsilon > 0$ for $t \geq T_1$, there exists a finite time $T_2(> T_1)$ such that $u_{1,j}(t) \geq \epsilon/2(1 \leq j \leq m)$ for $t \geq T_2$.*

Proof: By the proof of Lemma 5,

$$\lim_{t \to \infty} y_{1,j}(t) = c$$

where c is a constant. Since the communication graph is balanced, $\sum_{j=1}^{m} y_{1,j}/m$ is an invariant quantity of system (11) and

$$c = \frac{1}{m} \sum_{i=1}^{m} y_{1,i}(0).$$

Let

$$e = [e_1, \ldots, e_m]^T = y_1 - c\mathbf{1},$$

then

$$\dot{e} = -L_G e \tag{12}$$

and

$$\sum_{i=1}^{m} e_i = 0.$$

Let

$$V = \frac{1}{2} e^T e$$

differentiate it along (12), we have

$$\dot{V} = -e^T L_G e \leq -\lambda_2(L_{\hat{G}}) e^T e = 2\lambda_2(L_{\hat{G}}) V (\forall e \neq 0.)$$

Therefore, e exponentially converges to zero with the least rate $\lambda_2(L_{\hat{G}})$. Since

$$u_{1,j} = \dot{z}_{1,d} - \sum_{i \in J_j} (y_{1,j} - y_{1,i}),$$

$u_{1,j}$ exponentially converge to $\dot{z}_{1,d}$. Noting $\dot{z}_{1,d}(t) \geq \epsilon$ for $t \geq T_1$, there exists a finite time t_j such that $u_{1,j}(t) \geq \epsilon/2$ for $t \geq t_j$. Let $T_2 = \max\{t_j, 1 \leq j \leq m\}$, $u_{1,j}(t) \geq \epsilon/2(1 \leq j \leq m)$ for $t \geq T_2$. ∎

In Lemma 6, $z_{1,j}$ increase monotonically with time for $t > T_2$, which means

$$z_{1,j} \to \infty \Leftrightarrow t \to \infty.$$

Therefore, we can use the following time-scaling technique. Let

$$z_{i,j}^{(1)} = \frac{dz_{i,j}}{dz_{1,j}}, (2 \le i \le n, 1 \le j \le m),$$

$$v = [v_1, v_2, \ldots, v_m]^T = [u_{2,1}/u_{1,1}, \ldots, u_{2,m}/u_{1,m}]^T$$

we have

$$\bar{z}_j^{(1)} = A_1 \bar{z}_j + B_1 v_j, (1 \le j \le m) \tag{13}$$

where $\bar{z}_j = [z_{2,j}, \ldots, z_{n,j}]^T$,

$$A_1 = \begin{bmatrix} 0 & 1 & 0 & \cdots & 0 & 0 \\ 0 & 0 & 1 & \cdots & 0 & 0 \\ \vdots & \vdots & \vdots & \ddots & \vdots & \vdots \\ 0 & 0 & 0 & \cdots & 0 & 1 \\ 0 & 0 & 0 & \cdots & 0 & 0 \end{bmatrix}, B_1 = \begin{bmatrix} 0 \\ 0 \\ \vdots \\ 0 \\ 1 \end{bmatrix}.$$

Using Kronecker product, (13) can be written as

$$\bar{z}^{(1)} = A\bar{z} + Bv \tag{14}$$

where $\bar{z} = [\bar{z}_1^T, \ldots, \bar{z}_m^T]^T$, $A = I_m \otimes A_1$, $B = I_m \otimes B_1$, $v = [v_1, \ldots, v_m]^T$, and

$$\bar{z}^{(1)} = [\frac{d\bar{z}_1^T}{dz_{1,1}}, \ldots, \frac{d\bar{z}_m^T}{dz_{1,m}}]^T.$$

Define

$$h_j = [h_{j,y}, 0, \ldots, 0]^T \in \mathbb{R}^{n-1} (1 \le j \le m),$$

and let ω_j be as follows

$$\omega_j = (\bar{z}_j - h_j) - \frac{1}{|\mathbb{J}_j|} \sum_{i \in \mathbb{J}_j} (\bar{z}_i - h_i), (1 \le j \le m),$$

then

$$\omega = [\omega_1^T, \ldots, \omega_m^T]^T = L(\bar{z} - h)$$

where $L = L_G \otimes I_{(n-1)}$ and $h = [h_1^T, h_2^T, \ldots, h_m^T]^T$.

Since (5) is satisfied if ω tends to zero. Therefore, we only need to design v such that ω converges to zero. Let the control law

$$v = \Gamma \omega, \tag{15}$$

where $\Gamma = \text{diag}[\Gamma_1, \ldots, \Gamma_1]$, we have the following result.

Lemma 7. *Assume the communication graph is directed, balanced and strongly connected, let $\Gamma_1 = [g_1, g_2, \ldots, g_{n-1}]$, if the matrix $A_1 + \lambda B_1 \Gamma_1$ is stable (Hurwitz) for each nonzero eigenvalue λ of the Laplacian matrix L_G, then the control law (15) makes ω converge to zero.*

Proof: Along the proof of Lemma 3, it is omitted here. ∎

By Lemma 7, we can obtain the control law for time $t \geq T_2$. Next, we give the controller of system (1) for all time.

Theorem 2. *For system (1), if the communication graph is directed, balanced and strongly connected, the control law*

$$\begin{cases} u_1 = \dot{\alpha}\mathbf{1} - L_G y_1, \\ u_2 = \mathrm{diag}[u_{1,1}, \ldots, u_{1,m}]\Gamma L(\bar{z} - h) \end{cases} \tag{16}$$

makes (4)-(5) satisfied, where $\dot{\alpha} \geq \epsilon > 0$, $\Gamma_1 = [g_1, \ldots, g_{n-1}]$ and such that $A_1 + \lambda B_1 \Gamma_1$ is stable (Hurwitz) for each nonzero eigenvalue λ of the Laplacian matrix L_G.

Proof: By Lemma 5, (4) is satisfied. For $t \geq T_2$, (5) is satisfied by Lemma 7 if the state is bounded at time T_2. It is only needed to show $\bar{z}$ is bounded during time interval $[0, T_2]$. Noting the structure of system (1), $\bar{z}$ is bounded due to the boundedness of $u_{1,j}$ and $u_{2,j}$. ∎

Control law (16) solves the formation problem in this chapter. With this control law, the robots move in formation as time tends to infinite. In the control law, α is a design parameter. Since $\dot{\alpha} > 0$, the robots always move on the plane. It should be noted that in this controller there is no singular point. Most of the existing results on the cooperative control of mobile robots with graph theory are based on linear models. This chapter shows how the graph theory can be applied successfully to the cooperative control of nonholonomic mobile robots.

5 Illustrative Examples

Consider five wheeled mobile robots moving on a plane (Fig. 1). The five robots are the same and have the following kinematics (Fig. 2).

$$\begin{bmatrix} \dot{x} \\ \dot{y} \\ \dot{\theta} \\ \dot{\phi} \end{bmatrix} = \begin{bmatrix} \cos\theta \\ \sin\theta \\ \tan\phi/l \\ 0 \end{bmatrix} Rv_1 + \begin{bmatrix} 0 \\ 0 \\ 0 \\ 1 \end{bmatrix} v_2 \tag{17}$$

where $q = [x, y, \theta, \phi]^T$ is the system state, (x, y) represents the Cartesian coordinates of the middle point of the rear wheel axle, θ is the orientation of the robot body with respect to the X-axis, ϕ is the steering angle; l is the distance between the front and rear wheel-axle centers, R is the radius of rear

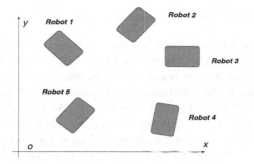

Fig. 1. Five mobile robots on a plane

driving wheel; v_1 is the angular velocity of the driving wheels, and v_2 is the steering velocity of the front wheels. $\phi \in (-\pi/2, \pi/2)$ due to the structure constraint of the robot.

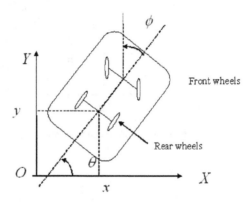

Fig. 2. Structure of a mobile robot

Let the state transformation

$$z_1 = x, z_2 = y, z_3 = \tan\theta, z_4 = \frac{\tan\phi}{l\cos^3\theta}, \tag{18}$$

and the input transformation

$$u_1 = v_1 R\cos\theta, u_2 = \frac{v_2 l\cos^2\theta + 3\sin\theta\sin^2\phi u_1}{l^2\cos^5\theta\cos^2\phi}, \tag{19}$$

system (17) is transformed into

$$\dot{z}_1 = u_1, \dot{z}_2 = z_3 u_1, \dot{z}_3 = z_4 u_1, \dot{z}_4 = u_2. \tag{20}$$

By the dynamic feedback linearization procedure, we can obtained a linear system (7) with $n = 4$. With the results in the last section, we can obtain the controller (6). By the inverse transformation, we can obtain the controller of the original system. In the simulation, the formation is defined by points $(0,0)$, $(15,0)$, $(0,15)$, $(8,25)$ and $(25,15)$ (see Fig. 3). Assume the communication link is shown as Fig. 4, Fig. 5 is the simulation result with this information graph. If the communication link is shown as Fig. 6, the group of mobile robots moves as in Fig. 7. The simulation results show that the proposed control law (6) is effective. In the simulation, the singular point never occurs since the robots always move forward.

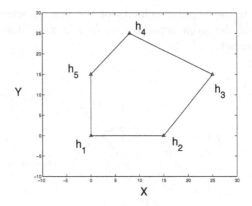

Fig. 3. Desired formation

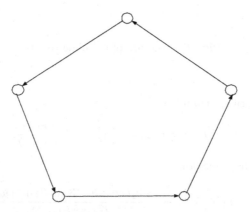

Fig. 4. Information link

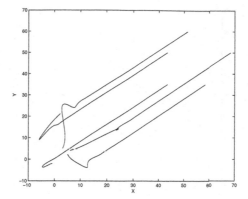

Fig. 5. Trajectories of the group of mobile robots in X-Y plane with controller (6)

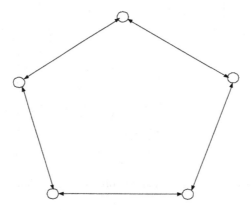

Fig. 6. Information link

We can easily obtain controller (16) too. In the controller, we choose $\alpha = 0.1t$. Fig. 8 is the simulation result with the information graph in Fig. 6. The simulation results show that the proposed control law (16) is effective if the communication graph is connected.

6 Conclusion

This chapter considers the formation control of mobile robots with nonlinear models. Two cooperative controllers are proposed based on the graph theory. Simulation results show the proposed controllers are effective. In this chapter, we do not consider collision and obstacles. How to consider them into controller design is an ongoing research work.

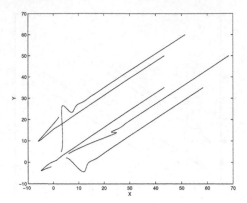

Fig. 7. Trajectories of the group of mobile robots in X-Y plane with controller (6)

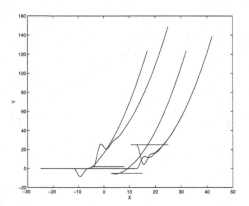

Fig. 8. Trajectories of the group of mobile robots in X-Y plane with controller (16)

References

1. R.C. Arkin, "Cooperation without coomunication: multiagent schemabased robot navigation," *J. of Robotic Systems,* vol.9, pp.351-364, 1992.
2. T. Balch and R. C. Arkin, "Behavior-based formation control for multirobot teams," *IEEE Trans. on Robotics and Automation,* vol.14, no.6, pp.926-939, 1998.
3. S.P. Bhattacharyya, H. Chapellet, L.H. Keel, Robust Control: The Parameter Approach. Prentice Hall, 1995.
4. Campion, G., Bastin, G., and d'Andrea-Novel, B., "Structure properties and classification of kinematic and dynamic models of wheeled mobile robots," *IEEE Trans. Robotics and Automation.* Vol.12, pp.47-62, 1991.
5. F.R.K. Chung. Spectral Graph Theory, volume 92 of Regional Conference Series in Mathe- matics. American Mathematical Soc., 1997.
6. R. Diestel. Graph Theory, volume 173 of Graduate Texts in Mathematics. Springer-Verlag, 1997.

7. A.K. Das, R. Fierro, and V. Kumar, Control Graphs for Robot Networks. Norwell, MA: Kluwer, 2002, Cooperative Control and Optimization of Applied Optimization, ch. 4, pp.5573.
8. J.P. Desai, J. Ostrowski, and V. Kumar, "Controlling formations of multiple mobile robots," *Proc. of the IEEE Int. Conf. on Robotics and Automation,* (Leuven, Belgium), pp.2864-2869, 1998.
9. J.P. Desai, J.P. Ostrowski, and V. Kumar, "Modeling and control of formations of nonholonomic mobile robots," *IEEE Trans. Robot. Automat.,* vol.17, pp.905908, Dec. 2001.
10. P.N.B.T. Eren and A.S. Morse, "Closing ranks in vehicle formations based on rigidity," *Proc. IEEE Conf. Decision and Control,* Las Vegas, NV, Dec. 2002, pp. 29592964.
11. J.A. Fax, and R. M. Murray, "Information Flow and Cooperative Control of Vehicle Formations," *IEEE Trans. on Automatic Control,* vol.49, pp.1465-1476, 2004.
12. R. Fierro, A. Das, V. Kumar, and J. Ostrowski, "Hybrid control of formations of robots," *Proc. of the IEEE Int. Conf. on Robotics and Automation,* (Seoul, Korea), pp.157-162, May 2001.
13. F. Giulietti, L. Pollini, and M. Innocenti, "Autonomous formation flight," *IEEE Control Systems Magazine,* vol.20, pp.34-44, 2000.
14. A. Jadbabaie, J. Lin, and A.S. Morse, "Coordination of Groups of Mobile Autonomous Agents Using Nearest Neighbor Rules," *IEEE Trans. on Automatic Control,* vol.48, pp.988-1001, 2003.
15. R.T. Jonathan J. Lawton, R.W. Beard, and B.J. Young, "A Decentralized Approach to Formation Maneuvers," *IEEE Trans. on Robotics and Automation,,* vol.19, pp.933-941, 2003.
16. W. Kang and H.-H. Yeh, "Co-ordinated attitude control of multi-satellite systems," *Int. J. of Robust and Nonlinear Control,* vol.12, pp.185-205, 2002.
17. G. Lafferriere, J. Caughman, and A. Williams, "Graph theoretic methods in the stability of veicle formations," *Proc. of Ameircan Control Conf.,* Boston, pp.3729-3734, 2004.
18. N.E. Leonard and E. Fiorelli, "Virtual leaders, artificial potentials and coordinated control of groups," *Proc. of the IEEE Conf. on Decision and Control,* (Orlando, Florida), pp.2968-2973, 2001.
19. M.A. Lewis and K.-H. Tan, "High precision formation control of mobile robots using virtual structures," *Autonomous Robots,* vol.4, pp.387-403, 1997.
20. Leroquais, W. and d'Andrea-Novel, B., "Transformation of the kinematic models of restricted mobility wheeled mobile robots with a single platform into chained forms," *Proc. of the IEEE Conf. Decision and Control,* 1995.
21. R. Merris, "A survey of graph Laplacians," *Linear and Multilinear Algebra,* Vol.39, pp.19-31, 1995.
22. M. Mesbahi and F.Y. Hadaegh, "Formation flying control of multiple spacecraft via graphs, matrix inequalities, and switching," *AIAA J. of Guidance, Control, and Dynamics,* vol.24, pp.369-377, 2001.
23. P. Ogren, E. Fiorelli, and N. E. Leonard, "Cooperative control of mobile sensor networks: adaptive gradient climbing in a distributed environment," *IEEE Trans. Automatic Control,* Vol.40, no.8, pp.1292-1302, 2004.
24. R. Olfati-Saber and R. M. Murray, "Distributed structural stabilization and tracking for formations of dynamic multiagents," *Proc. IEEE Conf. Decision and Control,* Las Vegas, NV, Dec. 2002, pp. 209215.

25. R. Olfati-Saber and R. M. Murray, "Consensus protocols for networks of dynamic agents," *Proc. of American Control conf.*, pp.951-956, 2003.
26. R. Olfati-Saber and R. M. Murray, "Consensus Problems in Networks of Agents with Switching Topology and Time-Delays," *IEEE Trans. on Automatic Control*, vol.49, pp.101-115, 2004.
27. L.E. Parker, "ALLIANCE: An architecture for fault tolerant multirobot cooperation," *IEEE Trans. on Robotics and Automation*, vol.14, pp.220-240, 1998.
28. W. Ren and R. W. Beard, "Consensus of information under dynamically changing interaction topologies," *Proc. of American Control Conf.*, pp.4939-4944, 2004.
29. W. Ren and R. W. Beard, "Formation Feedback Control for Multiple Spacecraft Via Virtual Structures," *submitted to IEE Proceedings - Control Theory and Applications*, 2004.
30. M. Schneider-Fontan and M. J. Mataric, "Territorial multi-robot task division," *IEEE Trans. on Robotics and Automation*, vol.14, pp.815-822, 1998.
31. D.J. Stilwell and B.E. Bishop, "Platoons of underwater vehicles," *IEEE Control Systems Magazine*, vol.20, pp.45-52, 2000.
32. K. Sugihara and I. Suzuki, "Distributed algorithms for formation of geometric patterns with many mobile robots," *J. Robot. Syst.*, vol.13, no.3, pp.127139, 1996.
33. H.G. Tanner, G.J. Pappas, and V. Kumar, "Leader-to-Formation Stability," *IEEE Trans. on Robotics and Automation*, vol.20, pp.443-455, 2004.
34. M. Veloso, P. Stone, and K. Han, "The CMUnited-97 robotic soccer team: Perception and multi-agent control," *Robot. Auton. Syst.*, vol.29, pp.133143, 1999.
35. P.K.C. Wang and F.Y. Hadaegh, "Coordination and control of multiple microspacecraft moving in formation," *The J. of the Astronautical Sciences*, vol.44, no.3, pp.315-355, 1996.
36. P. K. C. Wang, "Navigation strategies for multiple autonomous mobile robots moving in for- mation," *J. of Robotic Systems*, vol.8, no.2, pp.177-195, 1991.
37. H. Yamaguchi, "A cooperative hunting behavior by mobile robots troops," *Int. J. of Robotics Research*, vol.18, pp.921-940, 1999.
38. Y. Yang, A.A. Minai, and M.M. Polyearpou, "Decentralized cooperative search by networked UAVs in an uncertain environment," *Proc. American Control Conf.*, Boston, Massachusetts, 2004, pp.5558-5563.
39. Z. Lin, B. Francis, and M. Maggiore, "Necessary and sufficient graphical conditions for formation control of unicycles," *IEEE Trans. on Automatic Control*, vol.50, pp.121-127, 2005.

Comparison of Cooperative Search Algorithms for Mobile RF Targets Using Multiple Unmanned Aerial Vehicles

George W.P. York, Daniel J. Pack and Jens Harder

Department of Electrical and Computer Engineering
United States Air Force Academy USAF Academy, CO 80840-6236
E-mail: {george.york,daniel.pack}@usafa.af.mil

Summary. In this chapter, we compare two cooperative control algorithms for multiple Unmanned Aerial Vehicles (UAVs) to search, detect, and locate multiple mobile RF (Radio Frequency) emitting ground targets. We assume the UAVs are equipped with low-precision RF direction finding sensors with no ranging capability and the targets may emit signals randomly with variable duration. In the first algorithm the UAVs search a large area cooperatively until a target is detected. Once a target is detected, each UAV uses a cost function to determine whether to continue searching to minimize overall search time or to cooperate in localization of the target, joining in a proper orbit for precise triangulation to increase localization accuracy. In the second algorithm the UAVs fly in formations of three for both search and target localization. The first algorithm minimizes the total search time, while the second algorithm minimizes the time to localize targets after detection. Both algorithms combine a set of intentional cooperative rules with individual UAV behaviors optimizing a performance criterion to search a large area. This chapter will compare the total search time and localization accuracy generated by multiple UAVs using the two algorithms simulations as we vary ratios of the numbers of UAVs to the number of targets.

1 Introduction

Recently, military applications of UAVs have received considerable interest [1, 2, 3]. In most cases these UAVs, however, were individually controlled by human operators with significant training. We believe the frontier of UAV research lies in the autonomous control of multiple UAVs and their cooperative task completion. This chapter contributes toward that end by developing two cooperative search and localization algorithms for multiple UAVs and comparing their search time and localization accuracy for mobile RF emitting targets.

In this chapter we focus on a scenario where multiple homogeneous targets must be detected and localized by multiple homogeneous UAVs. Our targets are mobile and the signals are emitted randomly with arbitrary duration. We assume that each UAV has a limited communication range and a limited bandwidth. The communication range is assumed to be larger than the UAVs sensor range. Each UAV continually transmits to neighboring UAVs its location and the sensed direction toward any targets it has detected. Each UAV has sufficient memory to maintain a map showing the status of its neighboring UAVs over time. The UAV's RF sensor is assumed to only capable of direction finding (DF) with low precision (i.e., +/- 7 degrees). Finally, we assume the UAVs aprior knowledge of the search boundary and there are no objects to avoid within the area.

By fusing emergent swarm behaviors with intentional cooperation techniques among distributed systems, we demonstrate effective target search/localization with multiple UAVs. We compare two algorithms: (1) the UAVs search independently, then either cooperate in locating detected targets or continue searching independently; (2) the UAVs search and locate in triangular formations with each group working independently. Using MATLAB-based simulation, we compare the two approaches, varying the ratio of UAVs to targets to observe the trade-off in localization accuracy versus total task completion time. The task completion time includes search, detection, and localization of all targets in a predefined area. Our comparative analysis for the two cooperative control algorithms showed that the second algorithm, flying in formation, produced a higher accuracy in localization, but a much longer total search/localization time compared to the value obtained when the first algorithm was used. On the other hand, the first algorithm, in which the UAVs search independently and use a cost function to determine when to cooperate on localization, had a much reduced total search time at the cost of less accuracy in target localization.

2 Cooperative Control Algorithm I

Each UAV can operate in one of four stages: *(1) global search for targets, (2) approach detected target, (3) orbit and locate target, and (4) local search for lost mobile target.* A UAV can switch from one stage to another at anytime based on the cost involved in performing a particular task. Thus, a UAV that is engaged in the *approach detected target* phase can switch to any other three stages based on its current information of the environment.

2.1 Stage One: Global Search

To minimize the global search cost (time, fuel, distance traveled), each UAV keeps track of the flight trajectory history of all neighboring UAVs including its own. Each UAV attempts to fly a path as far away as possible from the paths of others, maximizing the search area. Since our targets are mobile, do not continually emit and can evade detection, we allow the UAVs to re-visit areas that were previously searched. This is accomplished by using the path history of UAVs. When a location is visited by a UAV, we set the history value within the UAV's sensor range to a maximum value and then decrease this value incrementally over time. The UAVs seek out locations with no information (not previously visited) or aged information. A UAV uses the following rules [5] to determine the next search point.

1. Fly to a point with the minimum explored history
2. Fly to a point farthest from other neighboring UAVs and the search boundaries
3. Fly straight (maximize fuel efficiency)

These rules are incorporated in the following cost function:

$$search = H(\sum \frac{1}{D_i} + \sum \frac{1}{D_j})(\sqrt{\frac{|\phi|}{\frac{\pi}{p}}} + 1) \tag{1}$$

where H corresponds to a numerical value representing the explored history of a location. Di represents the distance from the location to each known UAV i and Dj represents the shortest distance from the location to each search area boundaries j. Symbol ϕ is the turn angle required in radians and symbol p represents the number of discrete points used to determine a turning angle. Instead of exhaustively calculating this cost function for all potential points, only a finite number of points equally spaced on a 180 degree arc in front of the UAV are evaluated.

We have found the cost function in equation 1.1 sucessfully implements the above three rules, encouraging the UAVs to seek out unexplored locations but allowing them to return to previously search areas as our targets can temporarily disappear. On average the overall search time appears to be reduced (compared to random search and "lawn mower" search patterns [5]) due to the Di and Dj terms forcing the UAVs out over the search area. Finally the ϕ term prevents excessive turning (dampens oscillations) resulting in less fuel consumption or greater flight times, which is a critical concern for the smaller UAVs.

2.2 Stage Two: Approach Detected Target

When a UAV detects a target it then flies on a trajectory directed toward a tight orbit (minimum turn radius) around the target's estimated position. Its neighboring UAVs must decide whether to also cooperate with this UAV to improve the accuracy of target localization or to continue searching for other targets to minimize overall search time. The decision to switch between the first two stages for a UAV is governed by the overall emergent behavior of collective UAVs based on the following cost function computed by each UAV. Once a UAV learns that a target is detected, it computes the following cost

George W.P. York, Daniel J. Pack and Jens Harder

value and determines whether to switch from the *global search* stage to the *approach detected target* stage.

$$cooperate = w_1 \frac{D}{D_{max}} - w_2(n - s) + w_3(m - p) \tag{2}$$

D represents the estimated Euclidean distance from the current UAV location to the newly detected target. D_{max} corresponds to a normalized distance reflecting maximum distance possible in the search area. n is the preferred number of UAVs to accurately locate a target and s is the number of UAVs currently engaged in cooperatively locating the specified target, m stands for the total estimated number of targets in the search space, and symbol p represents the number of targets that have been detected or located. Finally, symbols w_i are the weights an operator can choose to influence the behavior of a UAV. When the cost function generates a positive number, the *global search* stage continues while a negative number indicates that the particular UAV should switch its operation to the *approach detected target* stage. The first term determines the relative distance of the current UAV to a detected target (if a new target is detected) and determines whether to approach the target based purely on the distance the UAV must travel to reach the target. The second term contributes to the decision by evaluating the number of UAVs that have already committed to locating the detected target. The final term contributes by evaluating the status of the global search task. If there is a large number of estimated targets that have not been detected, the cost function tends to become positive. If most of the targets have been detected, it influences the cost function to become negative, thus encouraging the current UAV to cooperate with others to locate the newly detected target.

We have found the cost function in equation 1.2 to be a simple and useful means for the UAV to quickly evaluate the tradeoff of whether it should help a neighboring UAV to locate a target (with greater precision) or instead continue the global search task to reduce the overall search time. The w_i terms allow the operator to aprior influence which side of this trade-off is more important for a given mission, reducing search time or reducing localization accuracy. One drawback to this cost function is it is most optimum if m is known aprior. If the operator's estimate for m is incorrect, the UAVs will still sucessfully perform the search/localization tasks, but in a less optimum manner.

Each UAV continually transmits its estimated location and, if available, its estimate of the angle to the target to neighboring UAVs. A UAV continues to approach a detected target unless a designated number of UAVs have already entered the orbit around the target.

2.3 Stage Three: Orbit and Locate Target

Once within the estimated target localization orbit, a UAV switches to the *orbit and locate target* stage. Since the sensor only detects the angle of arrival and has no ranging capability, the UAVs must triangulate to estimate the

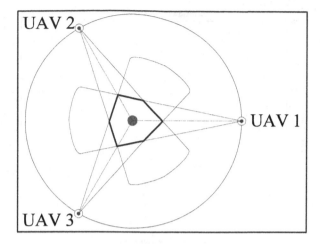

Fig. 1. Three UAVs in optimal geometry for triangulating. The target can lie any-where in the cone shaped areas determined by the accuracy of the sensor's DF angle and sensors maximum range.

target's location. The UAVs within the orbit adjust their velocities to form equi-angle distance among the UAVs (or 90 degree separation if only two UAVs) to improve the accuracy of the estimated target location, as shown in Fig. 1. The estimated localization error versus the number of UAVs in ideal geometry for triangulation is plotted in Fig. 2. This is used as a measure of merit to determine the optimal number of UAVs necessary to locate targets against all other requirements of the global search (i.e., finding other undiscov-ered emitters). For example, one could assign all UAVs to locate one emitter at the expense of hindering the global search. Fig. 2 indicates that accuracy improves little with more than three UAVs, thus we limit the number of UAVs cooperating in target localization to three.

When only one UAV is available and must localize on its own, it triangu-lates with its prior estimated angles as it flies around the target. The target can move during this process, increasing the localization error.

2.4 Stage Four: Local Search for Lost Mobile Target

The *local search* stage is initiated for UAVs (one to three) that have committed to locate a target when the target ceases to emit signals after it has been detected but before it is located. The committed UAVs then form an orbit with the radius greater than the one designated to locate a target. The radius continues to grow over time if the target is not re-detected. As the orbit radius grows, the UAVs continue to search for the target. Once the radius reaches a pre-defined maximum value, the UAVs engage in a local search pattern similar to the one in the *global search* stage, only in a smaller area with a maximum

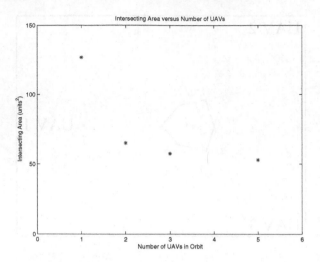

Fig. 2. Determining the optimal number of UAVs to locate a target. This figure shows how the location accuracy improves as the number of UAVs increases, assuming the UAVs are in ideal geometry.

of three UAVs. If a target is still not detected after a designated time interval, the UAVs return to the *global search* stage.

Currently, the radius of a local search orbit is a function of time, existence of signal, and two past trajectory points of the target emitter. As the time between emitter signals grow, the radius of the orbit will grow. At each point in time, UAVs involved in the local search uses the following two equations to determine the local orbit.

When no emitter signal is present, the local orbit equation is as follows.

$$(x - [e_x(t-1) + c_x(e_x(t-1) - e_x(t-2))])^2 +$$

$$(y - [e_y(t-1) + c_y(e_y(t-1) - e_y(t-2))])^2 = (r(t) + kt)^2 \qquad (3)$$

$e_x(t-1)$ and $e_y(t-1)$ are the last estimated x and y location values of the emitter; $e_x(t-2)$ and $e_y(t-2)$ are the second last estimated x and y location values of the emitter; $r(t)$ represents the radius of the current orbit; c_x and c_y are constants chosen to weigh the movement of the emitter location based on the past history; and kt represents the increase in the local orbit radius based on the elapsed time (t) and a constant (k) to accommodate the movement of the emitter location. When an emitter signal is detected, the local orbit equation yields to

$$(x - e_x(t))^2 + (y - e_y(t))^2 = r(t)^2 \qquad (4)$$

2.5 Simulation

Currently we have implemented our cooperative control algorithms in a MAT-LAB simulation. A simple example of three UAVs going through these four stages versus one target is shown in Fig. 3.

The initial locations of the three UAVs and three targets are randomly generated. Each UAV resides at the center of a large circle. The black dot at the center of a circle indicates the current UAV location. The radius of the circle represents the UAV sensor detection range as it sweeps 360 degrees. Similarly, each target is located at the center of the small circle, again represented with a black dot, and the radius of the small circle represents a circular orbit of the UAV's maximum turn radius around a target emitter where UAVs must fly to accurately locate the target. The trajectory history of UAVs and the targets are recorded and are shown in the figure by dotted lines within gray paths. Each UAV keeps track of the status of the search space and maintains a search map where each search location contains a numerical value which varies based on the search history of the particular location. Since the emitters we want to detect can be silent for an unspecified duration of time and are mobile, the UAVs not only must record the history of their past trajectories but also have a mechanism to slowly diminish numerical values for visited locations over time to realistically model the decaying intelligence gathered in the past. Such a scheme allows our UAVs to revisit the same spot and detect targets if those targets were 'silent' during previous flyovers.

3 Cooperative Control Algorithm II

The first algorithm performs well during the search stage, spreading the UAVs out to minimize search time. This approach has a disadvantage during target localization, however, as it takes time for the neighboring UAVs to join on localization orbit and the target may stop emitting in the process. This results in either a non-optimum localization due to poor geometry or a loss of the target resulting in more time spent in local search.

This motivated the second approach of having a set of three UAVs fly in an equilateral triangle formation as shown in Fig. 4. Instead of taking time to get on a proper orbit, the formation simply attempts to fly directly over the target, achieving the optimum geometry for triangulation when the centroid of the formation passes over the target. During the general search stage each formation leader follows the search strategy developed in the first algorithm, using the same cost function. This formation-based algorithm generally takes longer for the global search since the formations can not cover as much area as the individual UAVs when they are spread out [6].

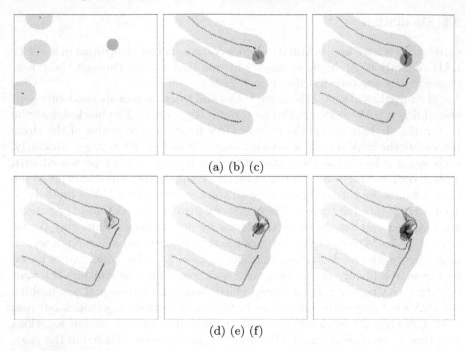

(a) (b) (c)

(d) (e) (f)

Fig. 3. Snapshots showing the progression of three UAVs searching, detecting, and localizing a moving target. Frame (a): This frame shows the initial locations of three UAVs (large circles) and one target (small circle). Frame (b): This frame shows the cooperative search mode where the three UAVs are spreading themselves across the area based on the search cost function. The top UAV detects a target (indicated by the black line showing the estimate angle to the target) and notifies the other two UAVs. Frame (c): Upon receiving the target discovery information, the nearby UAVs compute the 'search versus locate' cost function to determine whether or not to help localize the emitter. The frame shows that two UAVs have decided to join the efforts. The closest UAV starts an orbit, refining the location estimate on its own. Frame (d): The emitter turns off before the UAVs are in the proper geometry, so they switch to the local search mode. Frame (e): While the UAVs are in the local search mode, the target emits again within range of the UAVs, and they resume flying to the best geometry to locate the target. Frame (f): The three UAVs orbit around the detected emitter and adjust their positions to be equi-angle apart from each other for optimal triangulation.

3.1 Formation

We desire an algorithm to have a set of UAVs join in formation and maintain the formation without undue computation or communication requirements. Several authors have proposed simple methods such as Monteiro *et al's* attractor dynamics [8] and Spears' physics-based methods [7]. Our method is similar as illustrated in Fig. 5.

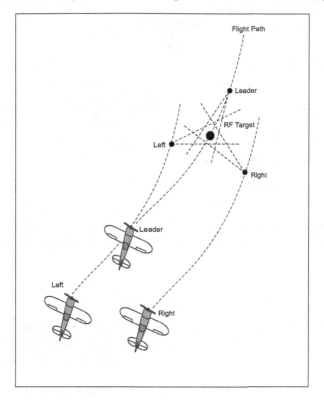

Fig. 4. Illustrating the triangular formation and the path required to ideally locate the RF target.

The UAVs first determine who their nearest two neighboring UAVs are. The one with the highest priority (known *a priori*) is designated the leader and the other two are the followers. The leader flies following the general search rules of Algorithm One. The two followers join in formation behind the leader by being attracted or repelled by their neighboring two UAVs as shown in Fig. 5. They are attracted up to a predefined range from the other UAVs and then repelled if they get too close. The UAVs continually adjust their velocity and trajectory incrementally as necessary to maintain this relationship. This simple method allows the UAVs to initially decide which formation to join and maintain the formation.

The follower first finds two attract/repel points (influenced by their two closest neighbors), then combine these to determine the resultant waypoint. An attract/repel point is found by

$$M = \sqrt{(x_1 - x_2)^2 + (y_1 - y_2)^2} \tag{5}$$

$$x_a = x_2 + r\frac{x_1 - x_2}{M} \tag{6}$$

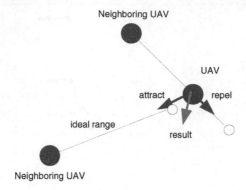

Fig. 5. Method to maintain formation. Following UAVs gravitate to a predefined range from their closest two neighboring UAVs using attractive and repelling forces.

$$y_a = y_2 + r\frac{y_1 - y_2}{M} \tag{7}$$

where r is the ideal formation range, (x_1, y_1) is the UAV's location, (x_2, y_2) is one of the neighboring UAV's location, and (x_a, y_a) is one of the attract/repel points. The other attract/repel point (x_b, y_b) is found similarly using the third UAV's location.

The waypoint goal (x_w, y_w) for the UAV is then computed

$$\Delta x = \frac{(x_a - x_1) + (x_b - x_1)}{2} \tag{8}$$

$$\Delta y = \frac{(y_a - y_1) + (y_b - y_1)}{2} \tag{9}$$

$$x_w = x_1 + \Delta x \tag{10}$$

$$y_w = y_1 + \Delta y \tag{11}$$

The UAV then attempts to fly to this goal waypoint, constrained by its flight dynamics (current trajectory, allowable velocity range, and the maximum bank angle for the given velocity). The following UAVs continually recompute the waypoint goals and effectively follow the leader in formation.

With this simple approach, there is no limit to the number of UAVs in a formation; however, the emergent property appears to be triangular formations of three. Occasionally a fourth UAV will temporarily join a group. When two formations pass closely to each other, occasionally a UAV will transfer to another formation due to the simple priority scheme.

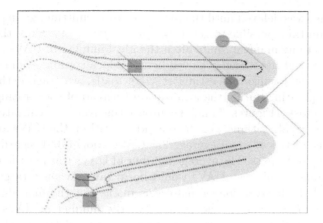

Fig. 6. Simulation using the formation algorithm. The squares indicate targets already located.

3.2 Formation Simulation

Fig. 6 is a snapshot from our MATLAB simulation using this formation approach. From the traces the snapshot shows that the two formations located three targets (squares) with four remaining. The upper-right formation is in the process of turning to fly over a target detected by one of the followers.

4 Results

In this section we present a comparative study to demonstrate how well the two algorithms compare against the competing requirements of minimizing the global search time and minimizing the target location error. We varied the ratio of the number of UAVs to the number of targets to get an indication of the scalability of the two approaches.

For our experiment, we changed the number of UAVs from 3 to 9, changed the number of targets from 1 to 9, and averaged the results from 100 simulations each. The initial target and UAV locations were randomly selected in a 50 km x 75 km area. The UAVs flew in a velocity range of 115 to 260 kph (cruise at 115 kph) with a minimum turn radius of 0.5 km. The target's maximum velocity was 37.5 kph, and a target traveled in random directions for random distances. A target emitted randomly, on for an average of 6.8 minutes and off for an average of 4.8 minutes. The sensor provided estimates every 12 seconds. We used both one degree and seven degree directional sensors. The maximum sensor range was assumed to be 4.3 km and the UAVs tried to fly an orbit of 2.2 km from the estimated target location for the first algorithm's localization. For the second algorithm, the formation tried to maintain an equal distance of 4.3 km between UAVs. The final localization estimate for a specific target was delayed until the target stopped emitting, giving the UAVs the maximum time possible to get in the proper geometry. Since the emitters can turn off at any moment, quite often the ideal number of UAVs and proper geometry may not have been achieved before a localization estimate is made.

When target localization estimates were made, we tracked the number of UAVs cooperating during the localization (instead of cooperating with the global search task). Figures 7 and 8 compare the average localization cooperation for the two algorithms. For the second algorithm, the UAVs are working in formations of three, so on average 2.5 UAVs cooperated regardless of the ratio of UAVs to targets. The average number of UAVs was not the ideal three since occasionally only two UAVs would be in range when a target stopped emitting and the localization estimate was made. For the first algorithm, as shown in Fig. 7, we can see the trade-off in the number of UAVs operating in the cooperative *target localization* stage and the *global search* stage. As expected, the best localization cooperation (> 2.5 UAVs) was achieved when 9 UAVs faced only 1 target, while the cooperation reduced (1 UAV) as the ratio changed down to 3 UAVs versus 9 targets. This amount of cooperation had a direct impact on the localization accuracy, as seen in Figures 9 through 12.

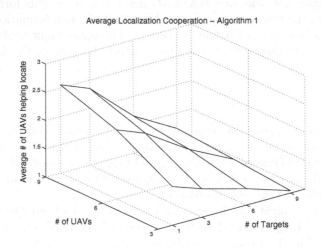

Fig. 7. Plot of the average number of UAVs cooperating for each localization estimate when the number of UAVs changes from 3 to 9 and the number of Targets changes from 1 to 9 for Algorithm I

Figures 9 through 12 compare the average localization error generated using the two algorithms with the one degree accurate sensor and the seven degree accurate sensor. For Algorithm II, flying in formation produced a more consistent localization error for all the UAV/target ratios. For the 1 degree sensor the average error was around 0.20 km while for the 7 degree sensor

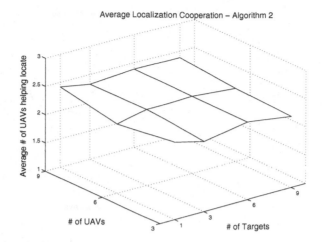

Fig. 8. Plot of the average number of UAVs cooperating for target localization when the number of UAVs change from 3 to 9 and the number of targets change from 1 to 9 for Algorithm II

it was around 0.36 km for Algorithm II. For Algorithm I, the localization error improved as the amount of localization cooperation improved. For the 1 degree sensor, the average error for 3 UAVs versus 9 targets was > 0.4 km, and the error reduced to 0.16 km for 9 UAVs versus 1 target. For the 7 degree sensor using the first algorithm with 3 UAVs versus 9 targets, the error was 0.57 km, and for the case with 9 UAVs versus 1 target, the error was 0.38 km. Thus, flying in formation appears to consistently improve localization accuracy. However, the cost is in the total search/localization time, as shown in Figures 13 and 14.

Figures 13 and 14 compare the total search and localization time for each of the cases for both algorithms. For Algorithm I, the best time (19 minutes) occurred as expected when 9 UAVs faced only one target; the time increased up to 174 minutes when 3 UAVs faced 9 targets. The total time for Algorithm II was much worse as the formations covered less area over time. The total times ranged from 60 minutes for 9 UAVs versus 1 target case to 417 minutes for the 3 UAVs versus 9 targets case.

5 Conclusion and Future Work

In this chapter, we introduced two algorithms for multiple UAVs to cooperatively search, detect, and locate RF mobile targets. The authors are not aware of any work that attempts to solve the current problem reported in the literature. We showed preliminary analysis work on the optimum number of UAVs required to locate targets; introduced a search cost function used to maximize

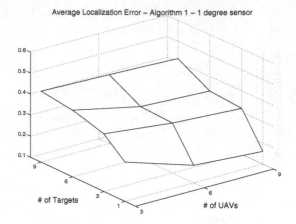

Fig. 9. Plot of the average localization error in kilometers when the number of UAVs changed from 3 to 9 and the number of Targets changed from 1 to 9 for Algorithm I using a 1 degree accurate sensor.

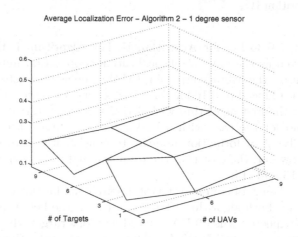

Fig. 10. Plot of the average localization error in kilometers when the number of UAVs changed from 3 to 9 and the number of Targets changed from 1 to 9 for Algorithm II using a 1 degree accurate sensor.

the use of multiple UAVs to individually and in formation search an area; proposed an algorithm to locate targets by cooperatively arranging multiple UAVs into the proper geometry, showed a quicker method of locating by flying a cooperative formation in the ideal geometry for localization; and illustrated a scheme to locally search targets that have stopped emitting but are expected to emit again. We demonstrated our proposed algorithms using simulated results. Our comparative analysis for the two cooperative control algorithms

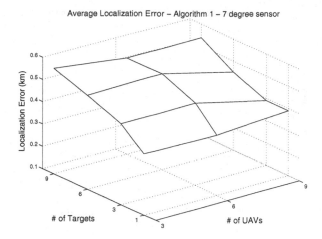

Fig. 11. Plot of the average localization error in kilometers when the number of UAVs changed from 3 to 9 and the number of Targets changed from 1 to 9 for Algorithm I using a 7 degree accurate sensor.

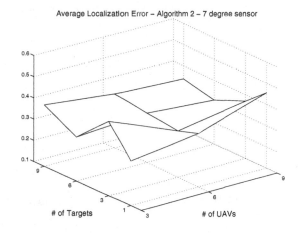

Fig. 12. Plot of the average localization error in kilometers when the number of UAVs changed from 3 to 9 and the number of Targets changed from 1 to 9 for Algorithm II using a 7 degree accurate sensor.

showed that Algorithm II, flying in formation, produced higher accuracy in target localization, but a much longer total search/localization time. On the other hand, Algorithm I, in which the UAVs search independently and use a cost function to determine when to cooperate on localization, had a much reduced total search time at the cost of less accuracy in target localization.

We have several plans to expand this research. Triangulating with directional sensors is known to occasionally produce very inaccurate results, partic-

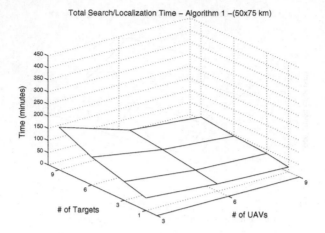

Fig. 13. Plot of the average total search/localization time in minutes when searching a 50x75 km area as the number of UAVs changed from 3 to 9 and the number of Targets changed from 1 to 9 for Algorithm I.

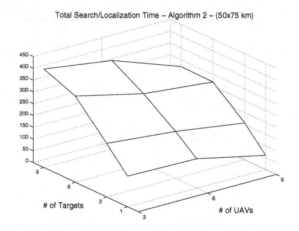

Fig. 14. Plot of the average total search/localization time in minutes when searching a 50x75 km area as the number of UAVs changed from 3 to 9 and the number of Targets changed from 1 to 9 for Algorithm II.

ularly when two UAVs are in an improper geometry close together in angle. We are working on improving triangular localization using a Kalman filter technique [9] as well as a rate-change-of-angle approach [10]. We plan to increase accuracy of simulator with 6 DOF UAV models and real RF DF sensor models. We also plan to experiment with hardware, first using robots in 2-D environment to compare the two cooperative control algorithms. We plan to follow this with experiments flying UAVs at the United States Air Force Academy test range.

References

1. Beard, R., Mclain, T., Goodrich, M., and Anderson, E. (2002). Coordinated Target Assignment of Intercept for Unmanned Air Vehicles, *IEEE Transactions on Robotics and Automation*, vol. 18, no. 6.
2. Chandler, P., Rasmussen, S., and Pachter, M. (2000). UAV Cooperative Path Planning, *AIAA Guidance, Navigation, and Control Conference and Exhibit*, pp. 1255-1265.
3. Dunbar, W. and Murray, R. (2002), Model Predictive Control of Coordinated Multi-vehicle Formations, *Proceedings of the 41st IEEE Conference on Decision and Control*, pp. 4631-4636.
4. Coffey, T. and Montgomery, J. (2002). The Emergence of Mini UAVs for Military Applications, *Defence Horizons*, No. 22, pp. 1 - 8.
5. Pack, D. and Mullins, B. (2003). Toward Finding an Universal Search Algorithm for Swarm Robots, *Proceedings of the 2003 IEEE/RJS Conference on Intelligent Robotic Systems (IROS)*, pp. 1945-1950.
6. York, G. W. P. and Pack, D. J. (2004). Minimal Formation Based Unmanned Aerial Vehicle Search Method to Detect RF Mobile Targets, *Proceedings of the 2nd Annual Swarming Conference: Networked Enabled C4ISR*.
7. Spears, W., Spears, D., Hamann, J., and Heil, R. (2004). Distributed, Physics-Based Control of Swarms of Vehicles, *Autonomous Robots*, Volume 17(2-3).
8. Montereiro, S., Vaz, M., and Bicho, E. (2004). Attractor Dynamics generates robot formations: from theory to implementation, *Proceedings of 2004 IEEE International Conference on Robotics and Automation*.
9. York, G. W. P., and Pack, D. J. (2005). Comparative Study on Time-Varying Target Localization Methods using Multiple Unmanned Aerial Vehicles: Kalman Estimation and Triangulation Techniques, *Proceedings of the 2005 IEEE International Conference On Networking, Sensing and Control*.
10. Gilbert, H. D., McGuirk, J. S., and Pack, D. J. (2005). A Comparative Study of Target Localization Methods for Large GDOP, *World Scientific*, publication pending.

Lecture Notes in Economics and Mathematical Systems

For information about Vols. 1–496
please contact your bookseller or Springer-Verlag